APPLIED ECONOMETRICS
USING THE SAS® SYSTEM

APPLIED ECONOMETRICS USING THE SAS® SYSTEM

VIVEK B. AJMANI, PHD
US Bank
St. Paul, MN

WILEY

A JOHN WILEY & SONS, INC., PUBLICATION

Published by John Wiley & Sons, Inc., Hoboken, New Jersey
Published simultaneously in Canada

For general information on our other products and services or for technical support, please contact our Customer Care Department within the United States at (800) 762-2974, outside the United States at (317) 572-3993 or fax (317) 572-4002.

Wiley also publishes its books in a variety of electronic formats. Some content that appears in print may not be available in electronic formats. For more information about Wiley products, visit our web site at www.wiley.com.

Library of Congress Cataloging-in-Publication Data:

Ajmani, Vivek B.
    Applied econometrics using the SAS system / Vivek B. Ajmani.
        p. cm.
    Includes bibliographical references and index.
    ISBN 978-0-470-12949-4 (cloth)
1.   Econometrics–Computer programs. 2.   SAS (Computer file)   I. Title.
    HB139.A46 2008
    330.0285'555–dc22
                                        2008004315

10 9 8 7 6 5 4

To My Wife, Preeti, and My Children, Pooja and Rohan

CONTENTS

PREFACE

The subject of econometrics involves the application of statistical methods to analyze data collected from economic studies. The goal may be to understand the factors influencing some economic phenomenon of interest, to validate a hypothesis proposed by theory, or to predict the future behavior of the economic phenomenon of interest based on underlying mechanisms or factors influencing it.

Although there are several well-known books that deal with econometric theory, I have found the books by Badi H. Baltagi, Jeffrey M. Wooldridge, Marno Verbeek, and William H. Greene to be very invaluable. These four texts have been heavily referenced in this book with respect to both the theory and the examples they have provided. I have also found the book by Ashenfelter, Levine, and Zimmerman to be invaluable in its ability to simplify some of the complex econometric theory into a form that can easily be understood by undergraduates who may not be well versed in advanced statistical methods involving matrix algebra.

When I embarked on this journey, many questioned me on why I wanted to write this book. After all, most economic departments use either Gauss or STATA to do empirical analysis. I used SAS Proc IML extensively when I took the econometric sequence at the University of Minnesota and personally found SAS to be on par with other packages that were being used. Furthermore, SAS is used extensively in industry to process large data sets, and I have found that economics graduate students entering the workforce go through a steep learning curve because of the lack of exposure to SAS in academia. Finally, after using SAS, Gauss, and STATA for my own personal work and research, I have found that the SAS software is as powerful or flexible compared to both Gauss and STATA.

There are several user-written books on how to use SAS to do statistical analysis. For instance, there are books that deal with regression analysis, logistic regression, survival analysis, mixed models, and so on. However, all these books deal with analyzing data collected from the applied or social sciences, and none deals with analyzing data collected from economic studies. I saw an opportunity to expand the SAS-by-user books library by writing this book.

I have attempted to incorporate some theory to lay the groundwork for the techniques covered in this book. I have found that a good understanding of the underlying theory makes a good data analyst even better. This book should therefore appeal to both students and practitioners, because it tries to balance the theory with the applications. ***However, this book should not be used as a substitute in place of the well-established texts that are being used in academia***. As mentioned above, the theory has been referenced from four main texts: Baltagi (2005), Greene (2003), Verbeek (2004), and Wooldridge (2002).

This book assumes that the reader is somewhat familiar with the SAS software and programming in general. The SAS help manuals from the SAS Institute, Inc. offer detailed explanation and syntax for all the SAS routines that were used in this book. Proc IML is a matrix programming language and is a component of the SAS software system. It is very similar to other matrix programming languages such as GAUSS and can be easily learned by running simple programs as starters. Appendixes A and B offer some basic code to help the inexperienced user get started. All the codes for the various examples used in this book were written in a very simple and direct manner to facilitate easy reading and usage by others. I have also provided detailed annotation with every program. The reader may contact me for electronic versions of the codes used in this book. The data sets used in this text are readily available over the Internet. Professors Greene and Wooldridge both have comprehensive web sites where the data are available for download. However, I have used data sets from other sources as well. The sources are listed with the examples provided in the text. All the data (except the credit card data from Greene (2003)) are in the public domain. The credit card data was used with permission from William H. Greene at New York University.

The reliance on Proc IML may be a bit confusing to some readers. After all, SAS has well-defined routines (Proc Reg, Proc Logistic, Proc Syslin, etc.) that easily perform many of the methods used within the econometric framework. I have found that using a matrix programming language to first program the methods reinforces our understanding of the underlying theory. Once the theory is well understood, there is no need for complex programming unless a well-defined routine does not exist.

It is assumed that the reader will have a good understanding of basic statistics including regression analysis. Chapter 1 gives a good overview of regression analysis and of related topics that are found in both introductory and advance econometric courses. This chapter forms the basis of the analysis progression through the book. That is, the basic OLS assumptions are explained in this chapter. Subsequent chapters deal with cases when these assumptions are violated. Most of the material in this chapter can be found in any statistics text that deals with regression analysis. The material in this chapter was adapted from both Greene (2003) and Meyers (1990).

Chapter 2 introduces regression analysis in SAS. I have provided detailed Proc IML code to analyze data using OLS regression. I have also provided detailed coverage of how to interpret the output resulting from the analysis. The chapter ends with a thorough treatment of multicollinearity. Readers are encouraged to refer to Freund and Littell (2000) for a thorough discussion on regression analysis using the SAS system.

Chapter 3 introduces hypothesis testing under the general linear hypothesis framework. Linear restrictions and the restricted least squares estimator are introduced in this chapter. This chapter then concludes with a section on detecting structural breaks in the data via the Chow and CUSUM tests. Both Greene (2003) and Meyers (1990) offer a thorough treatment of this topic.

Chapter 4 introduces instrumental variables analysis. There is a good amount of discussion on measurement errors, the assumptions that go into the analysis, specification tests, and proxy variables. Wooldridge (2002) offers excellent coverage of instrumental variables analysis.

Chapter 5 deals with the problem of heteroscedasticity. We discuss various ways of detecting whether the data suffer from heteroscedasticity and analyzing the data under heteroscedasticity. Both GLS and FGLS estimations are covered in detail. This chapter ends with a discussion of GARCH models. The material in this chapter was adapted from Greene (2003), Meyers (1990), and Verbeek (2004).

Chapter 6 extends the discussion from Chapter 5 to the case where the data suffer from serial correlation. This chapter offers a good introduction to autocorrelation. Brocklebank and Dickey (2003) is excellent in its treatment of how SAS can be used to analyze data that suffer from serial correlation. On the other hand, Greene (2003), Meyers (1990), and Verbeek (2004) offer a thorough treatment of the theory behind the detection and estimation techniques under the assumption of serial correlation.

Chapter 7 covers basic panel data models. The discussion starts with the inefficient OLS estimation and then moves on to fixed effects and random effects analysis. Baltagi (2005) is an excellent source for understanding the theory underlying panel data analysis while Greene (2003) offers an excellent coverage of the analytical methods and practical applications of panel data.

Seemingly unrelated equations (SUR) and simultaneous equations (SE) are covered in Chapters 8 and 9, respectively. The analysis of data in these chapters uses Proc Syslin and Proc Model, two SAS procedures that are very efficient in analyzing multiple equation models. The material in this chapter makes extensive use of Greene (2003) and Ashenfelter, Levine and Zimmerman (2003).

Chapter 10 deals with discrete choice models. The discussion starts with the Probit and Logit models and then moves on to Poisson regression. Agresti (1990) is the seminal reference for categorical data analysis and was referenced extensively in this chapter.

Chapter 11 is an introduction to duration analysis models. Meeker and Escobar (1998) is a very good reference for reliability analysis and offers a firm foundation for duration analysis techniques. Greene (2003) and Verbeek (2004) also offer a good introduction to this topic while Allison (1995) is an excellent guide on using SAS to analyze survival analysis/duration analysis studies.

Chapter 12 contains special topics in econometric analysis. I have included discussion on groupwise heterogeneity, Harvey's multiplicative heterogeneity, Hausman–Taylor estimators, and heterogeneity and autocorrelation in panel data.

Appendixes A and B discuss basic matrix algebra and how Proc IML can be used to perform matrix calculations. These two sections offer a good introduction to Proc IML and matrix algebra useful for econometric analysis. Searle (1982) is an outstanding reference for matrix algebra as it applies to the field of statistics.

Appendix C contains a brief discussion of the large sample properties of the OLS estimators. The discussion is based on a simple simulation using SAS.

Appendix D offers an overview of bootstrapping methods including their application to regression analysis. Efron and Tibshirani (1993) offer outstanding discussion on bootstrapping techniques and were heavily referenced in this section of the book.

Appendix E contains the complete code for some key programs used in this book.

St. Paul, MN   Vivek B. Ajmani

ACKNOWLEDGMENTS

I owe a great debt to Professors Paul Glewwe and Gerard McCullough (both from University of Minnesota) for teaching me everything I know about econometrics. Their instruction and detailed explanations formed the basis for this book. I am also grateful to Professor William Greene (New York University) for allowing me to access data from his text *Econometric Analysis*, 5th edition, 2003. The text by Greene is widely used to teach introductory graduate level classes in econometrics for the wealth of examples and theoretical foundations it provides. Professor Greene was also kind enough to nudge me in the right direction on a few occasions while I was having difficulties trying to program the many routines that have been used in this book.

I would also like to acknowledge the constant support I received from many friends and colleagues at Ameriprise Financials. In particular, I would like to thank Robert Moore, Ines Langrock, Micheal Wacker, and James Eells for reviewing portions of the book.

I am also grateful to an outside reviewer for critiquing the manuscript and for providing valuable feedback. These comments allowed me to make substantial improvements to the manuscript. Many thanks also go to Susanne Steitz-Filler for being patient with me throughout the completion of this book.

In writing this text, I have made substantial use of resources found on the World Wide Web. In particular, I would like to acknowledge Professors Jeffrey Wooldridge (Michigan State University) and Professor Marno Verbeek (RSM Erasmus University, the Netherlands) for making the data from their texts available on their homepages.

Although most of the SAS codes were created by me, I did make use of two programs from external sources. I would like to thank the SAS Institute for giving me permission to use the *% boot* macros. I would also like to acknowledge Thomas Fomby (Southern Methodist University) for writing code to perform duration analysis on the Strike data from Kennan (1984).

Finally, I would like to thank my wife, Preeti, for "holding the fort" while I was busy trying to crack some of the codes that were used in this book.

St. Paul, MN   Vivek B. Ajmani

1

INTRODUCTION TO REGRESSION ANALYSIS

1.1 INTRODUCTION

The general purpose of regression analysis is to study the relationship between one or more dependent variable(s) and one or more independent variable(s). The most basic form of a regression model is where there is one independent variable and one dependent variable. For instance, a model relating the log of wage of married women to their experience in the work force is a simple linear regression model given by $\log(\text{wage}) = \beta_0 + \beta_1 \text{exper} + \varepsilon$, where β_0 and β_1 are unknown coefficients and ε is random error. One objective here is to determine what effect (if any) the variable exper has on wage. In practice, most studies involve cases where there is more than one independent variable. As an example, we can extend the simple model relating log(wage) to exper by including the square of the experience (exper^2) in the work force, along with years of education (educ). The objective here may be to determine what effect (if any) the explanatory variables (exper, exper^2, educ) have on the response variable log(wage). The extended model can be written as

$$\log(wage) = \beta_0 + \beta_1 exper + \beta_2 exper^2 + \beta_3 educ + \varepsilon,$$

where β_0, β_1, β_2, and β_3 are the unknown coefficients that need to be estimated, and ε is random error.

An extension of the multiple regression model (with one dependent variable) is the multivariate regression model where there is more than one dependent variable. For instance, the well-known Grunfeld investment model deals with the relationship between investment (I_{it}) with the true market value of a firm (F_{it}) and the value of capital (C_{it}) (Greene, 2003). Here, i indexes the firms and t indexes time. The model is given by $I_{it} = \beta_{0i} + \beta_{1i}F_{it} + \beta_{2i}C_{it} + \varepsilon_{it}$. As before, β_{0i}, β_{1i}, and β_{2i} are unknown coefficients that need to be estimated and ε_{it} is random error. The objective here is to determine if the disturbance terms are involved in cross-equation correlation. Equation by equation ordinary least squares is used to estimate the model parameters if the disturbances are not involved in cross-equation correlations. A feasible generalized least squares method is used if there is evidence of cross-equation correlation. We will look at this model in more detail in our discussion of seemingly unrelated regression models (SUR) in Chapter 8.

Dependent variables can be continuous or discrete. In the Grunfeld investment model, the variable I_{it} is continuous. However, discrete responses are also very common. Consider an example where a credit card company solicits potential customers via mail. The response of the consumer can be classified as being equal to 1 or 0 depending on whether the consumer chooses to respond to the mail or not. Clearly, the outcome of the study (a consumer responds or not) is a discrete random variable. In this example, the response is a binary random variable. We will look at modeling discrete responses when we discuss discrete choice models in Chapter 10.

In general, a multiple regression model can be expressed as

$$y = \beta_0 + \beta_1 x_1 + \cdots + \beta_k x_k + \varepsilon = \beta_0 + \sum_{i=1}^{k} \beta_i x_i + \varepsilon, \tag{1.1}$$

Applied Econometrics Using the SAS® System, by Vivek B. Ajmani
Copyright © 2009 John Wiley & Sons, Inc.

where y is the dependent variable, $\beta_0, \ldots, \beta_k$ are the $k + 1$ unknown coefficients that need to be estimated, $x_1, \ldots, x_k$ are the k independent or explanatory variables, and ε is random error. Notice that the model is linear in parameters $\beta_0, \ldots, \beta_k$ and is therefore called a linear model. Linearity refers to how the parameters enter the model. For instance, the model $y = \beta_0 + \beta_1 x_1^2 + \cdots + \beta_k x_k^2 + \varepsilon$ is also a linear model. However, the exponential model $y = \beta_0 \exp(-x\beta_1)$ is a nonlinear model since the parameter β_1 enters the model in a nonlinear fashion through the exponential function.

1.1.1 Interpretation of the Parameters

One of the assumptions (to be discussed later) for the linear model is that the conditional expectation $E(\varepsilon|x_1, \ldots, x_k)$ equals zero. Under this assumption, the expectation, $E(y|x_1, \ldots, x_k)$ can be written as $E(y|x_1, \ldots, x_k) = \beta_0 + \sum_{i=1}^{k} \beta_i x_i$. That is, the regression model can be interpreted as the conditional expectation of y for given values of the explanatory variables $x_1, \ldots, x_k$. In the Grunfeld example, we could discuss the expected investment for a given firm for known values of the firm's true market value and value of its capital. The intercept term, β_0, gives the expected value of y when all the explanatory variables are set at zero. In practice, this rarely makes sense since it is very uncommon to observe values of all the explanatory variables equal to zero. Furthermore, the expected value of y under such a case will often yield impossible results. The coefficient β_k is interpreted as the expected change in y for a unit change in x_k holding all other explanatory variables constant. That is, $\partial E(y|x_1, \ldots, x_k)/\partial x_k = \beta_k$.

The requirement that all other explanatory variables be held constant when interpreting a coefficient of interest is called the *ceteris paribus condition*. The effect of x_k on the expected value of y is referred to as the marginal effect of x_k.

Economists are typically interested in *elasticities* rather than marginal effects. Elasticity is defined as the relative change in the dependent variable for a relative change in the independent variable. That is, elasticity measures the responsiveness of one variable to changes in another variable—the greater the elasticity, the greater the responsiveness.

There is a distinction between marginal effect and elasticity. As stated above, the marginal effect is simply $\partial E(y|\mathbf{x})/\partial x_k$ whereas elasticity is defined as the ratio of the percentage change in y to the percentage change in x. That is, $e = (\partial y/y)/(\partial x_k/x_k)$.

Consider calculating the elasticity of x_1 in the general regression model given by Eq. (1.1). According to the definition of elasticity, this is given by $e_{x_1} = (\partial y/\partial x_1)(x_1/y) = \beta_1(x_1/y) \neq \beta_1$. Notice that the marginal effect is constant whereas the elasticity is not. Next, consider calculating the elasticity in a log–log model given by $\log(y) = \beta_0 + \beta_1 \log(x) + \varepsilon$. In this case, elasticity of x is given by

$$\partial \log(y) = \beta_1 \partial \log(x) \Rightarrow \partial y \frac{1}{y} = \beta_1 \partial x \frac{1}{x} \Rightarrow \frac{\partial y}{\partial x} \frac{x}{y} = \beta_1.$$

The marginal effect for the log–log model is also β_1. Next, consider the semi-log model given by $y = \beta_1 + \beta_1 \log(x) + \varepsilon$. In this case, elasticity of x is given by

$$\partial y = \beta_1 \partial \log(x) \Rightarrow \partial y = \beta_1 \partial x \frac{1}{x} \Rightarrow \frac{\partial y}{\partial x} \frac{x}{y} = \beta_1 \frac{1}{y}.$$

On the other hand, the marginal effect in the semi-log model is given by $\beta_1(1/x)$.

For the semi-log model given by $\log(y) = \beta_0 + \beta_1 x + \varepsilon$, the elasticity of x is given by

$$\partial y \log(y) = \beta_1 \partial x \Rightarrow \partial y \frac{1}{y} = \beta_1 \partial x = \frac{\partial y}{\partial x} \frac{x}{y} = \beta_1 x.$$

On the other hand, the marginal effect in the semi-log model is given by $\beta_1 y$.

Most models that appear in this book have a log transformation on the dependent variable or the independent variable or both. It may be useful to clarify how the coefficients from these models are interpreted. For the semi-log model where the dependent variable has been transformed using the log transformation while the explanatory variables are in their original units, the coefficient β is interpreted as follows: For a one unit change in the explanatory variable, the dependent variable changes by $\beta \times 100\%$ holding all other explanatory variables constant.

In the semi-log model where the explanatory variable has been transformed by using the log transformation, the coefficient β is interpreted as follows: For a one unit change in the explanatory variable, the dependent variable increases (decreases) by $\beta/100$ units.

In the log–log model where both the dependent and independent variable have been transformed by using a log transformation, the coefficient β is interpreted as follows: A 1% change in the explanatory variable is associated with a $\beta\%$ change in the dependent variable.

1.1.2 Objectives and Assumptions in Regression Analysis

There are three main objectives in any regression analysis study. They are

  a. To estimate the unknown parameters in the model.
  b. To validate whether the functional form of the model is consistent with the hypothesized model that was dictated by theory.
  c. To use the model to predict future values of the response variable, y.

Most regression analysis in econometrics involves objectives (a) and (b). Econometric time series analysis involves all three. There are five key assumptions that need to be checked before the regression model can be used for the purposes outlined above.

  a. *Linearity:* The relationship between the dependent variable y and the independent variables $x_1, \ldots, x_k$ is linear.
  b. *Full Rank:* There is no linear relationship among any of the independent variables in the model. This assumption is often violated when the model suffers from multicollinearity.
  c. *Exogeneity of the Explanatory Variables:* This implies that the error term is independent of the explanatory variables. That is, $E(\varepsilon_i|x_{i1}, x_{i2}, \ldots, x_{ik}) = 0$. This assumption states that the underlying mechanism that generated the data is different from the mechanism that generated the errors. Chapter 4 deals with alternative methods of estimation when this assumption is violated.
  d. *Random Errors:* The errors are random, uncorrelated with each other, and have constant variance. This assumption is called the homoscedasticity and nonautocorrelation assumption. Chapters 5 and 6 deal with alternative methods of estimation when this assumption is violated. That is estimation methods when the model suffers from heteroscedasticity and serial correlation.
  e. *Normal Distribution:* The distribution of the random errors is normal. This assumption is used in making inference (hypothesis tests, confidence intervals) to the regression parameters but is not needed in estimating the parameters.

1.2 MATRIX FORM OF THE MULTIPLE REGRESSION MODEL

The multiple regression model in Eq. (1.1) can be expressed in matrix notation as $\mathbf{y} = \mathbf{X}\boldsymbol{\beta} + \boldsymbol{\varepsilon}$. Here, $\mathbf{y}$ is an $n \times 1$ vector of observations, $\mathbf{X}$ is a $n \times (k+1)$ matrix containing values of explanatory variables, $\boldsymbol{\beta}$ is a $(k+1) \times 1$ vector of coefficients, and $\boldsymbol{\varepsilon}$ is an $n \times 1$ vector of random errors. Note that $\mathbf{X}$ consists of a column of 1's for the intercept term $\boldsymbol{\beta}_0$. The regression analysis assumptions, in matrix notation, can be restated as follows:

  a. *Linearity*: $\mathbf{y} = \boldsymbol{\beta}_0 + \mathbf{x}_1\beta_1 + \cdots + \mathbf{x}_k\beta_k + \boldsymbol{\varepsilon}$ or $\mathbf{y} = \mathbf{X}\boldsymbol{\beta} + \boldsymbol{\varepsilon}$.
  b. *Full Rank*: $\mathbf{X}$ is an $n \times (k+1)$ matrix with rank $(k+1)$.
  c. *Exogeneity*: $E(\boldsymbol{\varepsilon}|\mathbf{X}) = \mathbf{0} - \mathbf{X}$ is uncorrelated with $\boldsymbol{\varepsilon}$ and is generated by a process that is independent of the process that generated the disturbance.
  d. *Spherical Disturbances*: $Var(\varepsilon_i|\mathbf{X}) = \sigma^2$ for all $i = 1, \ldots, n$ and $Cov(\varepsilon_i, \varepsilon_j|\mathbf{X}) = 0$ for all $i \neq j$. That is, $Var(\boldsymbol{\varepsilon}|\mathbf{X}) = \sigma^2\mathbf{I}$.
  e. *Normality*: $\boldsymbol{\varepsilon}|\mathbf{X} \sim N(\mathbf{0}, \sigma^2\mathbf{I})$.

1.3 BASIC THEORY OF LEAST SQUARES

Least squares estimation in the simple linear regression model involves finding estimators b_0 and b_1 that minimize the sums of squares $L = \Sigma_{i=1}^{n}(y_i - \beta_0 - \beta_1 x_i)^2$. Taking derivatives of L with respect to β_0 and β_1 gives

$$\frac{\partial L}{\partial \beta_0} = -2\sum_{i=1}^{n}(y_i - \beta_0 - \beta_1 x_i),$$

$$\frac{\partial L}{\partial \beta_1} = -2\sum_{i=1}^{n}(y_i - \beta_0 - \beta_1 x_i)x_i.$$

Equating the two equations to zero and solving for β_0 and β_1 gives

$$\sum_{i=1}^{n} y_i = n\hat{\beta}_0 + \hat{\beta}_1 \sum_{i=1}^{n} x_i,$$

$$\sum_{i=1}^{n} y_i x_i = \hat{\beta}_0 \sum_{i=1}^{n} x_i + \hat{\beta}_1 \sum_{i=1}^{n} x_i^2.$$

These two equations are known as normal equations. There are two normal equations and two unknowns. Therefore, we can solve these to get the ordinary least squares (OLS) estimators of β_0 and β_1. The first normal equation gives the estimator of the intercept, β_0, $\hat{\beta}_0 = \bar{y} - \hat{\beta}_1 \bar{x}$. Substituting this in the second normal equation and solving for $\hat{\beta}_1$ gives

$$\hat{\beta}_1 = \frac{n \sum_{i=1}^{n} y_i x_i - \sum_{i=1}^{n} y_i \sum_{i=1}^{n} x_i}{n \sum_{i=1}^{n} x_i^2 - \left(\sum_{i=1}^{n} x_i\right)^2}.$$

We can easily extend this to the multiple linear regression model in Eq. (1.1). In this case, least squares estimation involves finding an estimator $\mathbf{b}$ of $\boldsymbol{\beta}$ to minimize the error sums of squares $L = (\mathbf{y} - \mathbf{X}\boldsymbol{\beta})^T(\mathbf{y} - \mathbf{X}\boldsymbol{\beta})$. Taking the derivative of L with respect to $\boldsymbol{\beta}$ yields $k + 1$ normal equations with $k + 1$ unknowns (including the intercept) given by

$$\partial L/\partial \boldsymbol{\beta} = -2(\mathbf{X}^T \mathbf{y} - \mathbf{X}^T \mathbf{X}\hat{\boldsymbol{\beta}}).$$

Setting this equal to zero and solving for $\hat{\boldsymbol{\beta}}$ gives the least squares estimator of $\boldsymbol{\beta}$, $\mathbf{b} = (\mathbf{X}^T\mathbf{X})^{-1}\mathbf{X}^T\mathbf{y}$. A computational form for $\mathbf{b}$ is given by

$$\mathbf{b} = \left(\sum_{i=1}^{n} \mathbf{x}_i^T \mathbf{x}_i\right)^{-1} \left(\sum_{i=1}^{n} \mathbf{x}_i^T y_i\right).$$

The estimated regression model or predicted value of $\mathbf{y}$ is therefore given by $\hat{\mathbf{y}} = \mathbf{X}\mathbf{b}$. The residual vector $\mathbf{e}$ is defined as the difference between the observed and the predicted value of $\mathbf{y}$, that is, $\mathbf{e} = \mathbf{y} - \hat{\mathbf{y}}$.

The method of least squares produces unbiased estimates of $\boldsymbol{\beta}$. To see this, note that

$$
\begin{aligned}
E(\mathbf{b}|\mathbf{X}) &= E((\mathbf{X}^T\mathbf{X})^{-1}\mathbf{X}^T\mathbf{y}|\mathbf{X}) \\
&= (\mathbf{X}^T\mathbf{X})^{-1}\mathbf{X}^T E(\mathbf{y}|\mathbf{X}) \\
&= (\mathbf{X}^T\mathbf{X})^{-1}\mathbf{X}^T E(\mathbf{X}\boldsymbol{\beta} + \boldsymbol{\varepsilon}|\mathbf{X}) \\
&= (\mathbf{X}^T\mathbf{X})^{-1}\mathbf{X}^T\mathbf{X}\boldsymbol{\beta} E(\boldsymbol{\varepsilon}|\mathbf{X}) \\
&= \boldsymbol{\beta}.
\end{aligned}
$$

Here, we made use of the fact that $(\mathbf{X}^T\mathbf{X})^{-1} = (\mathbf{X}^T\mathbf{X}) = \mathbf{I}$, where $\mathbf{I}$ is the identity matrix and the assumption that $E(\boldsymbol{\varepsilon}|\mathbf{X}) = 0$.

1.3.1 Consistency of the Least Squares Estimator

First, note that a consistent estimator is an estimator that converges in probability to the parameter being estimated as the sample size increases. To say that a sequence of random variables X_n converges in probability to X implies that as $n \to \infty$ the probability that $|X_n - X| \geq \delta$ is zero for all δ (Casella and Berger, 1990). That is,

$$\lim_{n \to \infty} \Pr(|X_n - X| \geq \delta) = 0 \, \forall \, \delta.$$

Under the exogeneity assumption, the least squares estimator is a consistent estimator of $\boldsymbol{\beta}$. That is,

$$\lim_{n \to \infty} \Pr(|\mathbf{b}_n - \boldsymbol{\beta}| \geq \delta) = 0 \, \forall \, \delta.$$

To see this, let $\mathbf{x}_i$, $i = 1, \ldots, n$, be a sequence of independent observations and assume that $\mathbf{X}^T\mathbf{X}/n$ converges in probability to a positive definite matrix $\boldsymbol{\Psi}$. That is (using the probability limit notation),

$$p \lim_{n \to \infty} \frac{\mathbf{X}^T\mathbf{X}}{n} = \boldsymbol{\Psi}.$$

Note that this assumption allows the existence of the inverse of $\mathbf{X}^T\mathbf{X}$. The least squares estimator can then be written as

$$\mathbf{b} = \boldsymbol{\beta} + \left(\frac{\mathbf{X}^T\mathbf{X}}{n}\right)^{-1}\left(\frac{\mathbf{X}^T\boldsymbol{\varepsilon}}{n}\right).$$

Assuming that $\boldsymbol{\Psi}^{-1}$ exists, we have

$$p\lim \mathbf{b} = \boldsymbol{\beta} + \boldsymbol{\Psi}^{-1}p\lim\left(\frac{\mathbf{X}^T\boldsymbol{\varepsilon}}{n}\right).$$

In order to show consistency, we must show that the second term in this equation has expectation zero and a variance that converges to zero as the sample size increases. Under the exogeneity assumption, it is easy to show that $E(\mathbf{X}^T\boldsymbol{\varepsilon}|\mathbf{X}) = \mathbf{0}$ since $E(\boldsymbol{\varepsilon}|\mathbf{X}) = \mathbf{0}$. It can also be shown that the variance of $\mathbf{X}^T\boldsymbol{\varepsilon}/n$ is

$$Var\left(\frac{\mathbf{X}^T\boldsymbol{\varepsilon}}{n}\right) = \frac{\sigma^2}{n}\boldsymbol{\Psi}.$$

Therefore, as $n \rightarrow \infty$ the variance converges to zero and thus the least squares estimator is a consistent estimator for $\boldsymbol{\beta}$ (Greene, 2003, p. 66).

Moving on to the variance–covariance matrix of $\mathbf{b}$, it can be shown that this is given by

$$Var(\mathbf{b}|\mathbf{X}) = \sigma^2(\mathbf{X}^T\mathbf{X})^{-1}.$$

To see this, note that

$$\begin{aligned}
Var(\mathbf{b}|\mathbf{X}) &= Var((\mathbf{X}^T\mathbf{X})^{-1}\mathbf{X}^T\mathbf{y}|\mathbf{X})\\
&= Var((\mathbf{X}^T\mathbf{X})^{-1}\mathbf{X}^T(\mathbf{X}\boldsymbol{\beta} + \boldsymbol{\varepsilon})|\mathbf{X})\\
&= (\mathbf{X}^T\mathbf{X})^{-1}\mathbf{X}^T Var(\boldsymbol{\varepsilon}|\mathbf{X})\mathbf{X}(\mathbf{X}^T\mathbf{X})^{-1}\\
&= \sigma^2(\mathbf{X}^T\mathbf{X})^{-1}.
\end{aligned}$$

It can be shown that the least squares estimator is the best linear unbiased estimator of $\boldsymbol{\beta}$. This is based on the well-known result, called the Gauss–Markov Theorem, and implies that the least squares estimator has the smallest variance in the class of all unbiased estimators of $\boldsymbol{\beta}$ (Casella and Berger, 1990; Greene, 2003; Meyers, 1990).

An estimator of σ^2 can be obtained by considering the sums of squares of the residuals (SSE). Here, $SSE = (\mathbf{y} - \mathbf{Xb})^T(\mathbf{y} - \mathbf{Xb})$. Dividing SSE by its degrees of freedom, $n - k - 1$ yields $\hat{\sigma}^2$. That is, the mean square error is given by $\hat{\sigma}^2 = MSE = SSE/(n-k-1)$. Therefore, an estimate of the covariance matrix of $\mathbf{b}$ is given by $\hat{\sigma}^2(\mathbf{X}^T\mathbf{X})^{-1}$.

Using a similar argument as the one used to show consistency of the least squares estimator, it can be shown that $\hat{\sigma}^2$ is consistent for σ^2 and that the asymptotic covariance matrix of $\mathbf{b}$ is $\hat{\sigma}^2(\mathbf{X}^T\mathbf{X})^{-1}$ (see Greene, 2003, p. 69 for more details). The square root of the diagonal elements of this yields the standard errors of the individual coefficient estimates.

1.3.2 Asymptotic Normality of the Least Squares Estimator

Using the properties of the least squares estimator given in Section 1.3 and the Central Limit Theorem, it can be easily shown that the least squares estimator has an asymptotic normal distribution with mean $\boldsymbol{\beta}$ and variance–covariance matrix $\sigma^2(\mathbf{X}^T\mathbf{X})^{-1}$. That is, $\hat{\boldsymbol{\beta}} \sim asym.N\left(\boldsymbol{\beta}, \sigma^2(\mathbf{X}^T\mathbf{X})^{-1}\right)$.

1.4 ANALYSIS OF VARIANCE

The total variability in the data set (SST) can be partitioned into the sums of squares for error (SSE) and the sums of squares for regression (SSR). That is, $SST = SSE + SSR$. Here,

$$SST = \mathbf{y}^T\mathbf{y} - \frac{\left(\sum_{i=1}^{n} y_i\right)^2}{n},$$

$$SSE = \mathbf{y}^T\mathbf{y} - \mathbf{b}^T\mathbf{X}^T\mathbf{y},$$

$$SSR = \mathbf{b}^T\mathbf{X}^T\mathbf{y} - \frac{\left(\sum_{i=1}^{n} y_i\right)^2}{n}.$$

TABLE 1.1. Analysis of Variance Table

Source of Variation	Sums of Squares	Degrees of Freedom	Mean Square F_0
Regression	SSR	k	$MSR = SSR/k$
Error	SSE	$n - k - 1$	$MSE = SSE/(n - k - 1)$
Total	SST	$n - 1$	MSR/MSE

The mean square terms are simply the sums of square terms divided by their degrees of freedom. We can therefore write the analysis of variance (ANOVA) table as given in Table 1.1.

The F statistic is the ratio between the mean square for regression and the mean square for error. It tests the global hypotheses

$$H_0: \quad \beta_1 = \beta_2 = \ldots = \beta_k = 0,$$
$$H_1: \quad \text{At least one } \beta_i \neq 0 \quad \text{for } i = 1, \ldots, k.$$

The null hypothesis states that there is no relationship between the explanatory variables and the response variable. The alternative hypothesis states that at least one of the k explanatory variables has a significant effect on the response. Under the assumption that the null hypothesis is true, F_0 has an F distribution with k numerator and $n - k - 1$ denominator degrees of freedom, that is, under H_0, $F_0 \sim F_{k, n-k-1}$. The p value is defined as the probability that a random variable from the F distribution with k numerator and $n - k - 1$ denominator degrees of freedom exceeds the observed value of F, that is, $\Pr(F_{k, n-k-1} > F_0)$. The null hypothesis is rejected in favor of the alternative hypothesis if the p value is less than α, where α is the type I error.

1.5   THE FRISCH–WAUGH THEOREM

Often, we may be interested only in a subset of the full set of variables included in the model. Consider partitioning $\mathbf{X}$ into $\mathbf{X}_1$ and $\mathbf{X}_2$. That is, $\mathbf{X} = [\mathbf{X}_1 \ \mathbf{X}_2]$. The general linear model can therefore be written as $\mathbf{y} = \mathbf{X}\boldsymbol{\beta} + \boldsymbol{\varepsilon} = \mathbf{X}_1\boldsymbol{\beta}_1 + \mathbf{X}_2\boldsymbol{\beta}_2 + \boldsymbol{\varepsilon}$. The normal equations can be written as (Greene, 2003, pp. 26–27; Lovell, 2006)

$$\begin{bmatrix} \mathbf{X}_1^T\mathbf{X}_1 & \mathbf{X}_1^T\mathbf{X}_2 \\ \mathbf{X}_2^T\mathbf{X}_1 & \mathbf{X}_2^T\mathbf{X}_2 \end{bmatrix} \begin{bmatrix} \mathbf{b}_1 \\ \mathbf{b}_2 \end{bmatrix} = \begin{bmatrix} \mathbf{X}_1^T\mathbf{y} \\ \mathbf{X}_2^T\mathbf{y} \end{bmatrix}.$$

It can be shown that

$$\mathbf{b}_1 = (\mathbf{X}_1^T\mathbf{X}_1)^{-1}\mathbf{X}_1^T(\mathbf{y} - \mathbf{X}_2\mathbf{b}_2).$$

If $\mathbf{X}_1^T\mathbf{X}_2 = \mathbf{0}$, then $\mathbf{b}_1 = (\mathbf{X}_1^T\mathbf{X}_1)^{-1}\mathbf{X}_1^T\mathbf{y}$. That is, if the matrices $\mathbf{X}_1$ and $\mathbf{X}_2$ are orthogonal, then $\mathbf{b}_1$ can be obtained by regressing $\mathbf{y}$ on $\mathbf{X}_1$. Similarly, $\mathbf{b}_2$ can be obtained by regressing $\mathbf{y}$ on $\mathbf{X}_2$. It can easily be shown that

$$\mathbf{b}_2 = (\mathbf{X}_2^T\mathbf{M}_1\mathbf{X}_2)^{-1}(\mathbf{X}_2^T\mathbf{M}_1\mathbf{y}),$$

where $\mathbf{M}_1 = (\mathbf{I} - \mathbf{X}_1(\mathbf{X}_1^T\mathbf{X}_1)^{-1}\mathbf{X}_1^T)$ so that $\mathbf{M}_1\mathbf{y}$ is a vector of residuals from a regression of $\mathbf{y}$ on $\mathbf{X}_1$.

Note that $\mathbf{M}_1\mathbf{X}_2$ is a matrix of residuals obtained by regressing $\mathbf{X}_2$ on $\mathbf{X}_1$. The computations described here form the basis of the well-known Frisch–Waugh Theorem, which states that $\mathbf{b}_2$ can be obtained by regressing the residuals from a regression of $\mathbf{y}$ on $\mathbf{X}_1$ with the residuals obtained by regressing $\mathbf{X}_2$ on $\mathbf{X}_1$. One application of this result is in the derivation of the form of the least squares estimators in the fixed effects (LSDV) model, which will be discussed in Chapter 7.

1.6   GOODNESS OF FIT

Two commonly used goodness-of-fit statistics used are the coefficient of determination (R^2) and the adjusted coefficient of determination (R_A^2). R^2 is defined as

$$R^2 = \frac{SSR}{SST} = 1 - \frac{SSE}{SST}.$$

It measures the amount of variability in the response, y, that is explained by including the regressors $x_1, x_2, \ldots, x_k$ in the model. Due to the nature of its construction, we have $0 \leq R^2 \leq 1$. Although higher values (values closer to 1) are desired, a large value of R^2 does not necessarily imply that the regression model is a good one. Adding a variable to the model will always increase R^2 regardless of whether the additional variable is statistically significant or not. In other words, R^2 can be artificially inflated by overfitting the model.

To see this, consider the model $\mathbf{y} = \mathbf{X}_1 \boldsymbol{\beta}_1 + \mathbf{X}_2 \boldsymbol{\beta}_2 + \mathbf{U}$. Here, $\mathbf{y}$ is a $n \times 1$ vector of observations, $\mathbf{X}_1$ is the $n \times k_1$ data matrix $\boldsymbol{\beta}_1$ is a vector of k_1 coefficients, $\mathbf{X}_2$ is the $n \times k_2$ data matrix with k_2 added variables, $\boldsymbol{\beta}_2$ is a vector of k_2 coefficients, and $\mathbf{U}$ is a $n \times 1$ random vector. Using the Frisch–Waugh theorem, we can show that

$$\hat{\boldsymbol{\beta}}_2 = \left(\mathbf{X}_2^T \mathbf{M} \mathbf{X}_2\right)^{-1} \mathbf{X}_2^T \mathbf{M} \mathbf{y} = \left(\mathbf{X}_{2*}^T \mathbf{X}_{2*}\right)^{-1} \mathbf{X}_{2*}^T \mathbf{y}_*.$$

Here, $\mathbf{X}_{2*} = \mathbf{M} \mathbf{X}_2$, $\mathbf{y}_* = \mathbf{M} \mathbf{y}$, and $\mathbf{M} = \mathbf{I} - \mathbf{X}_1 (\mathbf{X}_1^T \mathbf{X}_1)^{-1} \mathbf{X}_1^T$. That is, $\mathbf{X}_{2*}$ and $\mathbf{y}_*$ are residual vectors of the regression of $\mathbf{X}_2$ and $\mathbf{y}$ on $\mathbf{X}_1$. We can invoke the Frisch–Waugh theorem again to get an expression for $\hat{\boldsymbol{\beta}}_1$. That is, $\hat{\boldsymbol{\beta}}_1 = \left(\mathbf{X}_1^T \mathbf{X}_1\right)^{-1} \mathbf{X}_1^T \left(\mathbf{y} - \mathbf{X}_2 \hat{\boldsymbol{\beta}}_2\right)$. Using elementary algebra, we can simplify this expression to get $\hat{\boldsymbol{\beta}}_1 = \mathbf{b} - \left(\mathbf{X}_1^T \mathbf{X}_1\right)^{-1} \mathbf{X}_1^T \mathbf{X}_2 \hat{\boldsymbol{\beta}}_2$, where $\mathbf{b} = \left(\mathbf{X}_1^T \mathbf{X}_1\right)^{-1} \mathbf{X}_1^T \mathbf{y}$. Next, note that $\mathbf{u} = \mathbf{y} - \mathbf{X}_1 \hat{\boldsymbol{\beta}}_1 - \mathbf{X}_2 \hat{\boldsymbol{\beta}}_2$. We can substitute the expression of $\hat{\boldsymbol{\beta}}_1$ in this to get $\hat{\mathbf{U}} = \mathbf{u} = \mathbf{e} - \mathbf{M} \mathbf{X}_2 \hat{\boldsymbol{\beta}}_2 = \mathbf{e} - \mathbf{X}_{2*} \hat{\boldsymbol{\beta}}_2$. The sums of squares of error for the extra variable model is therefore given by

$$\mathbf{u}^T \mathbf{u} = \mathbf{e}^T \mathbf{e} + \hat{\boldsymbol{\beta}}_2^T \left(\mathbf{X}_{2*}^T \mathbf{X}_{2*}\right) \hat{\boldsymbol{\beta}}_2 - 2\hat{\boldsymbol{\beta}}_2 \mathbf{X}_{2*}^T \mathbf{e} = \mathbf{e}^T \mathbf{e} + \hat{\boldsymbol{\beta}}_2^T \left(\mathbf{X}_{2*}^T \mathbf{X}_{2*}\right) \hat{\boldsymbol{\beta}}_2 - 2\hat{\boldsymbol{\beta}}_2^T \mathbf{X}_{2*}^T \mathbf{y}_*.$$

Here, $\mathbf{e}$ is the residual $\mathbf{y} - \mathbf{X}_1 \mathbf{b}$ or $\mathbf{M} \mathbf{y} = \mathbf{y}_*$. We can now, manipulate $\hat{\boldsymbol{\beta}}_2$ to get

$$\mathbf{X}_{2*}^T \mathbf{y}_* = \left(\mathbf{X}_{2*}^T \mathbf{X}_{2*}\right) \hat{\boldsymbol{\beta}}_2 \text{ and}$$

$$\mathbf{u}^T \mathbf{u} = \mathbf{e}^T \mathbf{e} - \hat{\boldsymbol{\beta}}_2^T \left(\mathbf{X}_{2*}^T \mathbf{X}_{2*}\right) \hat{\boldsymbol{\beta}}_2 \leq \mathbf{e}^T \mathbf{e}.$$

Dividing both sides by the total sums of squares, $\mathbf{y}^T \mathbf{M}^0 \mathbf{y}$, we get

$$\frac{\mathbf{u}^T \mathbf{u}}{\mathbf{y}^T \mathbf{M}^0 \mathbf{y}} \leq \frac{\mathbf{e}^T \mathbf{e}}{\mathbf{y}^T \mathbf{M}^0 \mathbf{y}} \Rightarrow R_{X_1, X_2}^2 \geq R_{X_1}^2,$$

where $\mathbf{M}^0 = \mathbf{I} - \mathbf{i}(\mathbf{i}^T \mathbf{i})^{-1} \mathbf{i}^T$. See Greene (2003, p. 30) for a proof for the case when a single variable is added to an existing model.

Thus, it is possible for models to have a high R^2 yet yield poor predictions of new observations for the mean response. It is for this reason that many practitioners also use the adjusted coefficient of variation, R_A^2, which adjusts R^2 with respect to the number of explanatory variables in the model. It is defined as

$$R_A^2 = 1 - \frac{SSE/(n-k-1)}{SST/(n-1)} = 1 - \left(\frac{n-1}{n-k-1}\right)(1 - R^2).$$

In general, it will increase only when significant terms that improve the model are added to the model. On the other hand, it will decrease with the addition of nonsignificant terms to the model. Therefore, it will always be less than or equal to R^2. When the two R^2 measures differ dramatically, there is a good chance that nonsignificant terms have been added to the model.

1.7   HYPOTHESIS TESTING AND CONFIDENCE INTERVALS

The global F test checks the hypothesis that at least one of the k regressors has a significant effect on the response. It does not indicate which explanatory variable has an effect. It is therefore essential to conduct hypothesis tests on the individual coefficients $\beta_j (j = 1, \ldots, k)$. The hypothesis statements are $H_0: \beta_j = 0$ and $H_1: \beta_j \neq 0$. The test statistic for testing this is the ratio of the least squares estimate and the standard error of the estimate. That is,

$$t_0 = \frac{b_j}{s.e.(b_j)}, \quad j = 1, \ldots, k,$$

where $s.e.(b_j)$ is the standard error associated with b_j and is defined as $s.e.(b_j) = \sqrt{\hat{\sigma}^2 C_{jj}}$, where C_{jj} is the jth diagonal element of $(\mathbf{X}^T\mathbf{X})^{-1}$ corresponding to b_j. Under the assumption that the null hypothesis is true, the test statistic t_0 is distributed as a t distribution with $n - k - 1$ degrees of freedom. That is, $t_0 \sim t_{n-k-1}$. The p value is defined as before. That is, $\Pr(|t_0| > t_{n-k-1})$. We reject the null hypothesis if the p value $< \alpha$, where α is the type I error. Note that this test is a marginal test since b_j depends on all the other regressors $x_i (i \neq j)$ that are in the model (see the earlier discussion on interpreting the coefficients).

Hypothesis tests are typically followed by the calculation of confidence intervals. A $100(1 - \alpha)\%$ confidence interval for the regression coefficient $\beta_j (j = 1, \ldots, k)$ is given by

$$b_j - t_{\alpha/2, n-k-1} s.e.(b_j) \leq \beta_j \leq b_j + t_{\alpha/2, n-k-1} s.e.(b_j).$$

Note that these confidence intervals can also be used to conduct the hypothesis tests. In particular, if the range of values for the confidence interval includes zero, then we would fail to reject the null hypothesis.

Two other confidence intervals of interest are the confidence interval for the mean response $E(\mathbf{y}|\mathbf{x}_0)$ and the prediction interval for an observation selected from the conditional distribution $f(\mathbf{y}|\mathbf{x}_0)$, where without loss of generality $f(\bullet)$ is assumed to be normally distributed. Also note that $\mathbf{x}_0$ is the setting of the explanatory variables at which the distribution of $\mathbf{y}$ needs to be evaluated. Notice that the mean of $\mathbf{y}$ at a given value of $\mathbf{x} = \mathbf{x}_0$ is given by $E(y|\mathbf{x}_0) = \mathbf{x}_0^T\boldsymbol{\beta}$.

An unbiased estimator for the mean response is $\mathbf{x}_0^T\mathbf{b}$. That is, $E(\mathbf{x}_0^T\mathbf{b}|\mathbf{X}) = \mathbf{x}_0^T\boldsymbol{\beta}$. It can be shown that the variance of this unbiased estimator is given by $\sigma^2 \mathbf{x}_0^T(\mathbf{X}^T\mathbf{X})^{-1}\mathbf{x}_0$. Using the previously defined estimator for σ^2 (see Section 1.3.1), we can construct a $100(1 - \alpha)\%$ confidence interval on the mean response as

$$\hat{y}(\mathbf{x}_0) - t_{\alpha/2, n-k-1}\sqrt{\hat{\sigma}^2 \mathbf{x}_0^T(\mathbf{X}^T\mathbf{X})^{-1}\mathbf{x}_0} \leq \mu_{y|\mathbf{x}_0} \leq \hat{y}(\mathbf{x}_0) + t_{\alpha/2, n-k-1}\sqrt{\hat{\sigma}^2 \mathbf{x}_0^T(\mathbf{X}^T\mathbf{X})^{-1}\mathbf{x}_0}.$$

Using a similar method, one can easily construct a $100(1 - \alpha)\%$ prediction interval for a future observation $\mathbf{x}_0$ as

$$\hat{y}(\mathbf{x}_0) - t_{\alpha/2, n-k-1}\sqrt{\hat{\sigma}^2(1 + \mathbf{x}_0^T(\mathbf{X}^T\mathbf{X})^{-1}\mathbf{x}_0)} \leq y(\mathbf{x}_0) \leq \hat{y}(\mathbf{x}_0) + t_{\alpha/2, n-k-1}\sqrt{\hat{\sigma}^2(1 + \mathbf{x}_0^T(\mathbf{X}^T\mathbf{X})^{-1}\mathbf{x}_0)}.$$

In both these cases, the observation vector $\mathbf{x}_0$ is defined as $\mathbf{x}_0 = (1, x_{01}, x_{02}, \ldots, x_{0k})$, where the "1" is added to account for the intercept term.

Notice that the width of the prediction interval at point $\mathbf{x}_0$ is wider than the width of the confidence interval for the mean response at $\mathbf{x}_0$. This is easy to see because the standard error used for the prediction interval is larger than the standard error used for the mean response interval. This should make intuitive sense also since it is easier to predict the mean of a distribution than it is to predict a future value from the same distribution.

1.8   SOME FURTHER NOTES

A key step in regression analysis is residual analysis to check the least squares assumptions. Violation of one or more assumptions can render the estimation and any subsequent hypothesis tests meaningless. As stated earlier, the least squares residuals can be computed as $\mathbf{e} = \mathbf{y} - \mathbf{Xb}$. Simple residual plots can be used to check a number of assumptions. Chapter 2 shows how these plots are constructed. Here, we simply outline the different types of residual plots that can be used.

1. A plot of the residuals in time order can be used to check for the presence of autocorrelation. This plot can also be used to check for outliers.
2. A plot of the residuals versus the predicted value can be used to check the assumption of random, independently distributed errors. This plot (and the residuals versus regressors plots) can be used to check for the presence of heteroscedasticity. This plot can also be used to check for outliers and influential observations.
3. The normal probability plot of the residuals can be used to check any violations from the assumption of normally distributed random errors.

2

REGRESSION ANALYSIS USING PROC IML AND PROC REG

2.1 INTRODUCTION

We discussed basic regression concepts and least squares theory in Chapter 1. This chapter deals with conducting regression analysis calculations in SAS. We will show the computations by using both Proc IML and Proc Reg. Even though the results from both procedures are identical, using Proc IML allows one to understand the mechanics behind the calculations that were discussed in the previous chapter. Freund and Littell (2000) offer an in-depth coverage of how SAS can be used to conduct regression analysis in SAS. This chapter discusses the basic elements of Proc Reg as it relates to conducting regression analysis.

To illustrate the computations in SAS, we will make use of the investment equation data set provided in Greene (2003). The source of the data is attributed to the *Economic Report of the President* published by the U.S. Government Printing Office in Washington, D.C. The author's description of the problem appears on page 21 of his text and is summarized here. The objective is to estimate an investment equation by using GNP (gross national product), and a time trend variable T. Note that T is not part of the original data set but is created in the data step statement in SAS. Initially, we ignore the variables Interest Rate and Inflation Rate since our purpose here is to illustrate how the computations can be carried out using SAS. Additional variables can be incorporated into the analysis with a few minor modifications of the program. We will first discuss conducting the analysis in Proc IML.

2.2 REGRESSION ANALYSIS USING PROC IML

2.2.1 Reading the Data

The source data can be read in a number of different ways. We decided to create temporary SAS data sets from the raw data stored in Excel. However, we could easily have entered the data directly within the data step statement since the size of data set is small. The Proc Import statement reads the raw data set and creates a SAS temporary data set named *invst_equation*. Using the approach taken by Greene (2003), the data step statement that follows creates a trend variable T, and it also converts the variables investment and GNP to real terms by dividing them by the CPI (consumer price index). These two variables are then scaled so that the measurements are now scaled in terms of trillions of dollars. In a subsequent example, we will make full use of the investment data set by regressing real investment against a constant, a trend variable, GNP, interest rate, and inflation rate that is computed as a percentage change in the CPI.

```
proc import out=invst_equation
    datafile="C:\Temp\Invest_Data"
    dbms=Excel
```

Applied Econometrics Using the SAS® System, by Vivek B. Ajmani
Copyright © 2009 John Wiley & Sons, Inc.

```
      replace;
      getnames=yes;
run;
data invst_equation;
      set invst_equation;
      T=_n_;
      Real_GNP=GNP/CPI*10);
      Real_Invest=Invest/(CPI*10);
run;
```

2.2.2   Analyzing the Data Using Proc IML

Proc IML begins with the statement "Proc IML;" and ends with the statement "Run;". The analysis statements are written between these two. The first step is to read the temporary SAS data set variables into a matrix. In our example, the data matrix **X** contains two columns: T and Real_GNP. Of course, we also need a column of 1's to account for the intercept term. The response vector **y** contains the variable Real_Invest. The following statements are needed to create the data matrix and the response vector.

```
use invst_equation;
read all var {'T' 'Real_GNP'} into X;
read all var {'Real_Invest'} into Y;
```

Note that the model degrees of freedom are the number of columns of **X** excluding the column of 1's. Therefore, it is a good idea to store the number of columns in **X** at this stage. The number of rows and columns of the data matrix are calculated as follows:

```
n=nrow(X);
k=ncol(X);
```

A column of 1's is now concatenated to the data matrix to get the matrix in analysis ready format.

```
X=J(n,1,1)||X;
```

The vector of coefficients can now easily be calculated by using the following set of commands:

```
C=inv(X'*X);
B_Hat=C*X'*Y;
```

Note that we decided to compute $(\mathbf{X}^T \mathbf{X})^{-1}$ separately since this matrix is used frequently in other computations, and it is convenient to have it calculated just once and ready to use.

With the coefficient vector computed, we can now focus our attention on creating the ANOVA table. The following commands compute the sums of squares (regression, error, total), the error degrees of freedom, the mean squares, and the F statistic.

```
SSE=y'*y-B_Hat'*X'*Y;
DFE=n-k-1;
MSE=SSE/DFE;
Mean_Y=Sum(Y)/n;
SSR=B_Hat'*X'*Y-n*Mean_Y**2;
MSR=SSR/k;
SST=SSR+SSE;
F=MSR/MSE;
```

Next, we calculate the coefficient of determination (R^2) and the adjusted coefficient of determination ($adj\ R^2$).

```
R_Square=SSR/SST;
Adj_R_Square=1-(n-1)/(n-k-1) * (1-R_Square);
```

We also need to calculate the standard errors of the regression estimates in order to compute the *t*-statistic values and the corresponding *p* values. The function PROBT will calculate the probability that a random variable from the *t* distribution with d*f* degrees of freedom will exceed a given *t* value. Since the function takes in only positive values of *t*, we need to use the absolute value function *abs*. The value obtained is multiplied by '2' to get the *p* value for a two-sided test.

```
SE=SQRT(vecdiag(C)#MSE);
T=B_Hat/SE;
PROBT=2*(1-CDF('T', ABS(T), DFE));
```

With the key statistics calculated, we can start focusing our attention on generating the output. We have found the following set of commands useful in creating a concise output.

```
ANOVA_Table=(k||SSR||MSR||F)//(DFE||SSE||MSE||{.});
STATS_Table=B_Hat||SE||T||PROBT;
Print 'Regression Results for the Investment
Equation';
Print ANOVA_Table (|Colname={DF SS MS F} rowname={Model
Error} format=8.4|);
Print 'Parameter Estimates';
Print STATS_Table (|Colname={BHAT SE T PROBT} rowname={INT
T Real_GNP} format=8.4|);
Print 'The value of R-Square is ' R_Square; (1 format = 8.41);
Print 'The value of Adj R-Square is ' Adj_R_Square;
(1 format = 8.41);
```

These statements produce the results given in Output 2.1. The results of the analysis will be discussed later.

Regression Results for the Investment Equation

ANOVA_TABLE				
	DF	SS	MS	F
MODEL	2.0000	0.0156	0.0078	143.6729
ERROR	12.0000	0.0007	0.0001	.

Parameter Estimates

STATS_TABLE				
	BHAT	SE	T	PROBT
INT	-0.5002	0.0603	-8.2909	0.0000
T	-0.0172	0.0021	-8.0305	0.0000
REAL_GNP	0.6536	0.0598	10.9294	0.0000

	R_SQUARE
The value of R-Square is	0.9599

	ADJ_R_SQUARE
The value of Adj R-Square is	0.9532

OUTPUT 2.1. Proc IML analysis of the investment data.

2.3 ANALYZING THE DATA USING PROC REG

This section deals with analyzing the investment data using Proc Reg. The general form of the statements for this procedure is

```
Proc Reg Data=dataset;
    Model Dependent Variable(s) = Independent Variable(s)/Model
    Options;
Run;
```

See Freund and Littell (2000) for details on other options for Proc Reg and their applications. We will make use of only a limited set of options that will help us achieve our objectives. The dependent variable in the investment data is Real Investment, and the independent variables are Real GNP and the time trend T. The SAS statements required to run the analysis are

```
Proc Reg Data=invst_equation;
    Model Real_Invest=Real_GNP T;
Run;
```

The analysis results are given in Output 2.2. Notice that the output from Proc Reg matches the output from Proc IML.

2.3.1 Interpretation of the Output (Freund and Littell, 2000, pp. 17–24)

The first few lines of the output display the name of the model (Model 1, which can be changed to a more appropriate name), the dependent variable, and the number of observations read and used. These two values will be equal unless there are missing observations in the data set for either the dependent or the independent variables or both. The investment equation data set has a total of 15 observations and there are no missing observations.

The analysis of variance table lists the standard output one would expect to find in an ANOVA table: the sources of variation, the degrees of freedom, the sums of squares for the different sources of variation, the mean squares associated with these, the

```
The REG Procedure
Model: MODEL1
Dependent Variable: Real_Invest
```

Number of Observations Read	15
Number of Observations Used	15

			Analysis of Variance		
Source	DF	Sum of Squares	Mean Square	F Value	Pr > F
Model	2	0.01564	0.00782	143.67	<0.0001
Error	12	0.00065315	0.00005443		
Corrected Total	14	0.01629			

Root MSE	0.00738	R-Square	0.9599
Dependent Mean	0.20343	Adj R-Sq	0.9532
Coeff Var	3.62655		

			Parameter Estimates		
Variable	DF	Parameter Estimate	Standard Error	t Value	Pr > |t|
Intercept	1	-0.50022	0.06033	-8.29	<0.0001
Real_GNP	1	0.65358	0.05980	10.93	<0.0001
T	1	-0.01721	0.00214	-8.03	<0.0001

OUTPUT 2.2. Proc Reg analysis of the investment data.

F-statistic value, and the *p* value. As discussed in Chapter 1, the degrees of freedom for the model are k, the number of independent variables, which in this example is 2. The degrees of freedom for the error sums of squares are $n - k - 1$, which is $15 - 2 - 1$ or 12. The total degrees of freedom are the sum of the model and error degrees of freedom or $n - 1$, the number of nonmissing observations minus one. In this example, the total degrees of freedom are 14.

  i. In Chapter 1, we saw that the total sums of squares can be partitioned into the model and the error sums of squares. That is, the Corrected Total Sums of Squares = Model Sums of Squares + Error Sums of Squares. From the ANOVA table, we see that $0.01564 + 0.00065$ equals 0.01629.
  ii. The mean squares are calculated by dividing the sums of squares by their corresponding degrees of freedom. If the model is correctly specified, then the mean square for error is an unbiased estimate of σ^2, the variance of ϵ, and the error term of the linear model. From the ANOVA table,

$$MSR = \frac{0.01564}{2} = 0.00782$$

and

$$MSE = \frac{0.00065315}{12} = 0.00005443.$$

  iii. The *F*-statistic value is the ratio of the mean square for regression and the mean square for error. From the ANOVA table,

$$F = \frac{0.00782}{0.00005443} = 143.67.$$

It tests the hypothesis that

$$H_0: \quad \beta_1 = \beta_2 = 0,$$
$$H_1: \quad \text{At least one of the } \beta\text{'s} \neq 0.$$

Here, β_1 and β_2 are the true regression coefficients for Real GNP and Trend. Under the assumption that the null hypothesis is true,

$$F = \frac{MSR}{MSE} \sim F_{2,12}$$

and the

$$p \text{ value} = \Pr(F > F_{2,12}) = \Pr(F > 143.67) \approx 0.$$

The *p* value indicates that there is almost no chance of obtaining an *F*-statistic value as high or higher than 143.67 under the null hypothesis. Therefore, the null hypothesis is rejected and we claim that the overall model is significant.

The root MSE is the square root of the mean square error and is an estimate of the standard deviation of $\epsilon(\sqrt{0.00005443} = 0.00738)$. The dependent mean is simply the mean of the dependent variable Real Invest. *Coeff Var* is the coefficient of variation and is defined as

$$\frac{root - mse}{dependent - mean} \times 100.$$

As discussed in Meyers (1990, p. 40), this statistic is scale free and can therefore be used in place of the root mean square error (which is not scale free) to assess the quality of the model fit. To see how this is interpreted, consider the investment data set example. In this example, the coefficient of variation is 3.63%, which implies that the dispersion around the least squares line as measured by the root mean square error is 3.63% of the overall mean of Real Invest.

The coefficient of determination (R^2) is 96%. This implies that the regression model explains 96% of the variation in the dependent variable. As explained in Chapter 1, it is calculated by dividing the model sums of squares by the total sums of squares and expressing the result as a percentage ($0.01564/0.01629 = 0.96$). The adjusted R^2 value is an alternative to the R^2 value and takes the number of parameters into account. In our example, the adjusted $R^2 = 95.32\%$. This is calculated as

$$R_A^2 = 1 - \frac{SSE/(n-k-1)}{SST/(n-1)} = 1 - \left(\frac{n-1}{n-k-1}\right)(1-R^2)$$

$$= 1 - \frac{14}{12} \times (1 - 0.96) = 0.9533.$$

Notice that the values of R^2 and adjusted R^2 are very close.

The parameter estimates table lists the intercept and the independent variables along with the estimated values of the coefficients, their standard errors, the t-statistic values, and the p values.

i. The first column gives the estimated values of the regression coefficients. From these, we can write the estimated model as

Estimated Real_Invest $= -0.50 + 0.65$ Real_GNP $- 0.017$ T.

The coefficient for Real_GNP is positive, indicating a positive correlation between it and Real_Invest. The coefficient value of 0.65 indicates that an increase of one trillion dollars of Real GNP would lead to an average of 0.65 trillion dollars of Real Investment (Greene, 2003). Here, we have to assume that Time (T) is held constant.

ii. The standard error column gives the standard errors for the coefficient estimates. These values are the square root of the diagonal elements of $\hat{\sigma}^2(\mathbf{X}^T\mathbf{X})^{-1}$. These are used to conduct hypothesis tests for the regression parameters and to construct confidence intervals.

iii. The t value column lists the t statistics used for testing

$$H_0: \quad \beta_i = 0,$$
$$H_1: \quad \beta_i \neq 0, \; i = 1, 2.$$

These are calculated by dividing the estimated coefficient values by their corresponding standard error values. For example, the t value corresponding to the coefficient for *Real_GNP* is $0.65358/0.05980 = 10.93$. The last column gives the p values associated with the t-test statistic values. As an example, the p value for *Real_GNP* is given by $P(|t| > 10.93)$. Using the t table with 12 degrees of freedom, we see that the p value for *Real_GNP* is zero, indicating high significance. In the real investment example, the p values for both independent variables offer strong evidence against the null hypothesis.

2.4   EXTENDING THE INVESTMENT EQUATION MODEL TO THE COMPLETE DATA SET

We will now extend this analysis by running a regression on the complete Investment Equation data set. Note that the CPI in 1967 was recorded as 79.06 (Greene, 2003, p. 947) and that Inflation Rate is defined as the percentage change in CPI. The following data step gets the data in analysis-ready format.

```
Data invst_equation;
    set invst_equation;
    T=_n_;
    Real_GNP=GNP/(CPI*10);
    Real_Invest=Invest/(CPI*10);
    CPI0=79.06;
    Inflation_Rate=100*((CPI-Lag(CPI))/Lag(CPI));
    if inflation_rate=. then inflation_rate=100*((CPI-
    79.06)/79.06);
    drop Year GNP Invest CPI CPI0;
run;
```

The REG Procedure
Model: MODEL1
Dependent Variable: Real_Invest

Number of Observations Read	15
Number of Observations Used	15

Analysis of Variance					
Source	DF	Sum of Squares	Mean Square	F Value	Pr > F
Model	4	0.01586	0.00397	91.83	<0.0001
Error	10	0.00043182	0.00004318		
Corrected Total	14	0.01629			

Root MSE	0.00657	R-Square	0.9735
Dependent Mean	0.20343	Adj R-Sq	0.9629
Coeff Var	3.23018		

Parameter Estimates						
Variable	Label	DF	Parameter Estimate	Standard Error	t Value	Pr > |t|
Intercept	Intercept	1	-0.50907	0.05393	-9.44	<0.0001
Real_GNP		1	0.67030	0.05380	12.46	<0.0001
T		1	-0.01659	0.00193	-8.60	<0.0001
Interest	Interest	1	-0.00243	0.00119	-2.03	0.0694
Inflation_Rate		1	0.00006320	0.00132	0.05	0.9627

OUTPUT 2.3. Proc Reg analysis of complete investment equation data.

The data can be analyzed using Proc IML or Proc Reg with only minor modifications of the code already presented. The following Proc Reg statements can be used. The analysis results are given in Output 2.3.

```
Proc reg data=invst_equation;
      model Real_Invest=Real_GNP T Interest Inflation_Rate;
Run;
```

The output indicates that both the Real_GNP and the time trend T are highly significant at the 0.05 type I error level. The variable Interest is significant at the 0.10 type I error rate, whereas Inflation Rate is not significant. The coefficients for both Real_GNP and T have the same signs as their signs in the model where they were used by themselves. The coefficient values for these variables are also very close to the values obtained in the earlier analysis. Notice that the values of the two coefficients of determination terms have now increased slightly.

2.5   PLOTTING THE DATA

Preliminary investigation into the nature of the correlation between the explanatory and dependent variables can easily be done by using simple scatter plots. In fact, we suggest that plotting the independent variables versus the dependent variable be the first step in any regression analysis project. A simple scatter plot offers a quick snapshot of the underlying relationship between the two variables and helps in determining the model terms that should be used. For instance, it will allow us to determine if a transformation of the independent variable or dependent variable or both should be used. SAS offers several techniques for producing bivariate plots. The simplest way of plotting two variables is by using the Proc Plot procedure. The general statements for this procedure are as follows:

```
Proc Plot data=dataset;
      Plot dependent_variable*independent_variable;
Run;
```

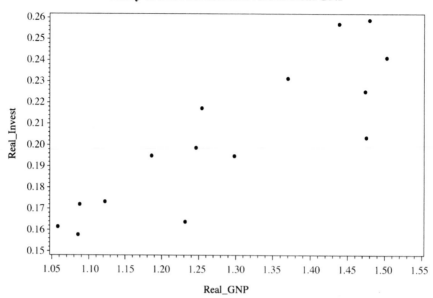

FIGURE 2.1. Plot of Real_Invest versus Real_GNP.

Proc Gplot is recommended if the intent is to generate high-resolution graphs. Explaining all possible features of Proc Gplot is beyond the scope of this book. However, we have found the following set of statements adequate for producing the basic high-resolution plots. The following statements produce a plot of Real_Invest versus Real_GNP (see Figure 2.1). Note that the size of the plotted points and the font size of the title can be adjusted by changing the "height=" and "h=" options.

```
proc gplot data=invst_equation;
      plot Real_Invest*Real_GNP
      /haxis=axis1
      vaxis=axis2;
      symbol1 value=dot c=black height=2;
      axis1 label=('Real_GNP');
      axis2 label=(angle=90 'Real_Invest');
      title2 h=4 'Study of Real Investment versus GNP';
run;
```

The statements can be modified to produce a similar plot for Real_Invest versus Time (*T*) (Figure 2.2).

Both plots indicate a positive correlation between the independent and dependent variables and also do not indicate any outliers or influential points. Later in this chapter, we will discuss constructing plots for the confidence intervals of the mean response and of predictions. We will also look at some key residual plots to help us validate the assumptions of the linear model.

2.6   CORRELATION BETWEEN VARIABLES

For models with several independent variables, it is often useful to examine relationships between the independent variables and between the independent and dependent variables. This is accomplished by using Proc Corr procedure. The general form of this procedure is

```
Proc Corr data=dataset;
      Var variables;
Run;
```

Study of Real Investment versus Time

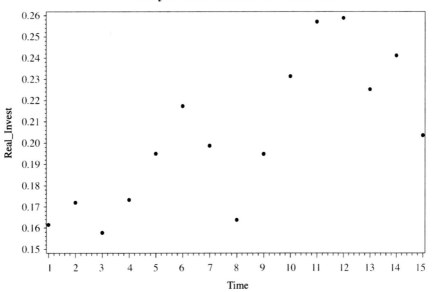

FIGURE 2.2. Plot of Real_Invest versus Time.

For example, if we want to study the correlation between all the variables in the investment equation model, we would use the statements

```
proc corr data=invst_equation;
    var Real_Invest Real_GNP T;
Run;
```

The analysis results are given in Output 2.4.

The first part of the output simply gives descriptive statistics of the variables in the model. The correlation coefficients along with their *p* values are given in the second part of the output. Notice that the estimated correlation between Real_Invest and Real_GNP is 0.86 with a highly significant *p* value. The correlation between Time Trend and Real_Invest is 0.75 and is also highly significant. Note that the correlation between the independent variables is 0.98, which points to multicollinearity problems with

The CORR Procedure

| 3 | Variables: | Real_Invest Real_GNP | T |

Simple Statistics						
Variable	N	Mean	Std Dev	Sum	Minimum	Maximum
Real_Invest	15	0.20343	0.03411	3.05151	0.15768	0.25884
Real_GNP	15	1.28731	0.16030	19.30969	1.05815	1.50258
T	15	8.00000	4.47214	120.00000	1.00000	15.00000

Pearson Correlation Coefficients, N = 15 Prob > |r| under H0: Rho=0			
	Real_Invest	Real_GNP	T
Real_Invest	1.00000	0.86283 <0.0001	0.74891 0.0013
Real_GNP	0.86283 <0.0001	1.00000	0.97862 <0.0001
T	0.74891 0.0013	0.97862 <0.0001	1.00000

OUTPUT 2.4. Correlation analysis of the investment equation data.

this data set. The problem of multicollinearity in regression analysis will be dealt with in later sections. However, notice that the scatter plot between Real_Invest and Time Trend indicated a positive relationship between the two (the Proc Corr output confirms this), but the regression coefficient associated with Time Trend is negative. Such contradictions sometimes occur because of multicollinearity.

2.7 PREDICTIONS OF THE DEPENDENT VARIABLE

One of the main objectives of regression analysis is to compute predicted values of the dependent variable at given values of the explanatory variables. It is also of interest to calculate the standard errors of these predicted values, confidence interval on the mean response, and prediction intervals. The following SAS statements can be used to generate these statistics (Freund and Littel, 2000, pp. 24–27).

```
Proc Reg Data=invst_equation;
    Model Real_Invest=Real_GNP T/p clm cli r;
Run;
```

The option 'p' calculates the predicted values and their standard errors, 'clm' calculates 95% confidence interval on the mean response, 'cli' generates 95% prediction intervals, and 'r' calculates the residuals and conducts basic residuals analysis. The above statements produce the results given in Output 2.5.

The first set of the output consists of the usual Proc Reg output seen earlier. The next set of output contains the analysis results of interest for this section. The column labeled Dependent Variable gives the observed values of the dependent variable, which is Real_Invest. The next column gives the predicted value of the dependent variable $\hat{y}$ and is the result of the 'p' option in Proc Reg. The next three columns are the result of using the 'clm' option. We get the standard error of the conditional mean at each observation, $E(\mathbf{y} \mid \mathbf{x}_0)$, and the 95% confidence interval for this. As explained in Chapter 1, the standard error of this conditional expectation is given by $\hat{\sigma}\sqrt{\mathbf{x}_0^T(\mathbf{X}^T\mathbf{X})^{-1}\mathbf{x}_0^T}$. Therefore, the 95% confidence interval is given by

$$\hat{y} \pm t_{0.025,\, n-k-1}\hat{\sigma}\sqrt{\mathbf{x}_o^T(\mathbf{X}^T\mathbf{X})^{-1}\mathbf{x}_0^T}.$$

The REG Procedure
Model: MODEL1
Dependent Variable: Real_Invest

Number of Observations Read	15
Number of Observations Used	15

Analysis of Variance					
Source	DF	Sum of Squares	Mean Square	F Value	Pr > F
Model	2	0.01564	0.00782	143.67	<0.0001
Error	12	0.00065315	0.00005443		
Corrected Total	14	0.01629			

Root MSE	0.00738	R-Square	0.9599
Dependent Mean	0.20343	Adj R-Sq	0.9532
Coeff Var	3.62655		

Parameter Estimates					
Variable	DF	Parameter Estimate	Standard Error	t Value	Pr > |t|
Intercept	1	-0.50022	0.06033	-8.29	<0.0001
Real_GNP	1	0.65358	0.05980	10.93	<0.0001
T	1	-0.01721	0.00214	-8.03	<0.0001

OUTPUT 2.5. Prediction and mean response intervals for the investment equation data.

				Output Statistics					
Obs	Dependent Variable	Predicted Value	Std Error Mean Predict	95% CL Mean		95% CL Predict		Residual	Std Error Residual	Student Residual
1	0.1615	0.1742	0.003757	0.1660	0.1823	0.1561	0.1922	-0.0127	0.00635	-1.993
2	0.1720	0.1762	0.003324	0.1690	0.1835	0.1586	0.1939	-0.004215	0.00659	-0.640
3	0.1577	0.1576	0.003314	0.1504	0.1648	0.1400	0.1752	0.0000746	0.00659	0.0113
4	0.1733	0.1645	0.002984	0.1580	0.1710	0.1472	0.1818	0.008823	0.00675	1.308
5	0.1950	0.1888	0.002330	0.1837	0.1939	0.1719	0.2056	0.006207	0.00700	0.887
6	0.2173	0.2163	0.003055	0.2096	0.2229	0.1989	0.2337	0.001035	0.00672	0.154
7	0.1987	0.1938	0.001988	0.1895	0.1981	0.1772	0.2105	0.004913	0.00710	0.692
8	0.1638	0.1670	0.003839	0.1586	0.1754	0.1489	0.1851	-0.003161	0.00630	-0.502
9	0.1949	0.1933	0.002433	0.1880	0.1986	0.1764	0.2102	0.001559	0.00696	0.224
10	0.2314	0.2229	0.002223	0.2180	0.2277	0.2061	0.2397	0.008547	0.00703	1.215
11	0.2570	0.2507	0.003599	0.2428	0.2585	0.2328	0.2685	0.006360	0.00644	0.988
12	0.2588	0.2602	0.004045	0.2514	0.2690	0.2419	0.2785	-0.001348	0.00617	-0.219
13	0.2252	0.2394	0.002990	0.2328	0.2459	0.2220	0.2567	-0.0142	0.00674	-2.100
14	0.2412	0.2409	0.003272	0.2337	0.2480	0.2233	0.2584	0.000314	0.00661	0.0475
15	0.2036	0.2059	0.004995	0.1950	0.2168	0.1865	0.2253	-0.002290	0.00543	-0.422

	Output Statistics	
Obs	-2-1 0 1 2	Cook's D
1	| ***| |	0.464
2	| *| |	0.035
3	| | |	0.000
4	| |** |	0.112
5	| |* |	0.029
6	| | |	0.002
7	| |* |	0.012
8	| *| |	0.031
9	| | |	0.002
10	| |** |	0.049
11	| |* |	0.102
12	| | |	0.007
13	| ****| |	0.289
14	| | |	0.000
15	| | |	0.050

Sum of Residuals	0
Sum of Squared Residuals	0.00065315
Predicted Residual SS (PRESS)	0.00099715

OUTPUT 2.5. (*Continued*)

Here, x_0 is the row vector of $\mathbf{X}$ corresponding to a single observation and $\hat{\sigma}$ is the root mean square error. The residual column is also produced by the 'p' option and is simply

$$observed\_value - predicted\_value.$$

The 'cli' option produces the 95% prediction intervals corresponding to each row x_0 of $\mathbf{X}$. As explained in Chapter 1, this is calculated by using the formula

$$\hat{y} \pm t_{0.025,\, n-k-1} \hat{\sigma} \sqrt{1 + \mathbf{x}_0^T (\mathbf{X}^T \mathbf{X})^{-1} \mathbf{x}_0^T}.$$

The 'r' option in Proc Reg does a residual analysis and produces the last five columns of the output. The actual residuals along with their corresponding standard errors are reported. This is followed by the standardized residual that is defined as e/σ_e. Here, e is the residual, and σ_e is the standard deviation of the residuals and is given by the square root of $\sigma_e = (1-h_{ii})\hat{\sigma}^2$ (Meyers, 1990 p. 220), where h_{ii} is the ith diagonal element of $\mathbf{X}(\mathbf{X}^T\mathbf{X})^{-1}\mathbf{X}^T$ and $\hat{\sigma}^2$ is an estimate of σ^2. Note that the standardized residuals corresponding to the 1st and 13th observations appear to be high. The graph columns of the output are followed by Cook's statistics, which measure how influential a point is. Cook's statistic or Cook's D is a measure of one change in the parameter estimate $\hat{\boldsymbol{\beta}}$ when the ith observation is deleted. If we define $\mathbf{d}_i = \hat{\boldsymbol{\beta}} - \hat{\boldsymbol{\beta}}_{(i)}$, where $\hat{\boldsymbol{\beta}}_{(i)}$ is the parameter estimate one without the ith observation, then Cook's D for the ith observation is defined as (Meyers, 1990, p. 260)

$$\text{Cook's } D_i = \frac{\mathbf{d}_i^T(\mathbf{X}^T\mathbf{X})^{-1}\mathbf{d}_i}{(k+1)\hat{\sigma}^2},$$

where k is the number of parameters in the model. A large value of the Cook's D statistic is typically used to declare a point influential.

Confidence intervals for the mean response and predicted values can be plotted fairly easily using Proc Gplot or by using the plotting features within Proc Reg. Here, we give an example of plotting the two confidence intervals within the Proc Reg statements. The following statements produce the plot for the mean interval (Freund and Littell, 2000, pp. 45–46).

```
Proc Reg Data=invst_equation;
    Model Real_Invest=Real_GNP T;
    plot p.*p. uclm.*p. lclm.*p./overlay;
run;
```

These statements produce the results given in Figure 2.3.
The prediction interval can be plotted by simply replacing the plot statements with

```
plot p.*p. ucl.*p. lcl.*p./overlay;
```

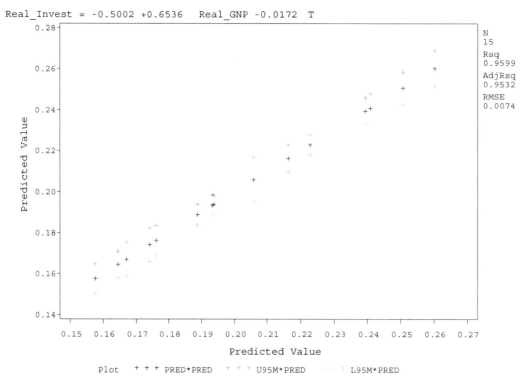

FIGURE 2.3. Proc Reg output with graphs of mean intervals.

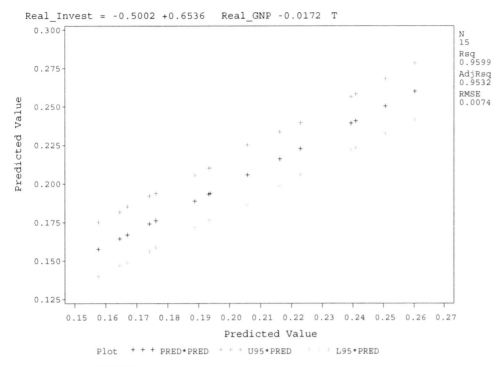

FIGURE 2.4. Proc Reg output with graphs of prediction intervals.

Of course, one can have both plot statements in the Proc Reg module to simultaneously create both plots. The prediction interval produced is given in Figure 2.4.

Notice that the prediction interval is wider than the confidence interval for the mean response since the variability in predicting a future observation is higher than the variability in predicting the mean response.

2.8 RESIDUAL ANALYSIS

Residual analysis is done to check the various assumptions underlying regression analysis. Failure of one or more assumptions may render a model useless for the purpose of hypothesis testing and predictions. The residual analysis is typically done by plotting the residuals. Commonly used residual graphs are

a. Residuals plotted in time order
b. Residuals versus the predicted value
c. Normal probability plot of the residuals.

We will use the investment equation regression analysis to illustrate creating these plots in SAS. To plot the residuals in time order, we have simply plotted the residuals versus the time trend variable since this captures the time order. The following statement added to the Proc Reg module will generate this plot (Freund and Littell, 2000, pp. 49–50).

```
plot r.*T;
```

Replacing "r." by "student." will create a trend chart of the standardized residuals (Figure 2.5).

Note that barring points 1 and 13, the residuals appear to be random over time. These two points were also highlighted in the influential point's analysis. To generate the residuals versus predicted response plot, use

```
plot student.*p.;
```

or

```
plot r.*p.;
```

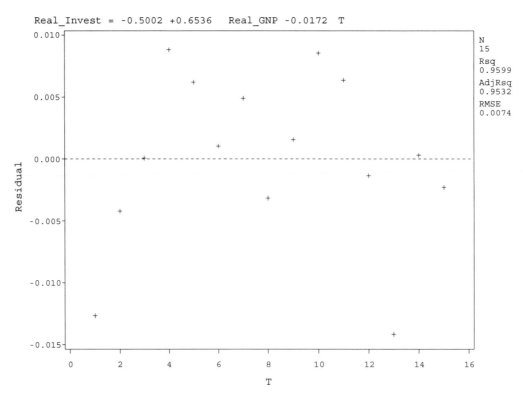

FIGURE 2.5. Plot of residuals versus time.

Note that two residual points appear to be anomalies and may need to be investigated further (Figure 2.6). It turns out that these points are data points 1 and 13. An ideal graph here is a random scatter of plotted points. A funnel-shaped graph here indicates heteroscedastic variance—that is, a model where the variance is dependent upon the conditional expectation $E(\mathbf{y} \mid \mathbf{X})$. Therefore, as $E(\mathbf{y} \mid \mathbf{X})$ changes, so does the variance.

To generate the normal probability plot of the residuals, we first create an output data set containing the residuals using the following code:

```
proc reg data=invst_equation;
    model Real_Invest=Real_GNP T;
    output out=resid r=resid;
Run;
```

The output data set then serves as an input to the Proc Univariate module in the following statements. The "probplot/normal(mu=0 sigma=1)" requests the calculated percentiles for the plots to be based on the standard normal distribution. It should be apparent that this option can be used to request probability plots based on other distributions. The plot is produced in Figure 2.7.

```
proc univariate data=resid;
    var resid;
    probplot/normal(mu=0 sigma=1);
run;
```

Note that barring the points around the 5th and 10th percentiles (which again are data points 1 and 13), the data appear to fall on a straight line and therefore we can be fairly confident that the residuals are distributed as a standard Normal distribution.

Real_Invest = -0.5002 +0.6536   Real_GNP -0.0172   T

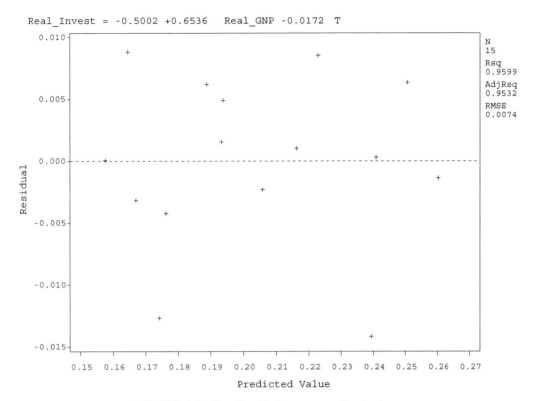

FIGURE 2.6. Plot of residuals versus predicted values.

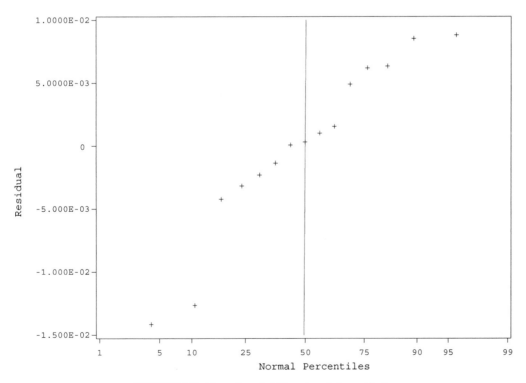

FIGURE 2.7. Normal probability plot of the residuals.

2.9   MULTICOLLINEARITY

The problem of multicollinearity was discussed earlier. We now provide more details about multicollinearity and discuss ways of detecting it using Proc Reg. Multicollinearity is a situation where there is a high degree of correlation between the explanatory variables in a model. This often arises in data mining projects (for example, in models to predict consumer behavior) where several hundred variables are screened to determine a subset that appears to best predict the response of interest. It may happen (and it often does) that many variables measure similar phenomena. As an example, consider modeling the attrition behavior of consumers with respect to Auto & Home insurance products. Three variables that could be studied are the number of premium changes, the number of positive premium changes, and the number of negative premium changes over the life of the policy holder's tenure with the company. We should expect the number of premium changes to be positively correlated with the number of positive (negative) premium changes. Including all three in the model will result in multicollinearity. So, what does multicollinearity do to our analysis results? First, note that the existence of multicollinearity does not lead to violations of any of the fundamental assumptions of regression analysis that were discussed in Chapter 1. That is, multicollinearity does not impact the estimation of the least squares estimator. However, it does limit the usefulness of the results. We can illustrate this by means of a simple example involving regression analysis with two explanatory variables. It is easy to show that the variance of the least squares estimator in this simple case is (Greene, 2003, p. 56)

$$Var(b_k) = \frac{\sigma^2}{(1-r_{12}^2)\sum_{i=1}^{n}(x_{ik}-\bar{x}_k)^2}, \qquad k=1,2.$$

Here, r_{12}^2 is the correlation between the two explanatory variables. It is clear that the higher the correlation between the two variables, the higher the variance of the estimator b_k. A consequence of the high variance is that explanatory variables that in reality are correlated with the response may appear insignificant. That is, the t-statistic value corresponding to the estimator will be underestimated. Another consequence are incorrect signs on the coefficients and/or really large coefficient values.

It can be shown that in a multiple regression with k explanatory variables, the variance of the least squares estimator b_k can be written as (Greene, 2003, p. 57)

$$Var(b_k) = \frac{\sigma^2}{(1-R_k^2)\sum_{i=1}^{n}(x_{ik}-\bar{x}_k)^2}, \qquad k=1,2,\ldots.$$

Here, R_k^2 is the R^2 value when x_k is regressed against the remaining $k-1$ explanatory variables. As discussed by the author,

a. The greater the correlation between x_k and other variables, the higher the variance of b_k.
b. The greater the variance in x_k, the lower the variance of b_k.
c. The better the model fit (the lower the σ^2), the lower the variance of b_k.

We will make use of the gasoline consumption data from Greene (2003) to illustrate how multicollinearity in the data is detected using SAS. The original source of this data set is the *Economic Report of the President* as published by the U.S. Government Printing Office in Washington, D.C. The objective is to conduct a regression of gasoline consumption on the price of gasoline, income, the price of new cars, and the price of used cars. All the variables were transformed using the log transformation. The hypothesized model and a general explanation of the problem are given in Greene (2003, p. 12).

There are three sets of statistics that may be used to determine the severity of multicollinearity problem. These statistics are as follows (Freund and Littell, 2000, p. 97; Meyer, 1990, pp. 369–370):

a. Comparing the significance of the overall model versus the significance of the individual parameter estimates.
b. Variance inflation factors (VIF) associated with each parameter estimate.
c. Analysis of the eigenvalues of the $\mathbf{X}^T\mathbf{X}$ matrix.

The following statements can be used to generate these statistics. The analysis results are given in Output 2.6.

```
proc reg data=clean_gas;
    model Ln_G_Pop=Ln_Pg Ln_Inc Ln_Pnc Ln_Puc/vif collinoint;
run;
```

The REG Procedure
Model: MODEL1
Dependent Variable: Ln_G_Pop

Number of Observations Read	36
Number of Observations Used	36

Analysis of Variance					
Source	DF	Sum of Squares	Mean Square	F Value	Pr > F
Model	4	0.78048	0.19512	243.18	<0.0001
Error	31	0.02487	0.00080237		
Corrected Total	35	0.80535			

Root MSE	0.02833	R-Square	0.9691
Dependent Mean	-0.00371	Adj R-Sq	0.9651
Coeff Var	-763.79427		

Parameter Estimates						
Variable	DF	Parameter Estimate	Standard Error	t Value	Pr > |t|	Variance Inflation
Intercept	1	-7.78916	0.35929	-21.68	<0.0001	0
Ln_Pg	1	-0.09788	0.02830	-3.46	0.0016	12.75251
Ln_Inc	1	2.11753	0.09875	21.44	<0.0001	4.20156
Ln_Pnc	1	0.12244	0.11208	1.09	0.2830	78.88071
Ln_Puc	1	-0.10220	0.06928	-1.48	0.1502	83.11980

Collinearity Diagnostics (intercept adjusted)						
			Proportion of Variation			
Number	Eigenvalue	Condition Index	Ln_Pg	Ln_Inc	Ln_Pnc	Ln_Puc
1	3.71316	1.00000	0.00541	0.01429	0.00088422	0.00084893
2	0.22345	4.07647	0.01482	0.81855	0.00704	0.00351
3	0.05701	8.07008	0.96334	0.01903	0.02941	0.03434
4	0.00638	24.11934	0.01644	0.14813	0.96266	0.96130

OUTPUT 2.6. Multicollinearity output for the gasoline consumption data.

The first two tables give the standard OLS regression statistics. The second table adds the variance inflation factor values for the regressors. The third table gives information about $\mathbf{X}^T\mathbf{X}$. From the first table, we see that the model is highly significant with an F-statistic value of 176.71 and a p value < 0.0001. However, examining the second table reveals p values of the regressors ranging from 0.032 to 0.126—much larger than the overall model significance. This is one problem associated with multicollinearity, that is, high model significance without any corresponding highly significant explanatory variables. However, notice that both R^2 values are high, indicating a good model fit—a contradiction. The correlation coefficients among the four regressors were created using Proc Corr and is given in Output 2.7.

The values below the coefficients are the p values associated with the null hypothesis of zero correlation. The regressors have strong correlations among them, with the price of used and new cars having the highest correlation—in fact, the price of used and new cars almost have a perfect correlation. It is not surprising, therefore, that the variation inflation factors associated with the two regressors is high (74.44, 84.22).

In general, variance inflation factors are useful in determining which variables contribute to multicollinearity. As given in Meyers (1990, p. 127) and Freund and Littell (2000, p. 98), the VIF associated with the kth regressor is given by $1/(1-R_k^2)$, where R_k^2 is the R^2 value when x_k is regressed against the other $k-1$ regressors. It can be shown (see Freund and Wilson, 1998) that the variance of b_k is inflated by a factor equal to the VIF of x_k in the presence of multicollinearity than in the absence of multicollinearity. Although there are no formal rules for determining what a cutoff should be for calling a VIF large, there are a few recommended approaches. As discussed in Freund and Littell (2000), many practitioners first compute $1/(1-R^2)$, where R^2 is the

The CORR Procedure

4	Variables:	Ln_Pg	Ln_Inc	Ln_Pnc	Ln_Puc

Simple Statistics						
Variable	N	Mean	Std Dev	Sum	Minimum	Maximum
Ln_Pg	36	0.67409	0.60423	24.26740	-0.08992	1.41318
Ln_Inc	36	3.71423	0.09938	133.71230	3.50487	3.82196
Ln_Pnc	36	0.44320	0.37942	15.95514	-0.00904	1.03496
Ln_Puc	36	0.66361	0.63011	23.89004	-0.17913	1.65326

Pearson Correlation Coefficients, N = 36 Prob > |r| under H0: Rho=0				
	Ln_Pg	Ln_Inc	Ln_Pnc	Ln_Puc
Ln_Pg	1.00000	0.84371 <0.0001	0.95477 <0.0001	0.95434 <0.0001
Ln_Inc	0.84371 <0.0001	1.00000	0.82502 <0.0001	0.84875 <0.0001
Ln_Pnc	0.95477 <0.0001	0.82502 <0.0001	1.00000	0.99255 <0.0001
Ln_Puc	0.95434 <0.0001	0.84875 <0.0001	0.99255 <0.0001	1.00000

OUTPUT 2.7. Proc Corr output of the independent variables in the gasoline consumption data.

coefficient of determination of the original model. In the example used, $1/(1 - R^2) = 23.81$. Regressors with VIF values greater than this are said to be more closely related to other independent variables than the dependent variable. In the gasoline consumption example, both *Ln_Pnc* and *Ln_Puc* have VIFs greater than 23.81. Furthermore, both have large *p* values and are therefore suspected of contributing to multicollinearity.

Let us now take a look at the output produced with the COOLINOINT option. The output produced contains the eigenvalues of the correlation matrix of the regressors along with the proportion of variation each regressor explains for the eigenvalues. The eigenvalues are ranked from highest to lowest. The extent or severity of the multicollinearity problem is evident by examining the size of the eigenvalues. For instance, big differences among the eigenvalues (large variability) indicate a higher degree of multicollinearity. Furthermore, small eigenvalues indicate near-perfect linear dependencies or high multicollinearity (Freund and Littell, 2000, pp. 100–101; Meyers, 1990, p. 370). In the example used, the eigenvalues corresponding to car prices are very small. The square root of the ratio of the largest eigenvalue to the smallest eigenvalue is given by the last element in the condition number column. In general, a large condition number indicates a high degree of multicollinearity. The condition number for the gasoline consumption analysis is 24.13 and indicates a high degree of multicollinearity. See Meyer (1990, p. 370) for a good discussion of condition numbers and how they are used to detect multicollinearity.

The Proportion of Variation output can be used to identify the variables which are highly correlated. The values measure the percentage contribution of the variance of the estimates toward the eigenvalues (Freund and Littell, 2000). As stated earlier, small eigenvalues indicate near-perfect correlations. As discussed in Meyer (2000, p. 372), a subset of explanatory variables with high contributions to the eigenvalues should be suspected of being highly correlated. As an example, the 4th eigenvalue is very small in magnitude (0.00638), and roughly 85% of the variation in both *Ln_Puc* and *Ln_Pnc* is associated with this eigenvalue. Therefore, these two are suspected (rightly so) of being highly correlated.

In reality, most econometric studies will be impacted by some correlation between the explanatory variables. In our experience, we have not found a clear and common fix to combat multicollinearity problems. An approach that we have found useful is to isolate the variables that are highly correlated and then prioritize the variables in terms of their importance to business needs. Variables that have a low priority are then dropped from further analysis. Of course, the prioritization of these variables is done after discussions with the business partners in marketing, finance, and so on. Arbitrarily dropping a variable from the model is not recommended (see Chapter 4) as it may lead to omitted variables bias.

3

HYPOTHESIS TESTING

3.1 INTRODUCTION

Chapters 1 and 2 introduced the concept of hypothesis testing in regression analysis. We looked at the "Global" F test, which tested the hypothesis of model significance. We also discussed the t tests for the individual coefficients in the model. We will now extend these to testing the joint hypothesis of the coefficients and also to hypothesis tests involving linear combinations of the coefficients. This chapter will conclude with a discussion on testing data for structural breaks and for stability over time.

3.1.1 The General Linear Hypothesis

Hypothesis testing on regression parameters, subsets of parameters, or a linear combination of the parameters can be done by considering a set of linear restrictions on the model $y = X\beta + \varepsilon$. These restrictions are of the form $C\beta = d$, where C is a $j \times k$ matrix of j restrictions on the k parameters $(j \leq k)$, β is the $k \times 1$ vector of coefficients, and d is a $j \times 1$ vector of constants. Note that here k is used to denote the number of parameters in the regression model. The ith restriction equation can be written as (Greene, 2003, p. 94; Meyers, 1990, p. 103)

$$c_{i1}\beta_1 + c_{i2}\beta_2 + \cdots + c_{ik}\beta_k = d_i \quad \text{for} \quad i = 1, \ldots, j$$

To see the general form of C, consider the following hypothetical model:

$$y = \beta_1 + \beta_2 X_1 + \beta_3 X_2 + \beta_4 X_3 + \beta_5 X_4 + \beta_6 X_5 + \varepsilon.$$

A linear restriction of the form $\beta_2 - \beta_3 = 0$ can be written as

$$0 \times \beta_1 + 1 \times \beta_2 - 1 \times \beta_3 + 0 \times \beta_4 + 0 \times \beta_5 + 0 \times \beta_6 = 0.$$

The C matrix is therefore given by $C = \begin{bmatrix} 0 & 1 & -1 & 0 & 0 & 0 \end{bmatrix}$ and the vector d is given by $d = [0]$.

Applied Econometrics Using the SAS® System, by Vivek B. Ajmani
Copyright © 2009 John Wiley & Sons, Inc.

3.1.2 Hypothesis Testing for the Linear Restrictions

We can very easily conduct a hypothesis test for a set of j linear restrictions on the linear model. The hypothesis statements are

$$H_0: \quad \mathbf{C\beta} - \mathbf{d} = \mathbf{0},$$
$$H_1: \quad \mathbf{C\beta} - \mathbf{d} \neq \mathbf{0}.$$

To see how the hypothesis test statements are written, consider the same general linear model as before. To test the hypothesis $H_0: \beta_3 = 0$, we need $\mathbf{C} = \begin{bmatrix} 0 & 0 & 1 & 0 & 0 & 0 \end{bmatrix}$ and $\mathbf{d} = [0]$. Note that this is equivalent to the t tests for the individual parameters that were discussed in Chapters 1 and 2. To test the hypothesis $H_0: \beta_4 = \beta_5$, we need $\mathbf{C} = \begin{bmatrix} 0 & 0 & 0 & 1 & -1 & 0 \end{bmatrix}$, and $\mathbf{d} = [0]$. To test several linear restrictions $H_0: \beta_2 + \beta_3 = 1, \beta_4 + \beta_6 = 1, \beta_5 + \beta_6 = 0$, we need

$$\mathbf{C} = \begin{bmatrix} 0 & 1 & 1 & 0 & 0 & 0 \\ 0 & 0 & 0 & 1 & 0 & 1 \\ 0 & 0 & 0 & 0 & 1 & 1 \end{bmatrix} \quad \text{and} \quad \mathbf{d} = \begin{bmatrix} 1 \\ 1 \\ 0 \end{bmatrix} \quad \text{(Greene, 2003, p. 96)}.$$

3.1.3 Testing the General Linear Hypothesis

We will now consider testing the general linear hypothesis. First note that the least squares estimator of $\mathbf{C\beta} - \mathbf{d}$ is given by $\mathbf{Cb} - \mathbf{d}$, where $\mathbf{b}$ is the least squares estimator of $\mathbf{\beta}$. It can be shown that this estimator is unbiased. That is, $E(\mathbf{Cb} - \mathbf{d} \mid \mathbf{X}) = CE(\mathbf{b} \mid \mathbf{X}) - \mathbf{d} = \mathbf{C\beta} - \mathbf{d}$. Its variance–covariance matrix is given by

$$
\begin{aligned}
Var(\mathbf{Cb} - \mathbf{d} | \mathbf{X}) &= Var(\mathbf{Cb} | \mathbf{X}) \\
&= \mathbf{C}Var(\mathbf{b} | \mathbf{X})\mathbf{C}^T \\
&= \sigma^2 \mathbf{C}(\mathbf{X}^T\mathbf{X})^{-1}\mathbf{C}^T.
\end{aligned}
$$

The test statistic for the linear restriction hypothesis is based on the F statistic given by (Greene, 2003, p. 97; Meyers, 1990, p. 105)

$$F = \frac{(\mathbf{Cb} - \mathbf{d})^T (\mathbf{C}(\mathbf{X}^T\mathbf{X})^{-1}\mathbf{C}^T)^{-1}(\mathbf{Cb} - \mathbf{d})}{s^2 j},$$

where s^2 is the mean square error and is estimated from the regression model. This test statistic can easily be derived by realizing that the F statistic is a ratio of two independent chi-squared random variables divided by their degrees of freedom. It is trivial to show that the distribution of χ_A^2 (defined below) has a chi-squared distribution with j degrees of freedom (Graybill, 2000). That is,

$$\chi_A^2 = \frac{(\mathbf{Cb} - \mathbf{d})^T (\mathbf{C}(\mathbf{X}^T\mathbf{X})^{-1}\mathbf{C}^T)^{-1}(\mathbf{Cb} - \mathbf{d})}{\sigma^2} \sim \chi_j^2.$$

Also note that the statistic $\chi_B^2 = (n-k)s^2/\sigma^2$ has a chi-squared distribution with $n-k$ degrees of freedom. Taking the ratio of χ_A^2 and χ_B^2 and dividing them by their degrees of freedom, we get the F statistic given above.

It is easy to show that $\mathbf{b}$ is independent of s^2, which in turn gives us the independence of the two chi-squared random variables, χ_A^2 and χ_B^2. It can also be shown that if the null hypothesis is true, the test statistic F is distributed as a F distribution with j numerator and $n-k$ denominator degrees of freedom. For testing the hypothesis of the ith linear restriction of the form

$$H_0: c_{i1}\beta_1 + c_{i2}\beta_2 + \cdots + c_{ik}\beta_k = \mathbf{c}_i^T \mathbf{\beta} = \mathbf{d}_i \quad (i = 1, \dots, j)$$

we can use the estimate $\hat{\mathbf{d}}_i = \mathbf{c}_i^T \mathbf{b}$ and the test statistic

$$t = \frac{\hat{\mathbf{d}}_i - \mathbf{d}}{s.e.(\hat{\mathbf{d}}_i)},$$

where

$$s.e.(\hat{\mathbf{d}}_i) = \sqrt{\mathbf{c}_i^T (s^2 (\mathbf{X}^T \mathbf{X})^{-1})^{-1} \mathbf{c}_i}.$$

Under the assumption that the null hypothesis is true, the test statistic t is distributed as a t distribution with $n-k$ degrees of freedom.

Note that we need not do anything special to test the hypothesis with a single linear restriction. That is, the F test can still be used for this since under the null hypothesis for the single restriction case $j = 1$ so that $F \sim F_{1,n-k}$. Also note that the relationship between the t and F statistic is given by $t^2 \sim F_{1,n-k}$.

3.2   USING SAS TO CONDUCT THE GENERAL LINEAR HYPOTHESIS

To illustrate the computations in SAS, consider the quarterly data on real investment, real GDP, an interest rate, and inflation measured by change in the log of CPI given in Greene (2003). The data are credited to the Department of Commerce, BEA. As discussed by Greene (2003, pp. 93 and 98), the model suggested for these data is a simple model of investment, I_t given by

$$\ln(I_t) = \beta_1 + \beta_2 i_t + \beta_3 \Delta p_t + \beta_4 \ln(Y_t) + \beta_5 t + \varepsilon_t,$$

which hypothesizes that investment depends upon the nominal interest rates, i_t; the rate of inflation, Δp_t; (the log of) real output, $\ln(Y_t)$; and the trend component, t. Next, consider the joint hypothesis

$$\beta_2 + \beta_3 = 0,$$
$$\beta_4 = 1,$$
$$\beta_5 = 0.$$

As discussed in Greene (2003), these restrictions test whether investments depend on the real interest rate, whether the marginal effect of real outputs equals 1, and whether there is a time trend. The $\mathbf{C}$ matrix and $\mathbf{d}$ vector are given by

$$\mathbf{C} = \begin{bmatrix} 0 & 1 & 1 & 0 & 0 \\ 0 & 0 & 0 & 1 & 0 \\ 0 & 0 & 0 & 0 & 1 \end{bmatrix} \quad \text{and} \quad \mathbf{d} = \begin{bmatrix} 0 \\ 1 \\ 0 \end{bmatrix}.$$

Proc IML can easily be used to compute the F-statistic value to test this hypothesis. The following statements show how to compute the F statistic and the associated p value. The data are first read into matrices as was shown in Chapter 2. The first set of statements is used to define the $\mathbf{C}$ matrix and $\mathbf{d}$ vector and to store the number of restrictions in the variable j; this is simply the number of rows in $\mathbf{C}$ (Note that the Proc IML statments make use of the notation given in Greene (2003).)

```
R={0 1 1 0 0,0 0 0 1 0,0 0 0 0 1};
q={0,1,0};
j=nrow(R);
```

The next set of statements is used to calculate the discrepancy vector $\mathbf{Cb} - \mathbf{d}$, the F-statistic value, and the corresponding p value. Note that $C = (\mathbf{X}^T \mathbf{X})^{-1}$ in the Proc IML code.

```
DisVec=R*B_Hat-q;
F=DisVec'*inv(R*MSE*C*R')*DisVec/j;
P=1-ProbF(F,J,n-k);
Print 'The value of the F Statistic is ' F;
Print 'The P-Value associated with this is ' P;
```

Notice that the least squares estimator B_Hat and the estimate for σ^2 are computed using the methods described in Chapter 2. These calculations yield an F value of 109.84 with a p value of 0. We can therefore reject the null hypothesis and claim that at least one of the three restrictions is false. We can then proceed with testing the individual restrictions by using the t test described earlier. Slight modification of the above code allows us to calculate the t-statistic values. A bit later, we will use the *restrict* statement in Proc Reg to calculate the t-statistic values. Recall that the t statistic is given by

$$t = \frac{\hat{\mathbf{d}}_i - \mathbf{d}}{s.e.(\hat{\mathbf{d}}_i)},$$

where $\hat{\mathbf{d}} = \mathbf{c}_i^T \mathbf{b}$, $\mathbf{d}_i = \mathbf{c}_i^T \boldsymbol{\beta}$ and $s.e.(\hat{\mathbf{d}}_i) = \sqrt{\mathbf{c}_i^T (s^2 (\mathbf{X}^T \mathbf{X})^{-1})^{-1} \mathbf{c}_i}$. Adding the following code to the code already provided will allow us to conduct the t test. The first step is to create the individual restrictions $\mathbf{c}$ and constant vectors $\mathbf{d}$ (Again, using the same notation as given in Greene (2003).)

```
R1=R[1,]; q1=q[1,];
R2=R[2,]; q2=q[2,];
R3=R[3,]; q3=q[3,];
```

We now calculate the individual estimates for the restrictions of interest and also calculate the standard errors for the estimated values. The t-statistic value is simply a ratio of these two values.

```
T_NUM1=R1*B_Hat-q1;
se1=sqrt(R1*MSE*C*R1');
T1=T_NUM1/se1;
p1=2*(1-CDF('T', abs(T1),n-k));
Print 'The value of the T Statistic for the first
restriction is ' t1;
Print 'The P-Value associated with this is ' P1;
T_NUM2=R2*B_Hat-q2;
se2=sqrt(R2*MSE*C*R2');
T2=T_NUM2/se2;
P2=2*(1-CDF('T', abs(T2),n-k));
Print 'The value of the T Statistic for the second
restriction is ' t2;
Print 'The P-Value associated with this is ' P2;
T_NUM3=R3*B_Hat-q3;
se3=sqrt(R3*MSE*C*R3');
T3=T_NUM3/se3;
P3=2*(1-CDF('T', abs(T3),n-k));
Print 'The value of the T Statistic for the third
restriction is ' t3;
Print 'The P-Value associated with this is ' P3;
```

The analysis results from the Proc IML statements are given in Output 3.1. Based on the results of the individual test statistic, we would expect the second and third hypotheses to be rejected.

	T1
The value of the T Statistic for the first restriction is	−1.843672

	P1
The P-Value associated with this is	0.0667179

	T2
The value of the T Statistic for the second restriction is	5.0752636

	P2
The P-Value associated with this is	8.855E-7

	T3
The value of the T Statistic for the third restriction is	−3.802964

	P3
The P-Value associated with this is	0.0001901

OUTPUT 3.1. Proc IML output of quarterly investment data.

3.3   THE RESTRICTED LEAST SQUARES ESTIMATOR

In this section, we discuss the computations involved in calculating the restricted least squares estimator, $\mathbf{b}^*$, given by (Greene, 2003, p.100)

$$\mathbf{b}^* = \mathbf{b} - (\mathbf{X}^T\mathbf{X})^{-1}\mathbf{C}^T(\mathbf{C}(\mathbf{X}^T\mathbf{X})^{-1}\mathbf{C}^T)^{-1}(\mathbf{C}\mathbf{b}-\mathbf{d}).$$

First note that the restricted least squares estimator is unbiased (under the null hypothesis, assuming that $\mathbf{C}\boldsymbol{\beta} = \mathbf{d}$), with variance–covariance matrix given by

$$Var(\mathbf{b}^*|\mathbf{X}) = \sigma^2(\mathbf{X}^T\mathbf{X})^{-1}[\mathbf{I} - \mathbf{C}^T(\mathbf{C}(\mathbf{X}^T\mathbf{X})^{-1}\mathbf{C}^T)^{-1}\mathbf{C}(\mathbf{X}^T\mathbf{X})^{-1}].$$

The unbiased property can easily be verified by noticing that $E(\mathbf{C}\mathbf{b}|\mathbf{X}) = \mathbf{C}\boldsymbol{\beta}$. Therefore, $E(\mathbf{b}^*|\mathbf{X}) = \boldsymbol{\beta}$ because the last term in the expression of $E(\mathbf{b}^*|\mathbf{X})$ is zero. To derive the expression for the variance, first write

$$\mathbf{b}^* - \boldsymbol{\beta} = \mathbf{b} - (\mathbf{X}^T\mathbf{X})^{-1}\mathbf{C}^T(\mathbf{C}(\mathbf{X}^T\mathbf{X})^{-1}\mathbf{C}^T)^{-1}(\mathbf{C}\mathbf{b}-\mathbf{d}) - \boldsymbol{\beta}.$$

Next, recall that the OLS estimator can be written as $\mathbf{b} = \boldsymbol{\beta} + (\mathbf{X}^T\mathbf{X})^{-1}\mathbf{X}^T\boldsymbol{\varepsilon}$. Substituting this in the above expression gives (after some algebra) $\mathbf{b}^* - \boldsymbol{\beta} = \mathbf{M}^*(\mathbf{X}^T\mathbf{X})^{-1}\mathbf{X}^T\boldsymbol{\varepsilon}$, where

$$\mathbf{M}^* = \mathbf{I} - (\mathbf{X}^T\mathbf{X})^{-1}\mathbf{C}^T[\mathbf{C}(\mathbf{X}^T\mathbf{X})^{-1}\mathbf{C}^T]^{-1}\mathbf{C}.$$

The variance of $\mathbf{b}^*$ is given by

$$\begin{aligned} Var(\mathbf{b}^*|\mathbf{X}) &= E\lfloor(\mathbf{b}^* - E(\mathbf{b}^*|\mathbf{X}))(\mathbf{b}^* - E(\mathbf{b}^*|\mathbf{X}))^T|\mathbf{X}\rfloor \\ &= E[\mathbf{M}^*(\mathbf{X}^T\mathbf{X})^{-1}\mathbf{X}^T\boldsymbol{\varepsilon}\boldsymbol{\varepsilon}^T\mathbf{X}(\mathbf{X}^T\mathbf{X})^{-1}\mathbf{M}^{*T}|\mathbf{X}] \\ &= \mathbf{M}^*(\mathbf{X}^T\mathbf{X})^{-1}\mathbf{X}^T E(\boldsymbol{\varepsilon}\boldsymbol{\varepsilon}^T|\mathbf{X})\mathbf{X}(\mathbf{X}^T\mathbf{X})^{-1}\mathbf{M}^{*T} \\ &= \sigma^2\mathbf{M}^*(\mathbf{X}^T\mathbf{X})^{-1}\mathbf{M}^{*T}. \end{aligned}$$

Substituting the expression for $\mathbf{M}^*$ in this equation gives the original expression for $Var(\mathbf{b}^*|\mathbf{X})$. We can easily calculate these using Proc IML. For simplicity, we will use the first restriction in the investment equation example ($\beta_2 + \beta_3 = 0$) to illustrate these computations. The following code will compute the restricted least squares estimates:

```
B_Star=B_Hat-C*R'*inv(R*C*R')*(R*B_Hat-q);
```

The restricted least squares estimators are $\beta_1 = -7.907$, $\beta_2 = -0.0044$, $\beta_3 = 0.0044$, $\beta_4 = 1.764$, and $\beta_5 = -0.0044$. Use the following code to compute the variance–covariance matrix of the restricted least squares estimator.

```
temp=I(5)-R'*inv(R*C*R')*R*C;
VarCov_Star=mse*c*Temp;
print VarCov_Star;
SE=J(5,1,0);
do i=1 to 5;
     SE[i,1]=sqrt(VarCov_Star[i,i]);
end;
```

Output 3.2 contains both the variance–covariance matrix and the standard errors of the restricted least squares estimator.

As we mentioned before, these computations are also available in the Proc Reg module. The following statements can be used (Freund and Littell, 2000, pp. 41–42). The analysis results are given in Output 3.3.

```
proc reg data=Rest_Invst_Eq;
     model Invest=interest delta_p output T;
     restrict interest+delta_p=0;
run;
```

Note that Proc Reg also provides the t test for the restriction of interest. If we use the 'Test' statement instead of the 'Restrict' statement, we will get the OLS estimates of the parameters followed by the F test on the restriction. The statement and output (Output 3.4) are given below. Also note that the p value for the test on the single restriction matches up to what was obtained using Proc IML. At the 5% type 1 error level, we would reject the null hypothesis that the sum of the coefficients equal 0.

```
test interest+delta_p=0;
```

VARCOV_STAR				
1.4243433	−0.000427	0.0004268	−0.190468	0.0015721
−0.000427	5.0923E-6	−5.092E-6	0.0000573	−5.556E-7
0.0004268	−5.092E-6	5.0923E-6	−0.000057	5.5561E-7
−0.190468	0.0000573	−0.000057	0.0254728	−0.00021
0.0015721	−5.556E-7	5.5561E-7	−0.00021	1.7499E-6

SE
1.1934585
0.0022566
0.0022566
0.159602
0.0013228

OUTPUT 3.2. Proc IML output of the variance–covariance matrix of the restricted least squares estimator.

The REG Procedure
Model: MODEL1
Dependent Variable: Invest

Note: Restrictions have been applied to parameter estimates.

Number of Observations Read	203
Number of Observations Used	203

Analysis of Variance					
Source	DF	Sum of Squares	Mean Square	F Value	Pr > F
Model	3	71.13325	23.71108	3154.48	<0.0001
Error	199	1.49581	0.00752		
Corrected Total	202	72.62906			

Root MSE	0.08670	R-Square	0.9794
Dependent Mean	6.30947	Adj R-Sq	0.9791
Coeff Var	1.37410		

Parameter Estimates					
Variable	DF	Parameter Estimate	Standard Error	t Value	Pr > |t|
Intercept	1	−7.90716	1.20063	−6.59	<0.0001
interest	1	−0.00443	0.00227	−1.95	0.0526
delta_p	1	0.00443	0.00227	1.95	0.0526
output	1	1.76406	0.16056	10.99	<0.0001
T	1	−0.00440	0.00133	−3.31	0.0011
RESTRICT	−1	−4.77085	2.60324	−1.83	0.0667*

* Probability computed using beta distribution.

OUTPUT 3.3. Proc Reg output for restricted least squares of the quarterly investment data.

3.4 ALTERNATIVE METHODS OF TESTING THE GENERAL LINEAR HYPOTHESIS

We now discuss two alternate ways of testing the linear restriction hypothesis presented in the previous section. First, note that by definition and construction of the OLS estimator, the restricted least squares estimator cannot be better than the ordinary least squares estimator in terms of the error sums of squares. This is because the OLS estimators are the best linear unbiased estimators of the parameters in the regression model. If we let $\mathbf{e}^{*T}\mathbf{e}^*$ denote the error sums of squares associated with the restricted least squares estimator and let $\mathbf{e}^T\mathbf{e}$ denote the error sums of squares of the ordinary least squares estimator, then $\mathbf{e}^{*T}\mathbf{e}^* \geq \mathbf{e}^T\mathbf{e}$. It can be shown that (Greene, 2003, p.102; Meyers, 1990, pp. 108–109)

$$\mathbf{e}^{*T}\mathbf{e}^* - \mathbf{e}^T\mathbf{e} = (\mathbf{Cb}-\mathbf{d})^T[\mathbf{C}(\mathbf{X}^T\mathbf{X})^{-1}\mathbf{C}^T](\mathbf{Cb}-\mathbf{d}),$$

so that the original F test can be restated as

$$F = \frac{(\mathbf{e}^{*T}\mathbf{e}^* - \mathbf{e}^T\mathbf{e})/j}{\mathbf{e}^T\mathbf{e}/(n-k)},$$

The REG Procedure
Model: MODEL1
Dependent Variable: Invest

Number of Observations Read	203
Number of Observations Used	203

Analysis of Variance					
Source	DF	Sum of Squares	Mean Square	F Value	Pr > F
Model	4	71.15850	17.78962	2395.23	<0.0001
Error	198	1.47057	0.00743		
Corrected Total	202	72.62906			

Root MSE	0.08618	R-Square	0.9798
Dependent Mean	6.30947	Adj R-Sq	0.9793
Coeff Var	1.36589		

Parameter Estimates					
Variable	DF	Parameter Estimate	Standard Error	t Value	Pr > |t|
Intercept	1	−9.13409	1.36646	−6.68	<0.0001
interest	1	−0.00860	0.00320	−2.69	0.0077
delta_p	1	0.00331	0.00234	1.41	0.1587
output	1	1.93016	0.18327	10.53	<0.0001
T	1	−0.00566	0.00149	−3.80	0.0002

Test 1 Results for Dependent Variable Invest				
Source	DF	Mean Square	F Value	Pr > F
Numerator	1	0.02525	3.40	0.0667
Denominator	198	0.00743		

OUTPUT 3.4. Proc Reg output using the test statement.

which under the null hypothesis is distributed with an F distribution with j numerator and $n - k$ denominator degrees of freedom. If we divide the numerator and denominator of this F statistic by the total uncorrected sums of squares, we get

$$F = \frac{(R^2 - R^{*2})/j}{(1 - R^2)/(n - k)}.$$

These three statistics give us the same value for testing the linear restriction hypothesis. That is,

$$F = \frac{(\mathbf{Cb} - \mathbf{d})^T (\mathbf{C}(s^2(\mathbf{X}^T\mathbf{X})^{-1})\mathbf{C}^T)^{-1}(\mathbf{Cb} - \mathbf{d})}{j} = \frac{(\mathbf{e}^{*T}\mathbf{e}^* - \mathbf{e}^T\mathbf{e})/j}{\mathbf{e}^T\mathbf{e}/(n - k)} = \frac{(R^2 - R^{*2})/j}{(1 - R^2)/(n - k)}.$$

We will illustrate these computations by analyzing the production function data given in Greene (2003). The data are credited to Aigner et al. (1977) and Hildebrand and Liu (1957). As discussed by the author on pages 102–103, the objective is to determine if the Cobb–Douglas model given by

$$\ln(Y) = \beta_1 + \beta_2\ln(L) + \beta_3\ln(K) + \varepsilon$$

is more appropriate for the data than the translog model given by

$$\ln(Y) = \beta_1 + \beta_2\ln(L) + \beta_3\ln(K) + \beta_4(\tfrac{1}{2}\ln^2 L) + \beta_5(\tfrac{1}{2}\ln^2 K) + \beta_6\ln(L)\ln(K) + \varepsilon.$$

Here, Y is the output produced, L is the labor and K is the capital involved. The Cobb–Douglas model is produced by the restriction $\beta_4 = \beta_5 = \beta_6 = 0$. The following SAS code will give results for both the translog and the Cobb–Douglas models. (The third Proc Reg module uses the restrict statement that gives us the Cobb–Douglas model also.) Outputs 3.5 through 3.7 contain the analysis results. Output 3.7 contains the Cobb–Douglas model results from Proc Reg using the restrict statement. (Note that the parameter estimates for the Cobb–Douglas portion of the output matches the output produced for the Cobb–Douglas model in Output 3.5.)

```
proc import out=Prod_Func
     datafile="C:\Temp\TableF61"
     dbms=Excel Replace;
     getnames=yes;
run;
Data Prod_Func;
     set Prod_Func;
     LnY=log(ValueAdd);
     LnL=Log(Labor);
     LnK=Log(Capital);
     LPrime=0.5*LnL*LnL;
     KPrime=0.5*LnK*LnK;
```

The REG Procedure
Model: MODEL1
Dependent Variable: LnY

Number of Observations Read	27
Number of Observations Used	27

Analysis of Variance					
Source	DF	Sum of Squares	Mean Square	F Value	Pr > F
Model	2	14.21156	7.10578	200.25	<0.0001
Error	24	0.85163	0.03548		
Corrected Total	26	15.06320			

Root MSE	0.18837	R-Square	0.9435
Dependent Mean	7.44363	Adj R-Sq	0.9388
Coeff Var	2.53067		

Parameter Estimates					
Variable	DF	Parameter Estimate	Standard Error	t Value	Pr > |t|
Intercept	1	1.17064	0.32678	3.58	0.0015
LnL	1	0.60300	0.12595	4.79	<0.0001
LnK	1	0.37571	0.08535	4.40	0.0002

OUTPUT 3.5. Regression analysis of production data Cobb–Douglas model.

```
        CProd=LnL*LnK;
run;
proc reg data=Prod_Func;
        model LnY=LnL LnK;
run;
proc reg data=Prod_Func;
        model LnY=LnL LnK LPrime KPrime CProd;
run;
proc reg data=Prod_Func;
        model LnY=LnL LnK LPrime KPrime CProd;
        restrict LPrime=0;
        restrict KPrime=0;
        restrict CProd=0;
run;
```

To test the hypothesis that the translog model is more appropriate, we can use the F test given by

$$F = \frac{(\mathbf{e}^{*T}\mathbf{e}^{*} - \mathbf{e}^{T}\mathbf{e})/j}{\mathbf{e}^{T}\mathbf{e}/(n-k)}.$$

The REG Procedure
Model: MODEL1
Dependent Variable: LnY

Number of Observations Read	27
Number of Observations Used	27

Analysis of Variance					
Source	DF	Sum of Squares	Mean Square	F Value	Pr > F
Model	5	14.38327	2.87665	88.85	<0.0001
Error	21	0.67993	0.03238		
Corrected Total	26	15.06320			

Root MSE	0.17994	R-Square	0.9549
Dependent Mean	7.44363	Adj R-Sq	0.9441
Coeff Var	2.41733		

Parameter Estimates					
Variable	DF	Parameter Estimate	Standard Error	t Value	Pr > |t|
Intercept	1	0.94420	2.91075	0.32	0.7489
LnL	1	3.61364	1.54807	2.33	0.0296
LnK	1	−1.89311	1.01626	−1.86	0.0765
LPrime	1	−0.96405	0.70738	−1.36	0.1874
KPrime	1	0.08529	0.29261	0.29	0.7735
CProd	1	0.31239	0.43893	0.71	0.4845

OUTPUT 3.6. Regression analysis of production data-translog model.

The REG Procedure
Model: MODEL1
Dependent Variable: LnY

Note: Restrictions have been applied to parameter estimates.

Number of Observations Read	27
Number of Observations Used	27

Analysis of Variance					
Source	DF	Sum of Squares	Mean Square	F Value	Pr > F
Model	2	14.21156	7.10578	200.25	<0.0001
Error	24	0.85163	0.03548		
Corrected Total	26	15.06320			

Root MSE	0.18837	R-Square	0.9435
Dependent Mean	7.44363	Adj R-Sq	0.9388
Coeff Var	2.53067		

Parameter Estimates					
Variable	DF	Parameter Estimate	Standard Error	t Value	Pr > |t|
Intercept	1	1.17064	0.32678	3.58	0.0015
LnL	1	0.60300	0.12595	4.79	<0.0001
LnK	1	0.37571	0.08535	4.40	0.0002
LPrime	1	8.94539E-17	0	Infty	<0.0001
KPrime	1	−1.7828E-18	0	−Infty	<0.0001
CProd	1	−1.0976E-16	0	−Infty	<0.0001
RESTRICT	−1	−0.04266	0.21750	−0.20	0.8493*
RESTRICT	−1	0.49041	0.45811	1.07	0.2940*
RESTRICT	−1	0.28409	0.57683	0.49	0.6325*

* Probability computed using beta distribution.

OUTPUT 3.7. Regression analysis of production data using the restrict statement Cobb–Douglas model.

The Cobb–Douglas model is the restricted model, and the error sums of squares for this is 0.85163. The error sums of squares for the unrestricted model (translog) is 0.67993. The number of restrictions, j, is 3. The error degrees of freedom for the unrestricted model, $n - k$, is 21. Substituting these values into the F-statistic formula, we get

$$F = \frac{(0.85163 - 0.67993)/3}{0.67993/21} = 1.768.$$

The critical value from the F table is 3.07, so we do not reject the restricted model. We can therefore use the Cobb–Douglas model for the production data set. Note that, using the F statistic given by

$$F = \frac{(R^2 - R^{*2})/j}{(1 - R^2)/(n - k)},$$

we get

$$F = \frac{(0.9549 - 0.9435)/3}{(1 - 0.9549)/21} = 1.768,$$

which is equivalent to the result received when we used the F test with the error sums of squares.

3.5   TESTING FOR STRUCTURAL BREAKS IN DATA

We can extend the linear restriction hypothesis test from the previous sections to test for structural breaks in the time period within which the data set was collected. We illustrate this by using the gasoline consumption data set given in Greene (2003). The data set is credited to the Council of Economic Advisors, Washington, D.C. The author gives a description of the model for the gasoline consumption data on page 12 of his text. The data consists of several variables, which includes the total U.S. gasoline consumption (G), computed as total expenditure divided by a price index from 1960 to 1995, the gasoline price index, disposable income, the price of used and new cars, and so on. A time series plot of G is given in Figure 3.1.

The plot clearly shows a break in the U.S. gasoline consumption behavior after 1973. As pointed out in Greene (2003, p.130), up to 1973, fuel was abundant with stable worldwide gasoline prices. An embargo in 1973 caused a shift marked by shortages and rising prices.

Consider then a model of the log of per capita gasoline consumption (G/Pop) with respect to the log of the price index of gasoline (Pg), the log of per capita disposable income (Y), the log of the price index of new cars (Pnc), and the log of the price index of used cars (Puc). The regression model is given by (Greene, 2003, p. 136)

$$\ln(G/Pop) = \beta_1 + \beta_2 \ln(Pg) + \beta_3 \ln(y) + \beta_4 \ln(Pnc) + \beta_5 \ln(Puc) + \varepsilon.$$

We would expect that the entire relationship described by this regression model was shifted starting 1974. Let us denote the first 14 years of the data in $\mathbf{y}$ and $\mathbf{X}$ as $\mathbf{y}_1$ and $\mathbf{X}_1$ and the remaining years as $\mathbf{y}_2$ and $\mathbf{X}_2$. An unrestricted regression that allows the coefficients to be different in the two time periods is given by

$$\begin{bmatrix} \mathbf{y}_1 \\ \mathbf{y}_2 \end{bmatrix} = \begin{bmatrix} \mathbf{X}_1 & \mathbf{0} \\ \mathbf{0} & \mathbf{X}_2 \end{bmatrix} \begin{bmatrix} \boldsymbol{\beta}_1 \\ \boldsymbol{\beta}_2 \end{bmatrix} + \begin{bmatrix} \boldsymbol{\varepsilon}_1 \\ \boldsymbol{\varepsilon}_2 \end{bmatrix}.$$

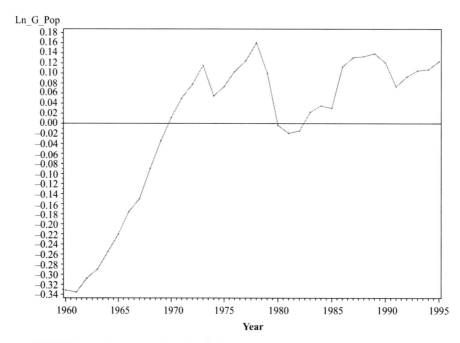

FIGURE 3.1. Time series plot of total U.S. gasoline consumption from 1960 to 1995.

The unrestricted least squares estimator is given by (Greene, 2003, pp. 129–130)

$$\mathbf{b} = (\mathbf{X}^T\mathbf{X})^{-1}\mathbf{X}^T\mathbf{y} = \begin{bmatrix} \mathbf{X}_1^T\mathbf{X}_1 & \mathbf{0} \\ \mathbf{0} & \mathbf{X}_2^T\mathbf{X}_2 \end{bmatrix}^{-1} \begin{bmatrix} \mathbf{X}_1^T\mathbf{y}_1 \\ \mathbf{X}_2^T\mathbf{y}_2 \end{bmatrix} = \begin{bmatrix} \mathbf{b}_1 \\ \mathbf{b}_2 \end{bmatrix},$$

which can be obtained by running two regressions separately. If we denote the two residual sums of squares as $\mathbf{e}_1^T\mathbf{e}_1$ and $\mathbf{e}_2^T\mathbf{e}_2$, then the unrestricted sums of squares is given by $\mathbf{e}^T\mathbf{e} = \mathbf{e}_1^T\mathbf{e}_1 + \mathbf{e}_2^T\mathbf{e}_2$. The restricted least squares estimator can be computed by simply stacking the data and running least squares regression on the stacked data. Let the restricted residual sums of squares be denoted by $\mathbf{e}^{*T}\mathbf{e}^*$. Then, the hypothesis test on whether the restricted model is more appropriate for the data can be conducted by using the F statistic discussed earlier. That is,

$$F = \frac{(\mathbf{e}^{*T}\mathbf{e}^* - \mathbf{e}^T\mathbf{e})/j}{\mathbf{e}^T\mathbf{e}/(n_1 + n_2 - 2k)}.$$

Note that here j is the number of restrictions (or simply the number of columns of $\mathbf{X}_2$), n_1 is the number of observations in the first data set, and n_2 is the number of observations in the second data set. This test is also called the Chow test (Chow, 1960). In the gasoline consumption example, $j = 5, n_1 = 14, n_2 = 22$, and $k = 5$. A bit later, we will discuss how to conduct the Chow test using the Proc Model. For now, let us run three regression analyses on the data set. The first two analyses are for the two separate data sets and the last one is for the combined data set. The output of the analysis on the data set before 1974 is given in Output 3.8. The output of the analysis for the data set after 1973 is given in Output 3.9. The output of the analysis of the combined data is given in Output 3.10.

The REG Procedure
Model: MODEL1
Dependent Variable: Ln_G_Pop

Number of Observations Read	14
Number of Observations Used	14

Analysis of Variance					
Source	DF	Sum of Squares	Mean Square	F Value	Pr > F
Model	4	0.33047	0.08262	647.56	<0.0001
Error	9	0.00115	0.00012758		
Corrected Total	13	0.33162			

Root MSE	0.01130	R-Square	0.9965
Dependent Mean	−0.13830	Adj R-Sq	0.9950
Coeff Var	−8.16742		

Parameter Estimates					
Variable	DF	Parameter Estimate	Standard Error	t Value	Pr > |t|
Intercept	1	−11.32637	1.15233	−9.83	<0.0001
Ln_pg	1	−0.07396	0.16776	−0.44	0.6697
Ln_Income	1	1.25341	0.12914	9.71	<0.0001
Ln_Pnc	1	0.80409	0.12076	6.66	<0.0001
Ln_Puc	1	−0.23754	0.10447	−2.27	0.0491

OUTPUT 3.8. Regression analysis of the gasoline consumption data prior to 1974.

The REG Procedure
Model: MODEL1
Dependent Variable: Ln_G_Pop

Number of Observations Read	22
Number of Observations Used	22

Analysis of Variance					
Source	DF	Sum of Squares	Mean Square	F Value	Pr > F
Model	4	0.05084	0.01271	27.28	<0.0001
Error	17	0.00792	0.00046586		
Corrected Total	21	0.05876			

Root MSE	0.02158	R-Square	0.8652
Dependent Mean	0.08194	Adj R-Sq	0.8335
Coeff Var	26.34150		

Parameter Estimates					
Variable	DF	Parameter Estimate	Standard Error	t Value	Pr > |t|
Intercept	1	–5.59999	3.00646	–1.86	0.0799
Ln_pg	1	–0.20862	0.04898	–4.26	0.0005
Ln_Income	1	0.63704	0.33190	1.92	0.0719
Ln_Pnc	1	0.06903	0.20662	0.33	0.7424
Ln_Puc	1	–0.02426	0.06311	–0.38	0.7054

OUTPUT 3.9. Regression analysis of the gasoline consumption data after 1973.

The sums of squares are as follows: $\mathbf{e}_1^T\mathbf{e}_1 = 0.00115$, $\mathbf{e}_2^T\mathbf{e}_2 = 0.00792$, and $\mathbf{e}^{*T}\mathbf{e}^* = 0.03384$. Therefore, the Chow test statistic is given by

$$F = \frac{(0.03384 - 0.00907)/5}{0.00907/26} = 14.20.$$

The p value associated with this is 9.71×10^{-7} and is highly significant. We can therefore reject the null hypothesis for the restricted model and conclude that there is significant evidence that the regression model changed after 1973.

The above computations are unnecessary since SAS Proc Model can be used to test the structural break hypothesis. Proc Model is typically used to analyze systems of simultaneous equations and seemingly unrelated regression equations where the equations can be linear or nonlinear. However, it can also be used to conduct basic OLS analysis. Since Proc Reg does not have an option to conduct the Chow test, we make use of Proc Model.

The following statements can be used to conduct the test. Output 3.11 contains the results of this analysis. Note the three main components of the Proc Model statements—the list of parameters (there are 5 in the gasoline consumption model), the actual model equation being estimated, and the response variable of interest highlighted with the *Fit* statement. The option "chow=15" requests a test to determine if the data sets before and after the 15th are significantly different.

```
proc model data=clean_gas;
    parm beta1 beta2 beta3 beta4 beta5;
    Ln_G_Pop=beta1 + beta2*Ln_Pg + beta3*Ln_Inc + beta4*Ln_Pnc
```

The REG Procedure
Model: MODEL1
Dependent Variable: Ln_G_Pop

Number of Observations Read	36
Number of Observations Used	36

Analysis of Variance					
Source	DF	Sum of Squares	Mean Square	F Value	Pr > F
Model	4	0.77152	0.19288	176.71	<0.0001
Error	31	0.03384	0.00109		
Corrected Total	35	0.80535			

Root MSE	0.03304	R-Square	0.9580
Dependent Mean	–0.00371	Adj R-Sq	0.9526
Coeff Var	–890.84882		

Parameter Estimates					
Variable	DF	Parameter Estimate	Standard Error	t Value	Pr > |t|
Intercept	1	–12.34184	0.67489	–18.29	<0.0001
Ln_pg	1	–0.05910	0.03248	–1.82	0.0786
Ln_Income	1	1.37340	0.07563	18.16	<0.0001
Ln_Pnc	1	–0.12680	0.12699	–1.00	0.3258
Ln_Puc	1	–0.11871	0.08134	–1.46	0.1545

OUTPUT 3.10. Regression analysis of the combined gasoline consumption data.

```
    +beta5*Ln_Puc;
    Fit Ln_G_Pop/chow=15;
run;
```

Note that the coefficients in this output are for the restricted model. The p value for the Chow test indicates that the models are indeed different between the two time periods.

3.6 THE CUSUM TEST

Model stability can also be tested by using the CUSUM test. See Montgomery (1991), Page (1954), and Woodall and Ncube (1985) for a discussion of the CUSUM procedure in quality control. See the documentation for Proc Autoreg from the SAS Institute, Brown et al. (1975), and Greene (2003) for a discussion of the CUSUM procedure to detect structural breaks in the data. This test is based on the cumulative sum of the least squares residuals. If we let $w_t = \frac{e_t}{\sqrt{v_t}}$ (e_t is the OLS residual at time t) with

$$v_t = 1 + \mathbf{x}_t^T \left[\sum_{i=1}^{t-1} \mathbf{x}_i \mathbf{x}_i^T \right]^{-1} \mathbf{x}_t,$$

The MODEL Procedure

Model Summary	
Model Variables	1
Parameters	5
Equations	1
Number of Statements	1

Model Variables	Ln_G_Pop
Parameters	beta1 beta2 beta3 beta4 beta5
Equations	Ln_G_Pop

The Equation to Estimate is	
Ln_G_Pop =	F(beta1(1), beta2(Ln_pg), beta3(Ln_Income), beta4(Ln_Pnc), beta5(Ln_Puc))

NOTE: At OLS Iteration 1 CONVERGE=0.001 Criteria Met.

The MODEL Procedure
OLS Estimation Summary

Data Set Options	
DATA=	GASOLINE

Minimization Summary	
Parameters Estimated	5
Method	Gauss
Iterations	1

Final Convergence Criteria	
R	2.34E-12
PPC	6.24E-12
RPC(beta1)	122197.4
Object	0.957917
Trace(S)	0.001092
Objective Value	0.00094

Observations Processed	
Read	36
Solved	36

OUTPUT 3.11. Chow test of structural break in gasoline data using Proc Model.

Nonlinear OLS Summary of Residual Errors							
Equation	DF Model	DF Error	SSE	MSE	Root MSE	R-Square	Adj R-Sq
Ln_G_Pop	5	31	0.0338	0.00109	0.0330	0.9580	0.9526

Nonlinear OLS Parameter Estimates				
Parameter	Estimate	Approx Std Err	t Value	Approx Pr > |t|
beta1	−12.3418	0.6749	−18.29	<0.0001
beta2	−0.0591	0.0325	−1.82	0.0786
beta3	1.373399	0.0756	18.16	<0.0001
beta4	−0.1268	0.1270	−1.00	0.3258
beta5	−0.11871	0.0813	−1.46	0.1545

Number of Observations		Statistics for System	
Used	36	Objective	0.000940
Missing	0	Objective*N	0.0338

Structural Change Test					
Test	Break Point	Num DF	Den DF	F Value	Pr > F
Chow	15	5	26	14.20	<0.0001

OUTPUT 3.11. (*Continued*)

then the CUSUM procedure can be defined as (Baltagi, 2008, p. 191; Greene, 2003, p. 135)

$$W_t = \sum_{r=K+1}^{r=t} \frac{w_r}{\hat{\sigma}},$$

where

$$\hat{\sigma}^2 = (T-k-1)^{-1} \sum_{r=k+1}^{T} (w_r - \bar{w})^2$$

and

$$\bar{w} = (T-k)^{-1} \sum_{r=k+1}^{T} w_r.$$

Here, T and k are the number of time periods and regressors, respectively. The critical values of the CUSUM at time t are given by

$$\pm a \left[\sqrt{T-k} + 2 \frac{(t-k)}{\sqrt{T-k}} \right].$$

The CUSUM values along with the lower- and upper-bound critical values are available in SAS through the Proc Autoreg procedure by using the "CUSUMLB" and "CUSUMUB" options (Proc Autoreg Documentation, SAS Institute, Inc.). More discussion on The Proc Autoreg procedure can be found in Chapter 6. However, note that this procedure is typically used to model data in the time series setting. It can handle autocorrelation in regression models and can also be used to fit autoregressive conditional heteroscedastic models (ARCH, GARCH, etc.). The "model" statement lists out the regression model of interest. This is followed by the "output" statement, which requests the lower and upper bounds along with the calculated values of the CUSUM. The analysis results are provided in Output 3.12.

Obs	lb	cusum	ub
1	.	.	.
2	.	.	.
3	.	.	.
4	.	.	.
5	.	.	.
6	−5.6188	−0.4207	5.6188
7	−5.9593	−0.0990	5.9593
8	−6.2998	−0.0107	6.2998
9	−6.6404	0.5628	6.6404
10	−6.9809	1.0289	6.9809
11	−7.3214	0.9724	7.3214
12	−7.6620	0.5926	7.6620
13	−8.0025	0.5352	8.0025
14	−8.3430	−0.5221	8.3430
15	−8.6836	−0.9944	8.6836
16	−9.0241	−0.9812	9.0241
17	−9.3646	−1.2040	9.3646
18	−9.7052	−1.7562	9.7052
19	−10.0457	−3.8517	10.0457
20	−10.3862	−6.4529	10.3862
21	−10.7267	−8.8309	10.7267
22	−11.0673	−9.7427	11.0673
23	−11.4078	−10.5631	11.4078
24	−11.7483	−11.0652	11.7483
25	−12.0889	−13.1237	12.0889
26	−12.4294	−15.7326	12.4294
27	−12.7699	−18.2970	12.7699
28	−13.1105	−19.0906	13.1105
29	−13.4510	−20.7915	13.4510
30	−13.7915	−21.9147	13.7915
31	−14.1321	−23.6449	14.1321
32	−14.4726	−26.4181	14.4726
33	−14.8131	−28.6394	14.8131
34	−15.1537	−29.5284	15.1537
35	−15.4942	−30.5195	15.4942
36	−15.8347	−31.4851	15.8347

OUTPUT 3.12. CUSUM values for the gasoline data using Proc Reg.

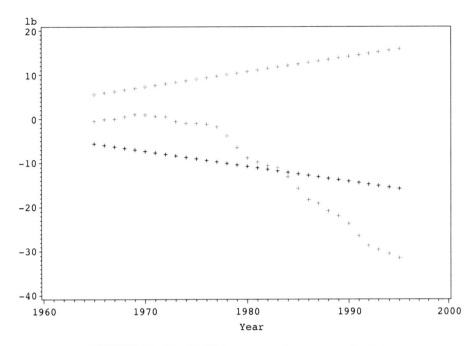

FIGURE 3.2. The CUSUM test on gasoline consumption data.

```
proc autoreg data=gasoline noprint;
    model Ln_G_Pop=Ln_Pg Ln_Income Ln_Pnc Ln_Puc;
    output out=cusum cusum=cusum cusumub=ub cusumlb=lb;
run;
proc print data=cusum;
    var lb cusum ub;
run;
```

The CUSUM plot can be created using the overlay option of Proc Gplot. The following statements are used. The results are provided in Figure 3.2.

```
proc gplot data=cusum;
    title 'The CUSUM Test on Gasoline Consumption Data';
    plot lb*year cusum*year ub*year/overlay;
run;
```

Note that a structural break is obvious around 1984. This is a bit contradictory to the results obtained from Chow's test, where the structural break was evident at 1974. However, there is evidence that the structural break started occurring around 1974.

3.7 MODELS WITH DUMMY VARIABLES

The models discussed so far consisted of quantitative explanatory variables. Often, the models of interest have explanatory variables that are discrete. A simple example is a variable recording the gender of a subject. Here gender may be set to 1 if the subject is a male and 0 if the subject is a female. As another example, consider a transportation study where one of the explanatory variables records the mode of public transportation used by the subjects in the study. Here, the values of the variable may be classified into three categories: Drive own car, Take Bus, and Take Metro. Practitioners familiar with Analysis of Variance (ANOVA) techniques may recall that this is analogous to the different levels of a treatment of interest. Here, the treatment is simply the mode of transportation. Greene (2003, p.116) discusses an example where the earning's of women is modeled as a function of age, a squared age term, level of education, and the presence or absence of children under 18. The data are credited to Mroz (1987). The model of interest is

Here,
$$\ln(earnings) = \beta_1 + \beta_2 \times age + \beta_3 \times age^2 + \beta_4 \times education + \beta_5 \times kids + \varepsilon.$$

$$kids = \begin{cases} 1 & \text{if there are kids} < 18, \\ 0 & \text{otherwise} \end{cases}$$

No special treatment is required in this case, and estimating the coefficients is straightforward using the techniques discussed so far. The analysis results are shown in Output 3.13. The estimated model for ln(*earnings*) is given by

$$\ln(earnings) = 3.24 + 0.20 \times age - 0.0023 \times age^2 + 0.067 \times education - 0.35 \times kids.$$

The earnings equation for women without children under 18 is given by

$$\ln(earnings) = 3.24 + 0.20 \times age - 0.0023 \times age^2 + 0.067 \times education$$

and the earnings equation for women with children under 18 is given by

$$\ln(earnings) = 2.89 + 0.20 \times age - 0.0023 \times age^2 + 0.067 \times education.$$

The REG Procedure
Model: MODEL1
Dependent Variable: L_WW

Number of Observations Read	428
Number of Observations Used	428

Analysis of Variance					
Source	DF	Sum of Squares	Mean Square	F Value	Pr > F
Model	4	25.62546	6.40637	4.52	0.0014
Error	423	599.45817	1.41716		.
Corrected Total	427	625.08363			

Root MSE	1.19044	R-Square	0.0410
Dependent Mean	8.05713	Adj R-Sq	0.0319
Coeff Var	14.77504		

Parameter Estimates						
Variable	Label	DF	Parameter Estimate	Standard Error	t Value	Pr > |t|
Intercept	Intercept	1	3.24010	1.76743	1.83	0.0675
WA	WA	1	0.20056	0.08386	2.39	0.0172
WA_SQ		1	−0.00231	0.00098688	−2.35	0.0195
WE	WE	1	0.06747	0.02525	2.67	0.0078
kids		1	−0.35120	0.14753	−2.38	0.0177

OUTPUT 3.13. Regression analysis of earnings data using Proc Reg.

Notice that the earnings model is a semi-log model (log in the dependent variable). We can therefore interpret the coefficient for the dummy variable *kids* as follows: *ceteris paribus*, the earnings of women with children under 18 are 35% less than the earnings of women without children under 18.

The general rule for creating dummy variables is to have one less variable than the number of categories present to avoid the "dummy variable trap". If a dummy variable category is not removed then the sum of the dummy variable category columns will equal "1" which will result in perfect multicollinearity in models with a constant term. Recall that the data matrix **X** for a model with an intercept term includes a column of 1's.

The earning's equation model had one dummy variable since we were dealing with two categories (kids under 18 or not). Suppose that we wanted to compare three groups: one with children under 10, another with children between the ages of 10 and 18, and a third group of children above 18. In this case we would need to include two dummy variables. Suppose the variables were labeled *Age_Grp*1 and *Age_Grp*2. Here,

$$Age\_Grp1 = \begin{cases} 1 & \text{if age} < 10 \\ 0 & \text{otherwise} \end{cases}$$

$$Age\_Grp2 = \begin{cases} 1 & \text{if } 10 \leq \text{age} < 18 \\ 0 & \text{otherwise.} \end{cases}$$

The third group with children above the age of 18 forms the "base". Of course, the coding scheme can be adjusted to make any of the groups the "base".

We will now look at a detailed example to illustrate the concept of dummy variables. Consider the airline data set in Greene (2003). Six airlines were followed for 15 years, and the following consumption model was proposed for the data set (Greene, 2003, pp. 118–119)

$$\ln(C_{i,t}) = \beta_1 + \beta_2 \ln(Q_{i,t}) + \beta_3 \ln^2(Q_{i,t}) + \beta_3 \ln(P_{fuel\ i,t}) + \beta_5 Loadfactor_{i,t} + \sum_{t=1}^{14} \theta_t D_{i,t} + \sum_{i=1}^{5} \delta_i F_{i,t} + \varepsilon_{i,t}.$$

The data set is credited to Christensen Associates of Madison, Wisconsin. The author provides a description of the data and model which is summarized here. In this model, the subscript i refers to airline, and the subscript t refers to year. The variable $C_{i,t}$ represents the total cost (in 1000's) for the ith firm in the tth year, Q is the output in revenue passenger miles, PF is the fuel price, and LF is the load factor or the average capacity utilization of the fleet. Note that the year 1984 is kept as a base, and we have 14 dummy variables for year. Similarly, one of the firms (Firm number 6) was kept as a base and we therefore have 5 dummy variables for firm. We will look at how the data are prepared for analysis in Proc IML and Proc Reg and then move on to analysis using Proc GLM. The following data statement will create the dummy variables. The array statement provides a convenient way of creating indicator variables for the six airlines and 14 time periods.

```
data airline;
    set airline;
    LnC=Log(C);
    LnQ=Log(Q);
    LnQ2=LnQ*LnQ;
    LnPF=Log(PF);
    array Firm{*} F1-F5;
    array time_y{*} D1-D14;
    do index1=1 to dim(Firm);
        if index1=i then Firm(index1)=1;else Firm(index1)=0;
    end;
    do index2=1 to dim(time_y);
        if index2=t then time_Y(index2)=1;else time_Y(index2)=0;
    end;
run;
```

The data can now be analyzed in Proc IML easily using the code provided in Chapter 2. Of course, an easier approach is to analyze the data using Proc Reg. The output of the analysis from Proc Reg is given in Output 3.14.

The REG Procedure
Model: MODEL1
Dependent Variable: LnC

Number of Observations Read	90
Number of Observations Used	90

Analysis of Variance					
Source	DF	Sum of Squares	Mean Square	F Value	Pr > F
Model	23	113.86832	4.95080	1893.45	<0.0001
Error	66	0.17257	0.00261		
Corrected Total	89	114.04089			

Root MSE	0.05113	R-Square	0.9985
Dependent Mean	13.36561	Adj R-Sq	0.9980
Coeff Var	0.38258		

Parameter Estimates						
Variable	Label	DF	Parameter Estimate	Standard Error	t Value	Pr > |t|
Intercept	Intercept	1	13.56249	2.26077	6.00	<0.0001
LnQ		1	0.88665	0.06284	14.11	<0.0001
LnQ2		1	0.01261	0.00986	1.28	0.2053
LnPF		1	0.12808	0.16576	0.77	0.4425
LF	LF	1	−0.88548	0.26051	−3.40	0.0012
D1		1	−0.73505	0.33784	−2.18	0.0332
D2		1	−0.67977	0.33209	−2.05	0.0447
D3		1	−0.64148	0.32983	−1.94	0.0561
D4		1	−0.58924	0.31954	−1.84	0.0697
D5		1	−0.49925	0.23176	−2.15	0.0349
D6		1	−0.44304	0.18796	−2.36	0.0214
D7		1	−0.41131	0.17290	−2.38	0.0203
D8		1	−0.35236	0.14972	−2.35	0.0216
D9		1	−0.28706	0.13470	−2.13	0.0368
D10		1	−0.23280	0.07611	−3.06	0.0032
D11		1	−0.09678	0.03385	−2.86	0.0057
D12		1	−0.01227	0.04585	−0.27	0.7899
D13		1	−0.00187	0.03816	−0.05	0.9611
D14		1	−0.01296	0.03069	−0.42	0.6742

Parameter Estimates						
Variable	Label	DF	Parameter Estimate	Standard Error	t Value	Pr > |t|
F1		1	0.05930	0.12421	0.48	0.6346
F2		1	0.02214	0.10438	0.21	0.8327
F3		1	−0.18000	0.05900	−3.05	0.0033
F4		1	0.16856	0.03326	5.07	<0.0001
F5		1	−0.04543	0.02238	−2.03	0.0464

OUTPUT 3.14. Dummy variables regression of the airlines data.

We will now analyze the data using Proc GLM. Note that in this case, we do not have to create dummy variables as the procedures automatically creates them for us. Proc GLM can be used to fit general linear models using the method of least squares. It can therefore be used for regression analysis, analysis of variance, and analysis of covariance. An advantage of Proc GLM over Proc Reg is that one can incorporate interaction terms directly into the model. Of course, one can also analyze nested and crossed effects designs very easily within Proc GLM. It also provides the flexibility of analyzing random and fixed effects in the model. The procedure also provides an overall significance of the classification variables that quickly helps the analyst gauge whether the variables are significant or not. In the following statements, the "class" statement is used to specify category variables. This is followed by the "model" statement that lists out the dependent variable and the explanatory variables in the model. The "solution" option in the "model" statement generates the parameter estimates for all the terms in the model.

The following statements should be used at the minimum.

```
proc glm data=airline;
    class I T;
    model LnC=LnQ LnQ2 LnPF LnLF I T/solution;
run;
```

The output from the analysis is given in Output 3.15. The first set of tables gives information on the classification variables and the sample size used in the analysis. In the gasoline example, we have six airlines and 15 time periods for a total of 90 observations.

The next table gives the analysis of variance table which lists out the sources of variation, the degrees of freedom, the sums of squares, the mean squares, the F statistic, and the corresponding p values. Note that the total degrees of freedom is $90 - 1 = 89$. The

The GLM Procedure

Class Level Information		
Class	Levels	Values
I	6	1 2 3 4 5 6
T	15	1 2 3 4 5 6 7 8 9 10 11 12 13 14 15

Number of Observations Read	90
Number of Observations Used	90

The GLM Procedure

Dependent Variable: LnC

Source	DF	Sum of Squares	Mean Square	F Value	Pr > F
Model	23	113.8683247	4.9507967	1893.45	<0.0001
Error	66	0.1725702	0.0026147		
Corrected Total	89	114.0408949			

R-Square	Coeff Var	Root MSE	LnC Mean
0.998487	0.382580	0.051134	13.36561

Source	DF	Type I SS	Mean Square	F Value	Pr > F
LnQ	1	103.6813479	103.6813479	39653.2	<0.0001
LnQ2	1	0.0618892	0.0618892	23.67	<0.0001
LnPF	1	8.7201792	8.7201792	3335.06	<0.0001
LF	1	0.3025573	0.3025573	115.71	<0.0001
I	5	1.0067672	0.2013534	77.01	<0.0001
T	14	0.0955840	0.0068274	2.61	0.0046

OUTPUT 3.15. Dummy variable regression of airlines data using Proc GLM.

Source	DF	Type III SS	Mean Square	F Value	Pr > F
LnQ	1	0.52059518	0.52059518	199.10	<0.0001
LnQ2	1	0.00427810	0.00427810	1.64	0.2053
LnPF	1	0.00156096	0.00156096	0.60	0.4425
LF	1	0.03020843	0.03020843	11.55	0.0012
I	5	0.86213348	0.17242670	65.95	<0.0001
T	14	0.09558402	0.00682743	2.61	0.0046

Parameter		Estimate		Standard Error	t Value	Pr > |t|
Intercept		13.56249268	B	2.26076834	6.00	<0.0001
LnQ		0.88664650		0.06283642	14.11	<0.0001
LnQ2		0.01261288		0.00986052	1.28	0.2053
LnPF		0.12807832		0.16576427	0.77	0.4425
LF		-0.88548260		0.26051152	-3.40	0.0012
I	1	0.05930014	B	0.12420645	0.48	0.6346
I	2	0.02213860	B	0.10437912	0.21	0.8327
I	3	-0.17999872	B	0.05900231	-3.05	0.0033
I	4	0.16855825	B	0.03326401	5.07	<0.0001

Parameter		Estimate		Standard Error	t Value	Pr > |t|
I	5	-0.04543271	B	0.02238459	-2.03	0.0464
I	6	0.00000000	B	.	.	.
T	1	-0.73505414	B	0.33783895	-2.18	0.0332
T	2	-0.67976508	B	0.33209031	-2.05	0.0447
T	3	-0.64147600	B	0.32983272	-1.94	0.0561
T	4	-0.58924090	B	0.31953812	-1.84	0.0697
T	5	-0.49924839	B	0.23175982	-2.15	0.0349
T	6	-0.44304017	B	0.18795980	-2.36	0.0214
T	7	-0.41130701	B	0.17289580	-2.38	0.0203
T	8	-0.35235703	B	0.14971845	-2.35	0.0216
T	9	-0.28705848	B	0.13470473	-2.13	0.0368
T	10	-0.23279711	B	0.07610725	-3.06	0.0032
T	11	-0.09678442	B	0.03385162	-2.86	0.0057
T	12	-0.01226693	B	0.04585181	-0.27	0.7899
T	13	-0.00186796	B	0.03815702	-0.05	0.9611
T	14	-0.01295665	B	0.03068805	-0.42	0.6742
T	15	0.00000000	B	.	.	.

Note: The X'X matrix has been found to be singular, and a generalized inverse was used to solve the normal equations. Terms whose estimates are followed by the letter 'B' are not uniquely estimable.

OUTPUT 3.15. (*Continued*)

model degrees of freedom is calculated as follows: one degree of freedom is used for each explanatory variable in the model and $a-1$ degrees of freedom are used for the classification variables where a is the number of levels of the classification variable. For the airlines example, we have four explanatory variables that contribute one degree of freedom, the six airlines contribute 5 degrees of freedom, and the 15 time periods contribute 14 degrees of freedom. Therefore, the total model degrees of freedom equals 24. Note that the p value indicates that the "global" F test for model significance is rejected.

This is followed by the R^2 value, the coefficient of variation (see Chapter 2), the root mean square error, and the mean of the dependent variable. Notice that the R^2 value is very high.

The next two tables gives the Type 1 and Type 3 sums of squares for each term in the model along with their F statistic and p values. The Type 1 sums of squares gives the amount of variation attributed to each term in the model assuming that the terms listed in the table before it has already been included in the model. This is also referred to as the sequential sums of squares. The Type 3 sums of squares gives the amount of variation attributed to each term after adjusting for the other terms. In other words, it measures the amount by which the error sums of squares is reduced if the term in question is added to a model already consisting of the other terms. Note that the p values for the four explanatory variables from the Type 3 sums of squares matches the p values of the variables from Proc Reg. Proc Reg uses Type 3 sums of squares since our objective is to determine if the variable in question is meaningful to the general model consisting of the other terms. The p values from the Type 3 table indicates that the load factor, the airlines, and the time periods are significantly different.

The "solution" option is used to list out the parameter estimates. Note that the coefficients are identical to the ones from Proc Reg where dummy variables were used. Also notice that the sixth airline and the 15th time period have been taken as a base. The interpretation of the models for the other firms will be relative to the sixth firm. Note that firms 3, 4, and 5 are significantly different from firm 6. Also note that the first 11 time periods are significantly different from the 15th time period.

4

INSTRUMENTAL VARIABLES

4.1 INTRODUCTION

The analysis methods presented in the previous chapters were based on the assumption that the independent variables are exogenous ($E(\varepsilon \mid X) = 0$). That is, the error terms in the linear regression model are uncorrelated or independent of the explanatory variables. In Chapter 1, we saw that under the exogeneity assumption, the least squares estimator $\mathbf{b}$ is an unbiased and consistent estimator of $\boldsymbol{\beta}$. This chapter explores the properties of $\mathbf{b}$ under departures from the exogenous assumption? Explanatory variables that are not exogenous are called endogenous variables.

Under departures from the exogeneity conditions, $\mathbf{b}$ is no longer an unbiased and consistent estimator of $\boldsymbol{\beta}$. To see this, in the simple linear regression case, consider the model $y_i = \beta_0 + \beta_1 x_i + \varepsilon_i$ where the disturbances are correlated with the explanatory variable. The least squares estimator of β_1 is given by (Chapter 1)

$$b_1 = \frac{\sum_{i=1}^{n}(x_i - \bar{x})(y_i - \bar{y})}{\sum_{i=1}^{n}(x_i - \bar{x})^2}$$

which upon simplifying can be written as

$$b_1 = \beta_1 + \frac{\sum_{i=1}^{n}(x_i - \bar{x})(\varepsilon_i - \bar{\varepsilon})}{\sum_{i=1}^{n}(x_i - \bar{x})^2}.$$

The second term gives the expression of the least squares estimator of a regression of ε on x. It should be obvious that the second term is not zero unless the disturbances are uncorrelated with the explanatory variable. Therefore, the least squares estimator of β_1 is biased. The inconsistency of the least squares estimator can be seen by dividing the numerator and denominator of the second term by n and then taking the probability limits to get

$$p \lim(b_1) = \beta_1 + \frac{Cov(\varepsilon, x)}{Var(x)}.$$

Applied Econometrics Using the SAS® System, by Vivek B. Ajmani
Copyright © 2009 John Wiley & Sons, Inc.

The least squares estimator is therefore not consistent. The magnitude and direction of the bias depend on the second term of the above expression.

To see this in the general case, consider the linear model $\mathbf{y} = \mathbf{X}\boldsymbol{\beta} + \boldsymbol{\varepsilon}$ with $E(\boldsymbol{\varepsilon}|\mathbf{X}) = \boldsymbol{\gamma}$. The OLS estimator is given by $\mathbf{b} = (\mathbf{X}^T\mathbf{X})^{-1}\mathbf{X}^T\mathbf{y}$ or $\mathbf{b} = \boldsymbol{\beta} + (\mathbf{X}^T\mathbf{X})^{-1}\mathbf{X}^T\boldsymbol{\varepsilon}$. Taking the conditional expectation $E(\mathbf{b}|\mathbf{X})$, we see that (Greene, 2003, p. 76)

$$
\begin{aligned}
E(\mathbf{b}|\mathbf{X}) &= \boldsymbol{\beta} + (\mathbf{X}^T\mathbf{X})^{-1}E(\mathbf{X}^T\boldsymbol{\varepsilon}|\mathbf{X}) \\
&= \boldsymbol{\beta} + (\mathbf{X}^T\mathbf{X})^{-1}\mathbf{X}^T E(\boldsymbol{\varepsilon}|\mathbf{X}) \\
&= \boldsymbol{\beta} + (\mathbf{X}^T\mathbf{X})^{-1}\boldsymbol{\gamma} \neq \boldsymbol{\beta}.
\end{aligned}
$$

Next, assume that $p \lim (\mathbf{X}^T\mathbf{X}/n) = \boldsymbol{\Psi}$ a positive definite matrix with inverse $\boldsymbol{\Psi}^{-1}$. Note that this assumption ensures that $(\mathbf{X}^T\mathbf{X})^{-1}$ exists (in the simple linear regression case, this assumption implied that $Var(x)$ existed and was finite). Taking the limit in the probability of OLS estimator, we get

$$
p \lim \mathbf{b} = \beta + \boldsymbol{\Psi}^{-1} p \lim \left(\frac{\mathbf{X}^T\boldsymbol{\varepsilon}}{n}\right) \neq \boldsymbol{\beta}
$$

because $E(\boldsymbol{\varepsilon} \mid \mathbf{X}) \neq \mathbf{0}$. Therefore, the OLS estimator is not a consistent estimator if the exogeneity assumption is violated.

Endogeneity occurs for several reasons. Missing variables and measurement errors in the independent variables are often cited as major causes of endogeneity in regression models (Ashenfelter et al., 2003; Greene, 2003; Wooldridge, 2002). We will now briefly discuss both the omitted variable bias and the bias emerging from measurement errors.

4.2 OMITTED VARIABLE BIAS

The analysis in Chapters 1 through 3 was also based on the assumption that the linear model $\mathbf{y} = \mathbf{X}\boldsymbol{\beta} + \boldsymbol{\varepsilon}$ was correctly specified. We will now relax this assumption and see the effect this has on the parameter estimates. Suppose that $\mathbf{X} = [\mathbf{X}_1, \mathbf{X}_2]$ and $\boldsymbol{\beta} = [\boldsymbol{\beta}_1, \boldsymbol{\beta}_2]^T$ so that the true model is $\mathbf{y} = \mathbf{X}_1\boldsymbol{\beta}_1 + \mathbf{X}_2\boldsymbol{\beta}_2 + \boldsymbol{\varepsilon}$. Assume that for some reason we have no information available for $\mathbf{X}_2$ and therefore omit it and fit the model $\mathbf{y} = \mathbf{X}_1\boldsymbol{\beta}_1 + \boldsymbol{\varepsilon}$. The least squares estimator of $\boldsymbol{\beta}_1$ in this case is given by (Greene, 2003, p. 148)

$$
\begin{aligned}
\mathbf{b}_1 &= (\mathbf{X}_1^T\mathbf{X}_1)^{-1}\mathbf{X}_1^T\mathbf{y} \\
&= \boldsymbol{\beta}_1 + (\mathbf{X}_1^T\mathbf{X}_1)^{-1}\mathbf{X}_1^T\mathbf{X}_2\boldsymbol{\beta}_2 + (\mathbf{X}_1^T\mathbf{X}_2)^{-1}\mathbf{X}_1^T\boldsymbol{\varepsilon}.
\end{aligned}
$$

It is easy to verify that the conditional expectation of $\mathbf{b}_1$ given $\mathbf{X}$ is

$$
E(\mathbf{b}_1|\mathbf{X}) = \boldsymbol{\beta}_1 + (\mathbf{X}_1^T\mathbf{X}_1)^{-1}\mathbf{X}_1^T\mathbf{X}_2\boldsymbol{\beta}_2.
$$

The bias is given by the second term and is zero only if $\mathbf{X}_1^T\mathbf{X}_2 = \mathbf{0}$ ($\mathbf{X}_1$ and $\mathbf{X}_2$ are orthogonal). That is, if the omitted variables are not correlated with the included variables. Under the omitted variable model, the term $\mathbf{X}_2\boldsymbol{\beta}_2$ is absorbed into the error term. If the omitted variable $\mathbf{X}_2$ is related to $\mathbf{X}_1$, then it can easily be shown that $Cov(\boldsymbol{\varepsilon}|\mathbf{X}) \neq \mathbf{0}$ and therefore the exogenous assumption is violated.

To see this, consider the simple linear model given by $y = \beta_0 + \beta_1 x_1 + \beta_2 x_2 + \varepsilon$ and assume that this is the true population model. Consider the case where x_2 is omitted so that the model used is $y = \beta_0 + \beta_1 x_1 + \varepsilon$. Assume that x_2 is correlated with x_1 with the reduced form for x_2 given by $x_2 = \alpha_0 + \alpha_1 x_1 + \upsilon$. As shown earlier, the OLS estimator of β_1 is given by

$$
\hat{\beta}_1 = \frac{\sum_{i=1}^{n}(x_{1i}-\bar{x}_1)(y_i-\bar{y})}{\sum_{i=1}^{n}(x_{1i}-\bar{x}_1)^2}.
$$

Substituting the true population model in the above expression and after some elementary algebraic manipulations, it can be shown that

$$\hat{\beta}_1 = \beta_1 + \beta_2 \frac{\sum_{i=1}^{n}(x_{1i}-\bar{x}_1)(x_{2i}-\bar{x}_2)}{\sum_{i=1}^{n}(x_{1i}-\bar{x}_1)^2} + \frac{\sum_{i=1}^{n}(x_{1i}-\bar{x}_1)(\varepsilon_i-\bar{\varepsilon})}{\sum_{i=1}^{n}(x_{1i}-\bar{x}_1)^2}.$$

The last term in the conditional expectation $E(\hat{\beta}_1|x_{1i}, x_{2i})$ drops out giving

$$\hat{\beta}_1 = \beta_1 + \beta_2 \frac{\sum_{i=1}^{n}(x_{1i}-\bar{x}_1)(x_{2i}-\bar{x}_2)}{\sum_{i=1}^{n}(x_{1i}-\bar{x}_1)^2}.$$

Again, the second term gives the least squares estimator of a regression of x_2 on x_1. It should be obvious that the OLS estimator of β_1 is biased unless $Cov(x_1, x_2) = 0$ with the magnitude and direction of the bias depending on the second term of the above expression.

4.3 MEASUREMENT ERRORS

We will now look at how measurement errors in the explanatory or dependent variables or both affect the least squares estimators. It turns out that the measurement error in the explanatory variables creates a correlation between the variables and the error term similar to the omitted variable case. On the other hand, measurement errors in the dependent variable may not be a problem (Wooldridge, 2002). We will illustrate the issue by using a simple linear regression model with just one explanatory variable—that is, a model of the form (Ashenfelter et al., 2003, p. 197):

$$y_i = \beta_0 + \beta_1 x_i + \varepsilon_i \ i = 1, \dots, n$$

where x_i is assumed to be exogenous. Suppose that we observe x_i with error. That is, we observe $x_i' = x_i + u_i$. Assume that x is independent of u_i and that the disturbances u_i and ε_i are independent of each other. Furthermore, assume that $u_i \sim i.i.d.(0, \sigma_u^2)$ and $\varepsilon_i \sim i.i.d.(0, \sigma_\varepsilon^2)$.

By substituting the observed value x_i' in the equation for y_i, we get $y_i = \beta_0 + \beta_1 x_i' + v_i$ where $v_i = \varepsilon_i - \beta_1 u_i$. Note that by construction $Cov(v_i, x_i') \neq 0$ since both x_i' and v_i are influenced by the random component u_i. Therefore, the OLS assumptions are violated and the least squares estimate for β_1 is biased and inconsistent. To see this, note that the OLS estimate for β_1 can be written as

$$\hat{\beta}_1 = \frac{\sum_{i=1}^{n}(x_i'-\bar{x}')(y_i-\bar{y})}{\sum_{i=1}^{n}(x_i'-\bar{x}')^2}.$$

Simple algebraic manipulation can be used to rewrite the above expression as

$$\hat{\beta}_1 = \beta_1 + \frac{\sum_{i=1}^{n}(x_i'-\bar{x}')(v_i-\bar{v})}{\sum_{i=1}^{n}(x_i'-\bar{x}')^2}.$$

Dividing the numerator and denominator of the second term by n and taking the probability limits gives

$$\hat{\beta}_1 = \beta_1 + \frac{Cov(x', v)}{Var(x')}.$$

Using the assumptions stated earlier, it can be easily shown that $Cov(x', v) = -\beta_1\sigma_u^2$ and that $Var(x') = \sigma_{x'}^2 + \sigma_u^2$. Therefore, the bias is given by

$$-\frac{\beta_1\sigma_u^2}{\sigma_{x'}^2 + \sigma_u^2}.$$

Therefore, measurement errors in the explanatory variables result in biased and inconsistent OLS estimates. As before, the magnitude and the direction of the bias depend on the second term of the expression of $\hat{\beta}_1$.

Ashenfelter et al. (2003, p. 197) gives an elegant derivation to show the behavior of the least squares estimator under measurement errors in both the dependent and the independent variables. As discussed by the authors, measurement errors in the dependent variable, in general, does not lead to violation of the least squares assumptions because the measurement error in the dependent variable is simply absorbed in the disturbance term of the model. However, errors in the dependent variable may inflate the standard errors of the least squares estimates. To see this, consider the simple linear regression model given earlier and assume that we do not observe y_i but observe $y_i' = y_i + v_i$. Substituting this in the original model, we get $y_i' = \beta_0 + \beta_1 x_i + u_i$ where $u_i = \varepsilon_i + v_i$. It should be obvious that unless $Cov(x_i, v_i) \neq 0$, the OLS assumptions are not violated. Furthermore, since (Meyers, 1990, p. 14)

$$Var\left(\hat{\beta}_1\right) = \frac{\sigma_u^2}{\sum_{i=1}^{n}\left(x_i - \bar{x}\right)^2}, \sigma_u^2 = \sigma_\varepsilon^2 + \sigma_v^2, \text{ and } \sigma_x^2 = n^{-1}\sum_{i=1}^{n}\left(x_i - \bar{x}\right)^2$$

then

$$Var\left(\hat{\beta}_1\right) = \frac{\sigma_\varepsilon^2 + \sigma_u^2}{n\sigma_x^2}.$$

Therefore, the measurement errors in the dependent variable tends to inflate the standard errors of the estimates.

4.4 INSTRUMENTAL VARIABLE ESTIMATION

We will now discuss an alternative method to get unbiased and consistent estimators of $\boldsymbol{\beta}$ under departures from the exogenous assumption. To motivate the discussion of instrumental variables, consider the least squares model given by

$$\mathbf{y} = \mathbf{X}^T\boldsymbol{\beta} + \boldsymbol{\varepsilon}.$$

Assume that one or more variables in $\mathbf{X}$ may be correlated with $\boldsymbol{\varepsilon}$. That is, assume that $E(\boldsymbol{\varepsilon}|\mathbf{X}) = \boldsymbol{\eta} \neq \mathbf{0}$. Next, assume that there exists a set of L variables in $\mathbf{W}$, with $L \geq k$, such that $Cov(\mathbf{W}, \mathbf{X}) \neq \mathbf{0}$ but $E(\mathbf{W}^T\boldsymbol{\varepsilon}) = \mathbf{0}$. That is, the L variables in $\mathbf{W}$ are exogenous but are correlated with the explanatory variables. Check this.

Note that the exogenous variables from the original set of variables may be part of $\mathbf{W}$. The variables in the set $\mathbf{W}$ are referred to as instrumental variables. We will first look at the instrumental variable estimator for the case when $L = k$. Premultiplying the linear model by $\mathbf{W}^T$ gives

$$\mathbf{W}^T\mathbf{y} = \mathbf{W}^T\mathbf{X}\boldsymbol{\beta} + \mathbf{W}^T\boldsymbol{\varepsilon}.$$

By rewriting this as

$$\mathbf{y}^* = \mathbf{X}^*\boldsymbol{\beta} + \boldsymbol{\varepsilon}^*$$

and using the method of least squares, we can write

$$\hat{\boldsymbol{\beta}}_{IV} = (\mathbf{X}^{*T}\mathbf{X}^*)^{-1}\mathbf{X}^{*T}\mathbf{y}^* = (\mathbf{X}^T\mathbf{W}\mathbf{W}^T\mathbf{X})^{-1}\mathbf{X}^T\mathbf{W}\mathbf{W}^T\mathbf{y},$$

which can be simplified to

$$\hat{\beta}_{IV} = \beta + (\mathbf{W}^T\mathbf{X})^{-1}\mathbf{W}^T\varepsilon.$$

Using the assumption that $(\mathbf{W}^T\mathbf{X})^{-1}$ exists and $E(\mathbf{W}^T\varepsilon) = \mathbf{0}$, it is easy to show that $\hat{\beta}_{IV}$ is unbiased for β. Using the discussion used to show the consistency of OLS estimators, it is trivial to show that $p \lim (\hat{\beta}_{IV}) = \beta$.

Therefore, the instrumental variable estimator for the case $L = k$ is

$$\hat{\beta}_{IV} = (\mathbf{X}^T\mathbf{W}\mathbf{W}^T\mathbf{X})^{-1}\mathbf{X}^T\mathbf{W}\mathbf{W}^T,$$

which can be simplified to

$$\mathbf{b}_{IV} = (\mathbf{W}^T\mathbf{X})^{-1}\mathbf{W}^T\mathbf{y}.$$

Greene (2003, pp. 76–77) gives a thorough description and the assumptions underlying the instrumental variables estimator. Also see Wooldridge (2002, pp. 85–86) and Ashenfelter et al. (2003, pp. 199–200).

It is easy to show that the asymptotic variance of $\mathbf{b}_{IV}$ is

$$\hat{\sigma}^2(\mathbf{W}^T\mathbf{X})^{-1}(\mathbf{W}^T\mathbf{W})(\mathbf{X}^T\mathbf{W})^{-1}$$

where,

$$\hat{\sigma}^2 = \frac{(\mathbf{y}-\mathbf{X}\mathbf{b}_{IV})^T(\mathbf{y}-\mathbf{X}\mathbf{b}_{IV})}{n} \quad \text{(Greene, 2003, p. 77).}$$

As shown in Greene (2003, p. 78 and Wooldridge (2002, pp. 90–91), instrumental variables estimation when $L > k$ is done in two steps. In Step 1, the data matrix $\mathbf{X}$ is regressed against the matrix containing the instrumental variables $\mathbf{W}$ to get $\hat{\mathbf{X}}$ which is defined as $\hat{\mathbf{X}} = (\mathbf{W}^T\mathbf{W})^{-1}\mathbf{W}^T\mathbf{X}$. In Step 2, $\mathbf{y}$ is regressed on $\hat{\mathbf{X}}$ to get the instrumental variables estimator, $\mathbf{b}_{IV}$ given by

$$\begin{aligned}
\mathbf{b}_{IV} &= (\hat{\mathbf{X}}T\,\hat{\mathbf{X}})^{-1}\hat{\mathbf{X}}^T\mathbf{y} \\
&= [\mathbf{X}^T\mathbf{W}(\mathbf{W}^T\mathbf{W})^{-1}\mathbf{W}^T\mathbf{X}]^{-1}\mathbf{X}^T\mathbf{W}(\mathbf{W}^T\mathbf{W})^{-1}\mathbf{W}^T\mathbf{y}.
\end{aligned}$$

This estimator is often referred to as the two-stage least squares estimator of β and is abbreviated as 2SLS. The $k \times k$ matrix $\hat{\sigma}^2(\hat{\mathbf{X}}^T\hat{\mathbf{X}})^{-1}$ with

$$\hat{\sigma}^2 = \frac{(\mathbf{y} - \hat{\mathbf{X}}\mathbf{b}_{IV})^T(\mathbf{y} - \hat{\mathbf{X}}\mathbf{b}_{IV})}{n}$$

is the estimated covariance matrix of the 2SLS estimator.

We will now illustrate the computations involved in estimation with instrumental variables by using the data on working, married women in the well-known labor supply data from Mroz (1987). As discussed in Wooldridge (2002, p. 87), the objective is to estimate the following wage equation:

$$\log(wage) = \beta_0 + \beta_1 exper + \beta_2 exper^2 + \beta_3 educ + \varepsilon.$$

The variable *educ* contains the actual number of years of education of each woman. As stated by the author, information on ability of the women, the quality of education received and their family background is missing. These variables are suspected of being correlated with the education variable and are therefore assumed to contribute to omitted variables bias. Suppose that information on mother's education, *motheduc*, is available so that it can be used as an instrument for *educ*. The instrumental variables matrix $\mathbf{W}$ therefore has three variables: *exper*, *exper*^{2}, and *motheduc*.

We will first analyze the data in Proc IML and then show the analysis in Proc Syslin. The following statements will read in the data from a text file called "mroz_raw.txt." Note that the variable *inlf* indicates whether a person was in the labor force in 1975. Since we are interested in working women, we need to select only those records where *inlf* = 1.

```
data mroz;
     infile 'C:\Temp\MROZ.txt' lrecl=234;
     input inlf hours kidslt6 kidsge6 age educ wage repwage
     hushrs husage huseduc huswage faminc mtr motheduc
     fatheduc unem city exper nwifeinc lwage expersq;
     if inlf=1;
run;
```

Next, we read the data into matrices by using Proc IML. Note that the data matrix **X** has data for the actual labor market experience (*exper*), the squared term for this (*exper*^{2}), and years of schooling (*educ*). Of course, we need to also add the constant column of 1's. The response variable is the log of wage (*log*(*wage*)). We use the mother's years of schooling (*motheduc*) as the instrument for *educ*. Therefore, the instrumental variables matrix **W** contains the column of 1's along with *exper*, *exper*^{2}, and *motheduc*. The following statements invoke Proc IML and read the data into the matrices.

```
Proc IML;
     use mroz;
     read all var {'exper' 'expersq' 'educ'} into X;
     read all var {'lwage'} into Y;
     read all var {'exper' 'expersq' 'motheduc'} into W;
     n=nrow(X);
     k=ncol(X);
     X=J(n,1,1)||X;
     W=J(n,1,1)||W;
```

As discussed earlier, when the number of columns of **W** equals the number of columns of **X**, the least squares instrumental variables estimator is given by $\mathbf{b}_{IV} = (\mathbf{W}^T\mathbf{X})^{-1}\mathbf{W}^T\mathbf{y}$. The following statement will calculate this.

```
bhat_IV=inv(W'*X)*W'*y;
```

We can calculate the asymptotic variance–covariance matrix of this estimator by using the formulas outlined in this section. The following statements can be used to calculate the standard errors of the instrumental variable estimator. Output 4.1 contains the analysis results.

```
variance=((y-X*bhat_IV)'*(y-X*bhat_IV)/n;
variance_matrix=inv(W'*X)*(W'*W)*inv(X'*W);
var_cov_IV=variance*variance_matrix;
SE=SQRT(vecdiag(var_cov_IV));
```

The model used is a semi-log model and can be interpreted as follows: Ceteris Paribus, the estimate of the return to education is about 5% and is not significant. The implication is that each additional year of school is predicted to increase earnings by about

TABLE				
	BHAT_IV	**SE**	**T**	**PROBT**
INTERCEPT	0.1982	0.4707	0.4211	0.6739
EXPER	0.0449	0.0135	3.3194	0.0010
EXPER_SQ	−0.0009	0.0004	−2.2797	0.0231
EDUC	0.0493	0.0373	1.3221	0.1868

OUTPUT 4.1. Instrumental variable estimates for the earning data.

5%. The estimate of return to schooling from OLS estimation is about 10.7% and is highly significant. The standard error from the instrumental variable estimation is 0.0373 versus 0.014 for the OLS model. The *t*-test value for the instrumental variable estimator is therefore smaller than that from the OLS estimation, which explains the difference between the lack of significance of the instrumental variables estimator to the OLS estimator. The OLS model is given in Output 4.3.

The Proc Syslin procedure in SAS can be used to conduct instrumental variable analysis on the earning data. This procedure is extremely powerful and can be used to estimate parameters in systems of seemingly unrelated regression and systems of simultaneous equations. It can also be used for single equation estimation using OLS and is very useful when conducting instrumental variables regression for both the single and multiple equations systems. The following statements can be used at the minimum for instrumental variables analysis of the earning data. Note that all we do here is specify the endogenous and exogenous variables followed by a specification of the linear model. The analysis results are given in Output 4.2.

```
proc syslin 2SLS data=mroz;
    endogenous educ;
    instruments exper expersq motheduc;
    model lwage=exper expersq educ;
run;
```

The *t*-tests indicate that both experience variables are significant. The output from Proc Syslin (using the options for the example used) is very similar to the output we have seen with Proc Reg and Proc GLM. That is, we get the ANOVA table, followed by the model statistics and the parameter estimates. The results from OLS analysis are given in Output 4.3 and indicates a significance of all three variables.

We now turn our attention to the case when the number of columns in **W** exceeds the number of columns in **X**. We will now use the information on both parents' education (*fatheduc and motheduc*) and husband's education (*huseduc*) as instruments for *educ*. As before, we will analyze this data using Proc IML followed by analysis using Proc Syslin.

The SYSLIN Procedure
Two-Stage Least Squares Estimation

Model	lwage
Dependent Variable	lwage

Analysis of Variance					
Source	DF	Sum of Squares	Mean Square	F Value	Pr > F
Model	3	10.18121	3.393735	7.35	<0.0001
Error	424	195.8291	0.461861		
Corrected Total	427	223.3275			

Root MSE	0.67960	R-Square	0.04942
Dependent Mean	1.19017	Adj R-Sq	0.04270
Coeff Var	57.10123		

Parameter Estimates					
Variable	DF	Parameter Estimate	Standard Error	t Value	Pr > |t|
Intercept	1	0.198186	0.472877	0.42	0.6754
exper	1	0.044856	0.013577	3.30	0.0010
expersq	1	−0.00092	0.000406	−2.27	0.0238
educ	1	0.049263	0.037436	1.32	0.1889

OUTPUT 4.2. Instrumental variables analysis of the earning data using Proc Syslin.

The REG Procedure
Model: MODEL1
Dependent Variable: lwage

Number of Observations Read	428
Number of Observations Used	428

Analysis of Variance					
Source	DF	Sum of Squares	Mean Square	F Value	Pr > F
Model	3	35.02230	11.67410	26.29	<0.0001
Error	424	188.30515	0.44412		
Corrected Total	427	223.32745			

Root MSE	0.66642	R-Square	0.1568
Dependent Mean	1.19017	Adj R-Sq	0.1509
Coeff Var	55.99354		

Parameter Estimates					
Variable	DF	Parameter Estimate	Standard Error	t Value	Pr > |t|
Intercept	1	−0.52204	0.19863	−2.63	0.0089
exper	1	0.04157	0.01318	3.15	0.0017
expersq	1	−0.00081119	0.00039324	−2.06	0.0397
educ	1	0.10749	0.01415	7.60	<0.0001

OUTPUT 4.3. Ordinary least squares analysis of the earning data.

The following statements will invoke Proc IML and read the data into appropriate matrices. Note that the instrumental variables matrix **W** now contains the two exogenous variables *exper* and *exper*^{2} along with the three instrumental variables. The analysis results are given in Output 4.4.

```
Proc IML;
    use mroz;
    read all var {'exper' 'expersq' 'educ'} into X;
    read all var {'lwage'} into Y;
    read all var {'exper' 'expersq' 'motheduc'
    'fatheduc''huseduc'} into W;
    n=nrow(X);
    k=ncol(X);
    X=J(n,1,1)||X;
    W=J(n,1,1)||W;
```

TABLE				
	BHAT_IV	SE	T	PROBT
INTERCEPT	−0.1869	0.2971	−0.6288	0.5298
EXPER	0.0431	0.0138	3.1205	0.0019
EXPER_SQ	−0.0009	0.0004	−2.0916	0.0371
EDUC	0.0804	0.0227	3.5461	0.0004

OUTPUT 4.4. Instrumental variables estimator for the earning data when $L > k$.

The following statements will calculate and print the least squares instrumental estimator. Note that the first step is to calculate the predicted data matrix $\hat{\mathbf{X}}$. This is then used to produce the instrumental variables estimator.

```
Xhat=W*inv(W'*Z)*W'*X;
bhat_IV=inv(Xhat'*Xhat)*Xhat'*y;
```

The standard errors of these estimates can also be easily calculated as before. The following statements will do this for us.

```
variance=((y-Xhat*bhat_IV)'*(y-Xhat*bhat_IV))/n;
variance_matrix=inv(Xhat'*Xhat);
var_cov_IV=variance*variance_matrix;
SE=SQRT(vecdiag(var_cov_IV));
```

Proc Syslin can easily be used to conduct the analysis. The following statements can be used. The analysis results are given in Output 4.5. Notice that all three explanatory variables are now significant and the returns to schooling has increased to about 8% and is highly significant. The interpretation is as before: Ceteris Paribus, each additional year of schooling is expected to increase earnings by about 8%.

```
proc syslin 2SLS data=mroz;
    endogenous educ;
    instruments exper expersq motheduc fatheduc huseduc;
    model lwage=exper expersq educ;
run;
```

The SYSLIN Procedure
Two-Stage Least Squares Estimation

Model	lwage
Dependent Variable	lwage

Analysis of Variance					
Source	DF	Sum of Squares	Mean Square	F Value	Pr > F
Model	3	15.48784	5.162612	11.52	<0.0001
Error	424	189.9347	0.447959		
Corrected Total	427	223.3275			

Root MSE	0.66930	R-Square	0.07540
Dependent Mean	1.19017	**Adj R-Sq**	0.06885
Coeff Var	56.23530		

Parameter Estimates					
Variable	DF	Parameter Estimate	Standard Error	t Value	Pr > |t|
Intercept	1	−0.18686	0.285396	−0.65	0.5130
exper	1	0.043097	0.013265	3.25	0.0013
expersq	1	−0.00086	0.000396	−2.18	0.0300
educ	1	0.080392	0.021774	3.69	0.0003

OUTPUT 4.5. Proc Syslin output of the earning data when $L > k$.

4.5   SPECIFICATION TESTS

In the discussion so far, we assumed that the regression models suffer from the presence of endogenous explanatory variable(s). We also presented techniques to estimate the model parameters under the presence of endogenous variables. This section introduces methods to determine if endogeneity is indeed a problem. The Hausman test for endogeneity is perhaps the most widely used test and is based on comparing the OLS and the 2SLS estimators. We will discuss this test a bit later. For now, we will look at simple regression-based tests that can be used as an alternative to Hausman's test.

The steps for conducting the test are given below (Wooldridge, 2002, pp. 118–119).

1. First, consider the linear model $\mathbf{y} = \mathbf{X}\boldsymbol{\beta} + \alpha\mathbf{x} + \boldsymbol{\varepsilon}$ where $\mathbf{y}$ is $n \times 1$, $\mathbf{X}$ is $n \times k$, $\boldsymbol{\beta}$ is $k \times 1$, α is the coefficient of the $n \times 1$ vector $\mathbf{x}$ that is suspected of being endogenous, and $\boldsymbol{\varepsilon}$ is the $n \times 1$ unobserved error. Let $\mathbf{W}_{n \times L}$ be the set of L exogenous variables including the variables in $\mathbf{X}$. Next, consider the hypothesis $H_0 : \mathbf{x}$ is exogenous versus $H_1 : \mathbf{x}$ is endogenous.
2. Consider the reduced form equation relating $\mathbf{x}$ to $\mathbf{W}$ given by $\mathbf{x} = \mathbf{W}\boldsymbol{\delta} + \boldsymbol{\gamma}$ with the assumption that $E(\mathbf{W}^T\boldsymbol{\gamma}) = \mathbf{0}$. Here, $\boldsymbol{\delta}$ is an $L \times 1$ vector of unknown coefficients and $\boldsymbol{\gamma}$ is the $n \times 1$ disturbance vector.
3. As shown in (Wooldridge, 2002, p. 119), the expectation $E(\boldsymbol{\varepsilon}^T\boldsymbol{\gamma})$ equals $E(\boldsymbol{\varepsilon}^T\mathbf{x})$. Therefore, we can test endogeneity of $\mathbf{x}$ by simply checking whether $E(\boldsymbol{\varepsilon}^T\boldsymbol{\gamma}) = 0$.
4. Write the equation relating $\boldsymbol{\varepsilon}$ to $\boldsymbol{\gamma}$ as $\boldsymbol{\varepsilon} = \rho_1\boldsymbol{\gamma} + \mathbf{e}$ and substitute this in the original equation to get $\mathbf{y} = \mathbf{X}\boldsymbol{\beta} + \alpha\mathbf{x} + \rho_1\boldsymbol{\gamma} + \mathbf{e}$.
5. It is trivial to show that $\mathbf{e}$ is independent of $\mathbf{X}$, $\mathbf{x}$, and $\boldsymbol{\gamma}$. Therefore, a test of $H_0 : \rho_1 = 0$ can be conducted by looking at the t-test results in the regression of y on $\mathbf{X}$, $\mathbf{x}$, and $\boldsymbol{\gamma}$. Endogeneity of $\mathbf{x}$ is implied if the null hypothesis is not rejected. Here, $\boldsymbol{\gamma}$ can be estimated by the residuals of the regression of $\mathbf{x}$ on $\mathbf{W}$.

We will revisit the education data set with parents and husband's education as instruments to illustrate the above approach to test for endogeneity. The objective here is to determine if the variable *educ* is endogenous. The first step is to regress *educ* on a constant, *exper*, *exper*,^{2} *motheduc*, *fatheduc*, and *huseduc*. The residuals from this (v) regression is saved and used as an explanatory variable in the regression of $log(wage)$ against a constant, *exper*, *exper*^{2}, *educ*, and v. If the t statistic corresponding to v is significant, then the null hypothesis is rejected and we conclude that the variable *educ* is endogenous. The following SAS statements can be used to do the analysis. Notice that the first Proc Reg statements save the residuals in a temporary SAS data set called mroz2. The analysis results are given in Output 4.6. The results indicate that we have evidence of endogeneity of *educ* at the 10% significance level (p-value $= 0.0991$).

```
proc reg data=mroz noprint;
    model educ=exper expersq motheduc fatheduc huseduc;
    output out=mroz2 residual=v;
run;
    proc reg data=mroz2;
    model lwage=exper expersq educ v;
run;
```

4.5.1   Testing Overidentifying Restrictions

We now turn our attention to addressing the problem of determining if the regression model has more instruments than is necessary. The question we address here is, "Are the extra instrument variables truly exogenous?" That is, are the extra instruments uncorrelated with the error term? Wooldridge (2002, p. 123) gives details on a simple regression-based Sargan's hypothesis test (1958) to determine whether the regression model has more instruments than is required. The steps are as follows:

1. Consider, the linear model given by $\mathbf{y} = \mathbf{X}\boldsymbol{\beta} + \boldsymbol{\Gamma}\boldsymbol{\delta} + \boldsymbol{\varepsilon}$ where $\mathbf{y}$ is $n \times 1$, $\mathbf{X}$ is $n \times L_1$, $\boldsymbol{\beta}$ is $L_1 \times 1$, $\boldsymbol{\Gamma}$ is $n \times G$, $\boldsymbol{\delta}$ is $G \times 1$, and $\boldsymbol{\varepsilon}$ is $n \times 1$. Here, $\boldsymbol{\Gamma}$ contains variables that are suspected of being endogenous. As before, let $\mathbf{W} = (\mathbf{X}, \mathbf{W}^*)$ be the set of all instrumental variables. Here, $\mathbf{W}^*$ is $n \times L_2$ so that $\mathbf{W}$ is $n \times L$ with $L = L_1 + L_2$ and $L_2 > G$.
2. Conduct a 2SLS and obtain $\hat{\boldsymbol{\varepsilon}}$.
3. Conduct an OLS of $\hat{\boldsymbol{\varepsilon}}$ on $\mathbf{W}$ and obtain R^2.

The REG Procedure
Model: MODEL1
Dependent Variable: lwage

Number of Observations Read	428
Number of Observations Used	428

Analysis of Variance					
Source	DF	Sum of Squares	Mean Square	F Value	Pr > F
Model	4	36.23050	9.05763	20.48	<0.0001
Error	423	187.09695	0.44231		
Corrected Total	427	223.32745			

Root MSE	0.66506	R-Square	0.1622
Dependent Mean	1.19017	Adj R-Sq	0.1543
Coeff Var	55.87956		

Parameter Estimates						
Variable	Label	DF	Parameter Estimate	Standard Error	t Value	Pr > |t|
Intercept	Intercept	1	–0.18686	0.28359	–0.66	0.5103
exper		1	0.04310	0.01318	3.27	0.0012
expersq		1	–0.00086280	0.00039368	–2.19	0.0290
educ		1	0.08039	0.02164	3.72	0.0002
v	Residual	1	0.04719	0.02855	1.65	0.0991

OUTPUT 4.6. Using Proc Reg to check for endogeneity.

4. Sargan's test statistic is nR_u^2. Under the null hypothesis of exogenous extra instruments, the test statistic is distributed as a chi-squared random variable with $L_2 - G$ degrees of freedom.

If the null hypothesis is rejected, then we need to reexamine the instruments that were selected for the analysis. The general idea is that if the instruments are truly exogenous, then they should not be correlated with the disturbance term.

We will now illustrate the computations by using the earning equation with parents and husband's education as instruments. The first step is to estimate the true model by using 2SLS and to store the residuals. The following SAS statements can be used. Note that the output has been suppressed because we are interested only in storing the residuals from this analysis.

```
proc syslin 2SLS noprint data=mroz out=step1_resid;
     endogenous educ;
     instruments exper expersq motheduc fatheduc huseduc;
     model lwage=exper expersq educ;
     output residual=out1_resid;
run;
```

The next step is to regress the residuals from the 2SLS analysis on all exogenous variables in the model. The following SAS statements can be used. The results of the analysis one given in Output 4.7.

```
proc reg data=step1_resid;
     model out1_resid=exper expersq motheduc
     fatheduc huseduc;
run;
```

The REG Procedure

Model: MODEL1

Dependent Variable: out1_resid Residual Values

Number of Observations Read	428
Number of Observations Used	428

Analysis of Variance					
Source	DF	Sum of Squares	Mean Square	F Value	Pr > F
Model	5	0.49483	0.09897	0.22	0.9537
Error	422	189.43989	0.44891		
Corrected Total	427	189.93471			

Root MSE	0.67001	R-Square	0.0026
Dependent Mean	−5.3125E-16	Adj R-Sq	−0.0092
Coeff Var	−1.2612E17		

Parameter Estimates						
Variable	Label	DF	Parameter Estimate	Standard Error	t Value	Pr > |t|
Intercept	Intercept	1	0.00861	0.17727	0.05	0.9613
exper		1	0.00005603	0.01323	0.00	0.9966
expersq		1	−0.00000888	0.00039562	−0.02	0.9821
motheduc		1	−0.01039	0.01187	−0.87	0.3821
fatheduc		1	0.00067344	0.01138	0.06	0.9528
huseduc		1	0.00678	0.01143	0.59	0.5532

OUTPUT 4.7. Testing overidentifying restrictions in the earning data.

There are 428 observations in the data set and $R^2 = 0.0026$. Therefore, the test statistic value is $NR^2 = 1.11$. The critical value is $\chi^2_{2,0.05} = 5.99$. The degrees of freedom were calculated using the formula $L_2 - G$, where $L_2 = 3$ because we used *motheduc*, *fatheduc*, and *huseduc* as instruments beyond *exper* and *expersq*. We suspect only one variable (*educ*) as being endogenous, $G = 1$. Thus, the degree of freedom is 2. The null hypothesis is not rejected because the test statistic value is smaller than the critical value. That is that we can use the "extra" instruments to identify the model for *y*.

4.5.2 Weak Instruments

We now turn our attention to the problem of weak instruments—that is, the case when the selected instrumental variables used in estimation have a poor correlation with the endogenous variable.

We will discuss a general method for determining if weak instruments have been used in the model. Consider the model $\mathbf{y} = \mathbf{X}\boldsymbol{\beta} + \alpha\mathbf{x} + \boldsymbol{\varepsilon}$ where $\mathbf{x}$ is suspected of being endogenous. Assume that we have a set of instrumental variables $\mathbf{W}$, which includes the explanatory variables in $\mathbf{X}$. The reduced form equation relating $\mathbf{x}$ to $\mathbf{X}$ and $\mathbf{W}$ is written as $\mathbf{x} = \mathbf{W}\boldsymbol{\delta} + \boldsymbol{\gamma}$ (Wooldridge, 2002).

If $\boldsymbol{\delta} = \mathbf{0}$, the instruments in $\mathbf{W}$ have no predictive power in explaining $\mathbf{x}$. A value of $\boldsymbol{\delta}$ close to zero implies that the instruments are weak. A rule of thumb proposed in the literature is that the weak instruments problem is a non-issue if the F statistic of the regression in the reduced form equation exceeds 10 (Glewwe, 2006). We will illustrate the computations by looking at the earning data set. The variable *educ* was suspected of being endogenous. The variables *motheduc*, *fatheduc*, and *huseduc* were considered

The REG Procedure
Model: MODEL1
Dependent Variable: educ

Number of Observations Read	428
Number of Observations Used	428

Analysis of Variance					
Source	DF	Sum of Squares	Mean Square	F Value	Pr > F
Model	5	955.83061	191.16612	63.30	<0.0001
Error	422	1274.36565	3.01982		
Corrected Total	427	2230.19626			

Root MSE	1.73776	R-Square	0.4286
Dependent Mean	12.65888	Adj R-Sq	0.4218
Coeff Var	13.72763		

Parameter Estimates					
Variable	DF	Parameter Estimate	Standard Error	t Value	Pr > |t|
Intercept	1	5.53831	0.45978	12.05	<0.0001
exper	1	0.03750	0.03431	1.09	0.2751
expersq	1	−0.00060020	0.00103	−0.58	0.5589
motheduc	1	0.11415	0.03078	3.71	0.0002
fatheduc	1	0.10608	0.02952	3.59	0.0004
huseduc	1	0.37525	0.02963	12.66	<0.0001

OUTPUT 4.8. Weak instruments analysis in the earning data.

as instruments and therefore the reduced regression equation for the wage equation is

$$educ = \alpha_0 + \alpha_1 exper + \alpha_2 exper^2 + \alpha_3 motheduc + \alpha_4 fatheduc + \alpha_5 huseduc + \gamma.$$

The reduced form parameters are estimated by OLS regression. The following SAS statements can be used. The analysis results are given in Output 4.8. Note that the *F* statistic value is very large (larger than 10) and therefore we cannot reject the hypothesis that we have weak instruments.

```
proc reg data=mroz;
    model educ=exper expersq motheduc fatheduc huseduc;
run;
```

4.5.3   Hausman's Specification Test

Hausman's specification test can be used to determine if there are significant differences between the OLS and the IV estimators. As discussed in Greene (2003, pp. 80–83), under the null hypothesis of no endogeneity, both the OLS and the IV estimators are consistent. Under the alternative hypothesis of endogeneity, only the IV estimator is consistent. Hausman's test is based on the

TABLE		
	OLS	**IV**
INTERCEPT	−0.0453	−0.0208
YT	0.1847	0.0892
IT	−0.0017	−0.0012
CTI	0.8205	0.9140

	H
The Hausman Test Statistic Value is	21.093095

OUTPUT 4.9. Hausman analysis using Proc IML for consumption data.

principle that if there are two estimators $(\boldsymbol{\beta}^1, \boldsymbol{\beta}^2)$ which converge to $\boldsymbol{\beta}$ under the null hypothesis and converge to different values under the alternative hypothesis then the null hypothesis can be tested by testing whether the two estimators are different. The test statistic is given by

$$H = \mathbf{d}^T [s_{IV}^2 (\hat{\mathbf{X}} T \hat{\mathbf{X}})^{-1} - s_{OLS}^2 (\mathbf{X}^T \mathbf{X})^{-1}]^{-1}.$$

Here, $\mathbf{d} = [\mathbf{b}_{IV} - \mathbf{b}_{OLS}]$, and $s_{IV}^2 (\hat{\mathbf{X}}^T \hat{\mathbf{X}})^{-1}$, $s_{OLS}^2 (\mathbf{X}^T \mathbf{X})^{-1}$ are the terms associated with the asymptotic covariance of the two estimators, respectively. Under the null hypothesis, H is distributed as a χ^2 with k^* degrees of freedom. The degree of freedom, k^*, is the number of variables in $\mathbf{X}$ that are suspected of being endogenous. We will use the consumption function data in Greene (2003) to illustrate the computations involved in SAS. The data is credited to the Department of Commerce, BEA. The author proposes estimating a model given by $c_t = \beta_1 + \beta_2 y_t + \beta_3 i_t + \beta_4 c_{t-1} + \varepsilon_t$, where c_t is the log of real consumption, y_t is the log of real disposable income, and i_t is the interest rate. We suspect a possible correlation between y_t and ε_t and consider y_{t-1}, c_{t-1}, and i_t as possible instruments (Greene, 2003, Example 5.3). The following Proc IML commands can be used to calculate Hausman's test statistic. The analysis results are given in Output 4.9. We assume that the data have been read into a temporary SAS data set called *hausman*. The names of the variables are self-explanatory. For instance, y_{t-1} and c_{t-1} are labeled as *yt1* and *ct1*. The first step is to read the data into appropriate matrices.

```
Proc IML;
    use hausman;
    read all var {'yt' 'it' 'ct1'} into X;
    read all var {'ct'} into Y;
    read all var {'it' 'ct1' 'yt1'} into W;
    n=nrow(X);
    k=ncol(X);
    X=J(n,1,1)||X;
    W=J(n,1,1)||W;
```

Next, we need to compute the OLS and IV estimators.

```
CX=inv(X'*X);
CW=inv(W'*W);
OLS_b=CX*X'*y;
Xhat=W*CW*W'*X;
b_IV=inv(Xhat'*X)*Xhat'*y;
```

Next, we need to compute the difference vector and calculate the consistent estimator of σ^2.

```
d=b_IV-OLS_b;
SSE1=y'*y-OLS_b'*X'*Y;
SSE2=y'*y-b_IV'*X'*Y;
DFE1=n-k;
DFE2=n;
MSE1=SSE1/DFE1;
MSE2=SSE/DFE2;
```

The last step is to calculate the test statistic, H, and print out the results.

```
diff=ginv(MSE2*inv(Xhat'*Xhat)-MSE1*CX);
H=d'*diff*d;
```

Since the 95% critical value from the chi-squared table is 3.84, we reject the null hypothesis of no correlation between y_t and ε_t. Therefore, the IV estimator is more appropriate to use for the consumption model. Hausman's test can also be performed in SAS by using the Proc model procedure. The following SAS statements can be used. Notice that we specify the endogenous variable in the "endo" statement, the instruments in the "instruments" statement, and then write down the linear model to be estimated. This is followed by the "fit" statement using the dependent variable with the option that the Hausman test be used to compare the OLS and the instrumental variable estimator.

```
proc model data=hausman;
    endo yt;
    instruments it ct1 yt1;
    ct=beta1+beta2*yt+beta3*it+beta4*ct1;
    fit ct/ols 2sls hausman;
run;
```

The procedure checks if the OLS estimates are more efficient than the 2SLS procedure. The degree of freedom used for the test is k, the number of columns of **X**. The analysis results are produced in Output 4.10.

The MODEL Procedure

Model Summary	
Model Variables	2
Endogenous	1
Parameters	4
Equations	1
Number of Statements	1

Model Variables	yt ct
Parameters	beta1 beta2 beta3 beta4
Equations	ct

The Equation to Estimate is	
ct =	F(beta1(1), beta2(yt), beta3(it), beta4(ct1))
Instruments	1 it ct1 yt1

NOTE: At OLS iteration 1 CONVERGE=0.001 criterion met.

OUTPUT 4.10. Hausman test for the consumption data using Proc model.

The MODEL Procedure
OLS Estimation Summary

Data Set Options	
DATA=	HAUSMAN

Minimization Summary	
Parameters Estimated	4
Method	Gauss
Iterations	1

Final Convergence Criteria	
R	6.19E-11
PPC	1.95E-10
RPC(beta4)	8122.857
Object	0.999999
Trace(S)	0.000066
Objective Value	0.000065

Observations Processed	
Read	202
Solved	202

The MODEL Procedure

Nonlinear OLS Summary of Residual Errors							
Equation	DF Model	DF Error	SSE	MSE	Root MSE	R-Square	Adj R-Sq
ct	4	198	0.0131	0.000066	0.00814	0.9997	0.9997

Nonlinear OLS Parameter Estimates				
Parameter	Estimate	Approx Std Err	t Value	Approx Pr > |t|
beta1	−0.04534	0.0130	−3.49	0.0006
beta2	0.18466	0.0330	5.60	<0.0001
beta3	−0.00165	0.000294	−5.62	<0.0001
beta4	0.820509	0.0323	25.38	<0.0001

Number of Observations		Statistics for System	
Used	202	**Objective**	0.0000650
Missing	0	**Objective*N**	0.0131

NOTE: At 2SLS Iteration 1 convergence assumed because OBJECTIVE=7.589157E-27 is almost zero (<1E-12).

OUTPUT 4.10. *(Continued)*

The MODEL Procedure
2SLS Estimation Summary

Data Set Options

DATA=	HAUSMAN

Minimization Summary

Parameters Estimated	4
Method	Gauss
Iterations	1

Final Convergence Criteria

R	1
PPC	6.41E-11
RPC(beta1)	0.541796
Object	0.676767
Trace(S)	0.000069
Objective Value	7.59E-27

Observations Processed

Read	202
Solved	202

The MODEL Procedure

Nonlinear 2SLS Summary of Residual Errors

Equation	DF Model	DF Error	SSE	MSE	Root MSE	R-Square	Adj R-Sq
ct	4	198	0.0137	0.000069	0.00831	0.9997	0.9997

Nonlinear 2SLS Parameter Estimates

Parameter	Estimate	Approx Std Err	t Value	Approx Pr > |t|
beta1	−0.02077	0.0141	−1.47	0.1429
beta2	0.089197	0.0386	2.31	0.0218
beta3	−0.00116	0.000315	−3.67	0.0003
beta4	0.913973	0.0378	24.18	<0.0001

Number of Observations / **Statistics for System**

Used	202	Objective	7.589E-27
Missing	0	Objective*N	1.533E-24

Hausman's Specification Test Results

Comparing	To	DF	Statistic	Pr > ChiSq
OLS	2SLS	4	22.74	0.0001

OUTPUT 4.10. (*Continued*)

The first few tables contain some information about the model including the number of variables, and the number of endogenous and exogenous variables. This is followed by output from both OLS and instrumental variable estimation. Note that convergence was achieved very quickly for both models. Also note that the OLS standard errors are smaller than the instrumental variables standard error. The parameter estimates for the intercept, y_i and i_t, are larger in magnitude than the ones obtained from instrumental variable estimation. The OLS estimate for the parameter value of c_{t-1} is smaller than the instrumental variable estimate. The value of the test statistic for Hausman's test is 22.74 with 4 degrees of freedom and is highly significant indicating that the instrumental variable estimator is more efficient than the OLS estimator.

Notice that the test statistic value from Proc Model is very close to the one obtained by using Proc IML. On the other hand, the degree of freedom used for the test in Proc Model is equal to the number of variables in **X**. The test procedure described in Greene (2003) uses as degrees of freedom the number of variables that are suspected of being endogenous. Greene (2003, pp. 81–82) provides a brief explanation that justifies the setting of the degrees of freedom to the number of suspected endogenous variables.

5

NONSPHERICAL DISTURBANCES AND HETEROSCEDASTICITY

5.1 INTRODUCTION

The discussion in the previous chapters was based on the assumption that the disturbance vector in the linear model $\mathbf{y} = \mathbf{X}\boldsymbol{\beta} + \boldsymbol{\varepsilon}$ is such that the conditional distribution $\varepsilon_i \mid \mathbf{X}$ is independently and identically distributed with zero mean and constant variance σ^2. The implication of this assumption is that the variance of $\boldsymbol{\varepsilon}$ does not change with changes in the conditional expectation, $E(\mathbf{y} \mid \mathbf{X})$. A plot of $\boldsymbol{\varepsilon}$ versus $E(\mathbf{y} \mid \mathbf{X})$ should therefore exhibit a random scatter of data points. The random disturbances under this assumption are referred to as spherical disturbances. This chapter deals with alternative methods of analysis under violations of this assumption. The implication here is that $Var(\boldsymbol{\varepsilon} \mid \mathbf{X}) = \sigma^2 \boldsymbol{\Omega} = \boldsymbol{\Sigma}$, where $\boldsymbol{\Omega}$ is a positive definite, symmetric matrix. The random disturbances under this assumption are referred to as nonspherical disturbances. Although the general method presented in this chapter can be extended to instrumental variables regression very easily, we will assume that the explanatory variables that are used in the model are exogenous. We will deal with two cases of nonspherical disturbances.

1. *Heteroscedasticity:* Here, the disturbances are assumed to have different variances. The variance of the disturbance may, for example, be dependent upon the conditional mean $E(\mathbf{y} \mid \mathbf{X})$. For example, this will happen in the case when the disturbances are assumed to follow the binomial distribution. Recall that if a random variable γ has a binomial distribution with parameters n (the number of trials) and success probability p, then the mean and variance of γ are np and $np(1 - p)$, respectively. Therefore the variance is dependent on the mean np; and as the mean changes, so does the variance. We will revisit this case in Chapter 10 in our discussion of discrete choice models using logistic regression. In our discussion of heteroscedasticity we will assume uncorrelated disturbances so that $\boldsymbol{\Sigma} = diag[\sigma_1^2, \sigma_2^2, \ldots, \sigma_n^2]$. Chapter 6 deals with the case of autocorrelation where the disturbances are correlated.

2. *Autocorrelation:* This often occurs in time series data where error terms between time periods are correlated. Here, we assume homoscedastic disturbances where the variances of the disturbances are equal but the disturbances are correlated. Therefore, $\boldsymbol{\Sigma}$ is no longer diagonal and is given by (Greene, 2003, p. 192)

$$\boldsymbol{\Sigma} = \sigma^2 \begin{bmatrix} 1 & \rho_1 & \cdots & \rho_{n-1} \\ \rho_1 & 1 & \cdots & \rho_{n-2} \\ \vdots & \vdots & \ddots & \vdots \\ \rho_{n-1} & \rho_{n-2} & \cdots & 1 \end{bmatrix}.$$

Applied Econometrics Using the SAS® System, by Vivek B. Ajmani
Copyright © 2009 John Wiley & Sons, Inc.

In this chapter and the next, we will discuss analysis techniques when the disturbances are either heteroscedastic or autocorrelated.

5.2 NONSPHERICAL DISTURBANCES

We start our discussion with estimation techniques when the disturbances are heteroscedastic. As seen in Chapter 1, under the assumptions $E(\varepsilon|\mathbf{X}) = \mathbf{0}$ and $V(\varepsilon|\mathbf{X}) = \sigma^2\mathbf{I}$, the least squares estimator of $\boldsymbol{\beta}$ is given by $\mathbf{b} = (\mathbf{X}^T\mathbf{X})^{-1}\mathbf{X}^T\mathbf{y}$. We also saw that the least squares estimator is the best linear unbiased estimator and that it is consistent and asymptotically normal (if the disturbance vector is normally distributed). It is easy to show that the unbiased property of the least squares estimator is unaffected under departures from the spherical disturbances assumption and that the variance–covariance matrix of $\mathbf{b}$ under heteroscedasticity is given by

$$Var(\mathbf{b}|\mathbf{X}) = \sigma^2(\mathbf{X}^T\mathbf{X})^{-1}\mathbf{X}^T\boldsymbol{\Omega}\mathbf{X}(\mathbf{X}^T\mathbf{X})^{-1}.$$

If we assume that the disturbances are normally distributed, then it is easily shown that the conditional distribution of $\mathbf{b}$ is also normal. That is,

$$\mathbf{b}|\mathbf{X} \sim N(\boldsymbol{\beta}, \sigma^2(\mathbf{X}^T\mathbf{X})^{-1}\mathbf{X}^T\boldsymbol{\Omega}\mathbf{X}(\mathbf{X}^T\mathbf{X})^{-1}).$$

In the previous chapters, all inferences regarding the least squares estimator were done using the estimated covariance matrix $s^2(\mathbf{X}^T\mathbf{X})^{-1}$. However, under departures from the spherical disturbance assumption, $Var(\mathbf{b}|\mathbf{X}) \neq \sigma^2(\mathbf{X}^T\mathbf{X})^{-1}$ and therefore any inference with $s^2(\mathbf{X}^T\mathbf{X})^{-1}$ will be incorrect. That is, the hypothesis tests and confidence intervals using the t, F, χ^2 distributions will not be valid (Greene, 2003, p. 194).

5.2.1 Estimation of $\boldsymbol{\beta}$

There are two methods for estimating $\boldsymbol{\beta}$ under the assumption of nonspherical disturbances. The first case assumes that the structure of $\boldsymbol{\Omega}$ is known and the other when it is assumed unknown. Estimators obtained under the assumption that $\boldsymbol{\Omega}$ is known are called the *Generalized Least Squares* (GLS) estimators, while those obtained under the assumption that $\boldsymbol{\Omega}$ is unknown are called the *Feasible Generalized Least Squares* (FGLS) estimators.

To start with, assume that $\boldsymbol{\Omega}$ is a known positive definite symmetric matrix. Premultiplying the linear model by $\boldsymbol{\Omega}^{-1/2}$ gives

$$\boldsymbol{\Omega}^{-1/2}\mathbf{y} = \boldsymbol{\Omega}^{-1/2}\mathbf{X}\boldsymbol{\beta} + \boldsymbol{\Omega}^{-1/2}\varepsilon \quad \text{or} \quad \mathbf{y}^* = \mathbf{X}^*\boldsymbol{\beta} + \varepsilon^*.$$

It is easy to show that $E(\varepsilon^*|\mathbf{X}) = \mathbf{0}$ and that $Var(\varepsilon^*|\mathbf{X}) = \sigma^2\mathbf{I}$. See both Greene (2003, p. 207) and Verbeek (2004, p. 81) for more details.

Therefore, the classical regression model assumptions of spherical disturbances are satisfied and the analysis techniques from the previous chapters can be used for estimation on the transformed variables. It can be shown that under the transformation used, the GLS estimator is consistent and unbiased and is given by

$$\mathbf{b}_{GLS} = (\mathbf{X}^T\boldsymbol{\Omega}^{-1}\mathbf{X})^{-1}\mathbf{X}^T\boldsymbol{\Omega}^{-1}\mathbf{y}.$$

with variance–covariance matrix

$$Var(\mathbf{b}_{GLS}|\mathbf{X}) = \sigma^2(\mathbf{X}^T\boldsymbol{\Omega}^{-1}\mathbf{X})^{-1}.$$

The GLS estimator is asymptotically normally distributed if the disturbances are normally distributed. That is, $\mathbf{b}_{GLS} \sim N(\boldsymbol{\beta}, \sigma^2(\mathbf{X}^T\boldsymbol{\Omega}^{-1}\mathbf{X})^{-1})$.

In reality, $\boldsymbol{\Omega}$ is unknown and has to be estimated. As discussed in Greene (2003, p. 209), if $\boldsymbol{\Omega}$ is allowed to be unrestricted as in the case of autocorrelation, then there are $n(n + 1)/2$ additional parameters in $\sigma^2\boldsymbol{\Omega}$ that need to be estimated. This is impossible, given that we have a total of n observations. In such cases, calculating $\mathbf{b}_{GLS}$ is not possible and a FGLS estimator has to be used. The FGLS estimation involves putting a restriction on the number of parameters that needs to be estimated. For instance, in heteroscedasticity, we restrict $\sigma^2\boldsymbol{\Omega}$ to one new parameter, θ, defined as $\sigma_i^2 = \sigma^2 z_i^\theta$ (Greene, 2003, p. 210).

Details on estimating θ will be provided a bit later in this chapter. For now, assume that θ can be estimated by $\hat{\theta}$. We can then estimate $\boldsymbol{\Omega}$ with $\boldsymbol{\Omega}(\hat{\theta})$ and use this in the formula for the GLS estimate to obtain the FGLS estimator. Illustration of the computation methods involved in computing the GLS and the FGLS estimators will be discussed in the following sections. For the moment, we will shift our attention to the task of detecting whether the regression model suffers from the heteroscedasticity.

5.3 DETECTING HETEROSCEDASTICITY

As mentioned earlier, heteroscedasticity implies that the variances of the disturbances are not constant across observations. Therefore, an easy way of detecting heteroscedasticity is to plot the least squares residuals, $\hat{\varepsilon}_i$, against the predicted values of the dependent variable and against all the independent variables in the model. Heteroscedasticity should be suspected if any of the graphs indicate a funnel-shaped (or some other nonrandom) pattern. That is, the graph gets more scattered as the predicted value of the dependent or independent variables change. As an example, consider the credit card data in Greene (2003). The data consist of monthly credit card expenses for 100 individuals. This data set was used with permission from William H. Greene (New York University) and is credited to Greene (1992). The author conducted a linear regression of monthly expenses on a constant, age, a dummy variable indicating ownership of a house, income, and the square of income using 72 observations where the average expense is nonzero. The following statements can be used (note that the square of income was calculated in the data step statement) to conduct the analysis. The analysis results are given in Output 5.1. We will not discuss the output results as we are interested in the residual plots.

```
proc reg data=Expense;
      model AvgExp = Age OwnRent Income IncomeSq;
      output out=for_graphs student=r_s;
run;
```

The option '*output out=for_graph student=r_s*' creates a SAS data set with standardized residuals along with the variables used in the model. The following GPLOT statements can now be used to create the residual plots. Note that the statements are set to create a plot of the standardized residuals versus *income*.

```
proc gplot data=for_graphs;
      plot r_s*income;
run;
```

The REG Procedure
Model: MODEL1
Dependent Variable: AvgExp AvgExp

Number of Observations Read	72
Number of Observations Used	72

Analysis of Variance					
Source	DF	Sum of Squares	Mean Square	F Value	Pr > F
Model	4	1749357	437339	5.39	0.0008
Error	67	5432562	81083		
Corrected Total	71	7181919			

Root MSE	284.75080	R-Square	0.2436
Dependent Mean	262.53208	Adj R-Sq	0.1984
Coeff Var	108.46324		

Parameter Estimates						
Variable	Label	DF	Parameter Estimate	Standard Error	t Value	Pr > |t|
Intercept	Intercept	1	−237.14651	199.35166	−1.19	0.2384
Age	Age	1	−3.08181	5.51472	−0.56	0.5781
OwnRent	OwnRent	1	27.94091	82.92232	0.34	0.7372
Income	Income	1	234.34703	80.36595	2.92	0.0048
incomesq		1	−14.99684	7.46934	−2.01	0.0487

OUTPUT 5.1. Ordinary least squares analysis of credit card expenses data.

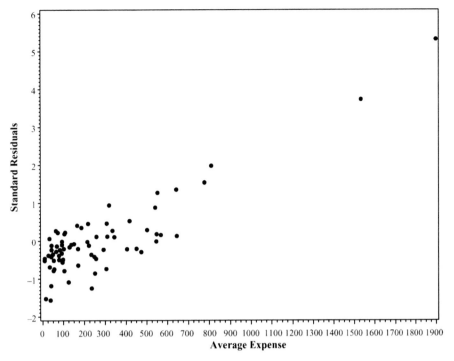

FIGURE 5.1. Plot of residuals versus average expense.

Three residual plots were created (Figures 5.1–5.3): one versus average expense, one versus age, and the last versus income. The residual plots with age and income show a funnel-shaped pattern with the residuals "ballooning" up with increases in age and income. We should therefore suspect that the regression model used for the credit card data suffers from heteroscedasticity. More specifically, it appears that the variance of the residuals is correlated to some function of the explanatory variable income.

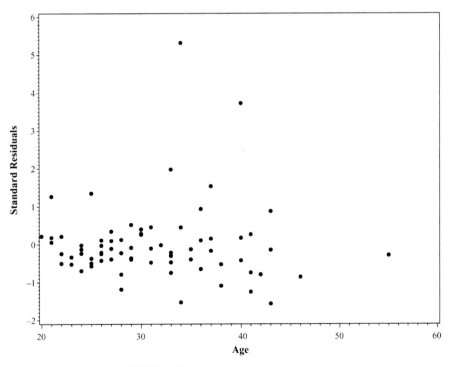

FIGURE 5.2. Plot of residuals versus age.

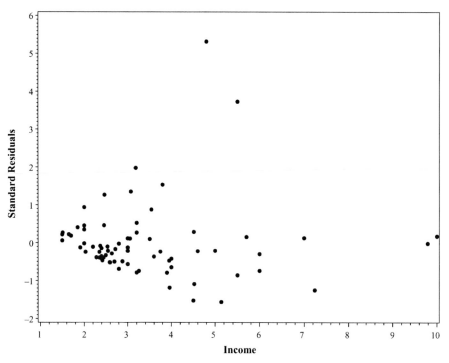

FIGURE 5.3. Plot of residuals versus income.

5.4   FORMAL HYPOTHESIS TESTS TO DETECT HETEROSCEDASTICITY

We prefer the simple residual plots analysis to detect heteroscedasticity. However, as discussed in Greene (2003, pp. 222–225) and Verbeek (2004, pp. 90–92), there are three formal tests for detecting the presence of nonspherical disturbances. They are

1. White's general test
2. The Goldfeld–Quandt test
3. The Breusch–Pagan test

5.4.1   White's Test

White's general test can be conducted via the hypothesis

$$H_0: \quad \sigma_i^2 = \sigma^2 \qquad \forall i,$$
$$H_1: \quad \text{Not } H_0.$$

It is a general test that looks for evidence of a relationship between the variance of the disturbance and the regressors (and functions of the regressors) without assuming any form of the relationship. This test can be conducted as follows:

1. Estimate ε_i via OLS to obtain $\hat{\varepsilon}_i$ and calculate $\hat{\varepsilon}_i^2$.
2. Conduct an OLS regression of $\hat{\varepsilon}_i^2$ on all unique variables in **X** along with all the squares and cross products of the unique variables in **X**.
3. The test statistic is given by nR^2, which under the null hypothesis is asymptotically distributed as chi-squared with $p - 1$ degrees of freedom, where p is the number of explanatory variables in the OLS regression in step 2.

See Greene (2003), Verbeek (2004), and the SAS/ETS User's Guide 9.1 for more details on this test. White's test can be programmed into Proc IML as follows:

1. First, read the data including the unique variables formed by using cross products into appropriate matrices. Using the credit card expense data, note that there are only 13 unique columns in the cross-product matrix since *ownrent²* = *ownrent*

and $income \times income = income^2$, which is already part of $\mathbf{X}$. The original list of explanatory variables include *age*, *ownrent*, *income*, and *incomesq*. The cross products are therefore *age*age*, *age*ownrent*, *age*income*, *age*incomesq*, *ownrent*income*, *ownrent*incomesq*, *incomesq*incomesq*, *income*incomesq*.

```
proc IML;
     use Expense;
     read all var {'age' 'ownrent' 'income'
     'incomesq'} into X;
     read all var {'age' 'ownrent' 'income'
     'incomesq' 'age_sq' 'incomefth' 'age_or'
     'age_inc' 'age_incsq' 'or_income' 'or_incomesq'
     'incomecube'} into XP;
     read all var {'avgexp'} into Y;
     n=nrow(X);
     np=nrow(XP);
     X=J(n,1,1)||X;
     XP=J(np,1,1)||XP;
     k=ncol(X);
     kp=ncol(XP);
```

Next, calculate the OLS residuals using the techniques from the previous chapters.

```
C=inv(X'*X);
beta_hat=C*X'*y;
resid=y-X*beta_hat;
```

2. The square of the OLS residuals is then regressed against the unique variables including the cross-product variables.

```
resid_sq=resid#resid;
C_E=inv(XP'*XP);
b_hat_e=C_E*XP'*resid_sq;
```

3. The R^2 value is then calculated from this regression.

```
Mean_Y=Sum(resid_sq)/np;
SSR=b_hat_e'*XP'*resid_sq-np*Mean_Y**2;
SSE=resid_sq'*resid_sq-b_hat_e'*XP'*resid_sq;
SST=SSR+SSE;
R_Square=SSR/SST;
```

4. Finally, the test statistic value is calculated.

```
     White=np*R_Square;
     pvalue= 1 - probchi(White, kp);
run;
```

The results of the analysis are given in Output 5.2.

The *p* value indicates that there is insufficient evidence to claim that the disturbances are heteroscedastic. White's test can also be done by using the Proc Model procedure. We will again use the credit card data to illustrate this. As opposed to the complexity involved in Proc IML of determining the number of unique columns in the cross product $\mathbf{X} \otimes \mathbf{X}$, the technicalities do not matter in

R_SQUARE
0.1990132

	WHITE
The test statistic value for Whites Test is	14.328953

	PVALUE
The p-value associated with this test is	0.3510922

OUTPUT 5.2. White's test for the credit card expense data.

the Proc Model. The following statements can be used. Note that we have also included an option to conduct the Breusch–Pagan test, which will be discussed later.

The Proc Model procedure used here contains four main parts. First, we define the parameters of the model using the "parms" option. Here, we chose the names Const (for the intercept), C_Age for Age, C_OwnRent for OwnRent, C_Income for Income, and C_IncomeSq for Income*Income. The next part is the actual layout of the model of interest. It should be obvious that we are regressing Average Expense against Age, Income, Income*Income, and OwnRent. The next part is used to define the squared Income term which will be used in the Breusch–Pagan test using the "breusch" option. We could have eliminated the definition had we chosen to simply conduct White's test. The final part uses the "fit" option on the dependent variable of interest to fit the model.

```
proc model data=Expense;
    parms Const C_Age C_OwnRent C_Income C_IncomeSq;
    AvgExp = Const + C_Age*Age + C_OwnRent*OwnRent
    + C_Income*Income + C_IncomeSq*Income*Income;
    income_sq = income * income;
    fit AvgExp/white breusch=(1 Income Income_Sq);
run;
```

Output 5.3 reveals that the test statistic value for White's test is 14.33 with p value equal to 0.28. Therefore, we do not reject the null hypothesis of homoscedastic disturbances. This is startling since the residual plots did indicate that the disturbances were nonspherical. This contradiction points to the nature of White's test. The generality of the test leads to a "poor power" of detecting heteroscedasticity when it may exist in reality. On the other hand, rejecting the null hypothesis leads to no indication of what should be done in terms of adjusting for heteroscedasticity since it offers no insight on the problematic variable(s).

The MODEL Procedure

Model Summary	
Model Variables	1
Parameters	5
Equations	1
Number of Statements	2

Model Variables	AvgExp
Parameters	Const C_Age C_OwnRent C_Income C_IncomeSq
Equations	AvgExp

The Equation to Estimate is	
AvgExp =	F(Const(1), C_Age(Age), C_OwnRent(OwnRent), C_Income(Income), C_IncomeSq)

NOTE: At OLS Iteration 1 CONVERGE=0.001 Criteria Met.

OUTPUT 5.3. White's test on credit card expense data using the Proc Model.

The MODEL Procedure
OLS Estimation Summary

Data Set Options	
DATA=	EXPENSE

Minimization Summary	
Parameters Estimated	5
Method	Gauss
Iterations	1

Final Convergence Criteria	
R	0
PPC	0
RPC(Const)	2347986
Object	0.55266
Trace(S)	81083.02
Objective Value	75452.25

Observations Processed	
Read	72
Solved	72

The MODEL Procedure

			Nonlinear OLS Summary of Residual Errors					
Equation	DF Model	DF Error	SSE	MSE	Root MSE	R-Square	Adj R-Sq	Label
AvgExp	5	67	5432562	81083.0	284.8	0.2436	0.1984	AvgExp

Nonlinear OLS Parameter Estimates				
Parameter	Estimate	Approx Std Err	t Value	Approx Pr > |t|
Const	-237.147	199.4	-1.19	0.2384
C_Age	-3.08181	5.5147	-0.56	0.5781
C_OwnRent	27.94091	82.9223	0.34	0.7372
C_Income	234.347	80.3660	2.92	0.0048
C_IncomeSq	-14.9968	7.4693	-2.01	0.0487

Number of Observations		Statistics for System	
Used	72	Objective	75452
Missing	0	Objective*N	5432562

Heteroscedasticity Test					
Equation	Test	Statistic	DF	Pr > ChiSq	Variables
AvgExp	White's Test	14.33	12	0.2802	Cross of all vars
	Breusch-Pagan	6.19	2	0.0453	1, Income, income_sq

OUTPUT 5.3. (*Continued*)

5.4.2   The Goldfeld–Quandt and Breusch–Pagan Tests

Most often, heteroscedasticity is caused by a relationship of the variance of the disturbances with one or more explanatory variables or their functions. The Goldfeld–Quandt and Breusch–Pagan tests are more powerful and therefore preferred over the White's test because they restrict attention on explanatory variables that appear to cause heteroscedasticity (Greene, 2003, p. 223; Verbeek, 2004, p. 90). The residual plots clearly indicated that heteroscedasticity may have been caused by *income*. We can therefore focus on this variable to determine whether there is evidence that the disturbances are heteroscedastic with respect to it.

For the Goldfeld–Quandt test, we assume that the data set can be split into two groups based on the explanatory variable that appears to be causing heteroscedasticity. The method involves first ranking the data with respect to the "problematic" explanatory variable. The hope is that this separation will split the data set into two groups with high and low variances. Regression analysis is then conducted on the two groups separately. The hypothesis being tested is

$$H_0: \quad \sigma_1^2 = \sigma_2^2$$
$$H_1: \quad \sigma_1^2 > \sigma_2^2$$

and the test statistic is

$$F = \frac{\mathbf{e}_1^T \mathbf{e}_1 / (n_1 - k - 1)}{\mathbf{e}_2^T \mathbf{e}_2 / (n_2 - k - 1)}.$$

Here, $\mathbf{e}_1^T \mathbf{e}_1$ and $\mathbf{e}_2^T \mathbf{e}_2$ are the error sums of squares from the two independent regressions where $\mathbf{e}_1^T \mathbf{e}_1 \geq \mathbf{e}_2^T \mathbf{e}_2$. Notice that this is simply the F test for comparing two variances. Under the null hypothesis of no heteroscedasticity, this test statistic has an F distribution with $n_1 - k - 1$ and $n_2 - k - 1$ degrees of freedom (Greene, 2003, p. 223).

As an example of implementing the test in SAS, we will use the credit card data again. The following statements can be used.

```
proc import out=Expense
     datafile="C:\TempTableF91"
     dbms=Excel Replace;
     getnames=yes;
run;
Data Expense;
     set Expense;
     incomesq=income*income;
     if avgexp > 0;
run;
proc sort data=Expense;
     by income;
run;
data Expense;
     set Expense;
     if _n_ < 37 then id=1;else id=2;
run;
proc reg data=Expense outest=est noprint;
     model avgexp=age ownrent income incomesq;
     by id;
run;
proc print data=est;
run;
```

Output 5.4 contains the results of the analysis. Note that we have suppressed all other output resulting from the above statements since we are interested in obtaining the mean square for error for the two splits of the data set. This is simply the square of the root mean square error (_RMSE_).

Obs	id	_MODEL_	_TYPE_	_DEPVAR_	_RMSE_	Intercept	Age
1	1	MODEL1	PARMS	AvgExp	102.587	153.130	-4.13740
2	2	MODEL1	PARMS	AvgExp	397.335	-259.108	-1.94040

Obs	OwnRent	Income	incomesq	AvgExp
1	108.872	16.886	3.6934	-1
2	-52.828	250.135	-16.1141	-1

OUTPUT 5.4. Regression summary statistics of the two splits of the credit card expenditure data.

The test statistic value is given by

$$F = \frac{397.335^2}{102.587^2} = 15.$$

Note that the numerator and denominator degrees of freedom are $36 - 4 - 1 = 31$ so that the critical value from the F table with type 1 error rate 5% is 1.822. The test statistic value exceeds the critical value, and we therefore reject the null hypothesis of homoscedasticity and state that there is evidence of heteroscedasticity caused by income.

As discussed in Greene (2003, p. 223), even though the Goldfeld–Quandt test has a higher power than White's test for detecting heteroscedasticity, there is a major criticism of the test. The test requires knowledge of the regressor that will be used to separate the data set and there may be instances where more than one regressor is involved. In the case of the credit card data, the residuals showed a heteroscedastic behavior with respect to income. A plot of residuals versus the square of income reveals heteroscedasticity also (we chose not to include the plot here). The Goldfeld–Quandt test therefore has limitations.

5.4.3   The Breusch–Pagan Test

The Lagrange Multiplier test designed by Breusch and Pagan takes into account the possibility of several "problematic" regressors. This test is based on the hypothesis that the variance of the disturbances is a function of one or more explanatory variables. That is, $\sigma_i^2 = \sigma^2 F(\alpha_0 + \boldsymbol{\alpha}^T \mathbf{z}_i)$, where $\mathbf{z}_i$ is a vector of independent variables (Greene, 2003, p. 224; Verbeek, 2004, p. 91). A test for homoscedasticity can therefore be conducted by testing

$$H_0 : \quad \boldsymbol{\alpha} = \mathbf{0},$$
$$H_1 : \quad \boldsymbol{\alpha} \neq \mathbf{0}.$$

The test statistic for this test is given by

$$LM = \frac{1}{2}\left[\mathbf{g}^T \mathbf{Z}(\mathbf{Z}^T\mathbf{Z})^{-1}\mathbf{Z}^T\mathbf{g}\right],$$

where $\mathbf{g}_i = n\,\varepsilon_i^2\,/\mathbf{e}^T\mathbf{e} - 1$, ε_i^2 is the square of the residuals and $\mathbf{e}^T\mathbf{e}$ is the OLS, residuals sums of squares, respectively (Breusch and Pagan, 1979; Greene, 2003, p. 224). From Chapter 1, it should be clear that the term within brackets of the LM statistic formula is the regression sums of squares when $\mathbf{g}$ is regressed on $\mathbf{Z}$. Under the null hypothesis, the LM statistic has a chi-squared distribution with k degrees of freedom, where k is the number of variables in $\mathbf{Z}$. SAS does not have a procedure that computes this version of the Breusch–Pagan test.

The version of the Breusch–Pagan test provided by SAS is the modification suggested by Koenker (1981) and Koenker and Bassett (1982). The authors showed that the LM test is not robust under departures from the normality assumption. They suggested a more robust estimate of the variance of the residuals. Details of their test can be found in Greene (2003, p. 224) and the SAS/ETS User's Guide 9.1.

We illustrate the computations involved by again making use of the credit card data. The vector $\mathbf{z}_i$ contains the constant term, income, and the square of income. The following statements in Proc IML will compute the test statistic. Note that we are assuming that the reader will have no problems reading the data into matrices and getting the results printed. We therefore just give the code that computes the test statistic value.

	LM
The Breusch Pagan Test Statistic Value is	41.920303

	PVAL
The p value associated with this is	7.891E-10

The null hypothesis of homoscedasticity is rejected

OUTPUT 5.5. The Breusch–Pagan test for the credit card expenditure data.

```
bhat_OLS=inv(X'*X)*X'*y;
SSE=(y-X*bhat_OLS)'*(y-X*bhat_OLS);
resid=y-X*bhat_OLS;
g=J(n,1,0);
fudge=SSE/n;
do index=1 to n;
     temp1=resid[index,1]*resid[index,1];
     g[index,1]=temp1/fudge - 1;
end;
LM=0.5*g'*Z*inv(Z'*Z)*Z'*g;
```

The analysis results are given in Output 5.5. The null hypothesis of homoscedasticity is rejected since the p value is almost zero.

The SAS output for White's test contained the test statistic and p value associated with the modified Breusch–Pagan test. The test statistic value is 6.19 with a p value of 0.0453, which leads us to reject the null hypothesis of homoscedasticity.

5.5   ESTIMATION OF β REVISITED

We now turn our attention back to estimation of the least parameters under heteroscedasticity. Our discussion starts with estimating a robust version of the variance–covariance matrix of the ordinary least squares estimator. These robust estimators will then be used to calculate the standard errors of the least squares estimators and to perform hypothesis tests. We will then move to weighted least squares estimation and estimation of the parameters using one-step and two-step FGLS.

Earlier in this chapter, we showed that under heteroscedasticity, the variance–covariance matrix of the least squares estimator is

$$Var(\mathbf{b}|\mathbf{X}) = (\mathbf{X}^T\mathbf{X})^{-1}(\mathbf{X}^T\sigma^2\mathbf{\Omega}\mathbf{X})(\mathbf{X}^T\mathbf{X})^{-1}.$$

In practice, $\mathbf{\Omega}$ is almost always unknown and therefore has to be estimated. White (1980) suggested the following estimator for $Var(\mathbf{b}|\mathbf{X})$ (Greene, 2003, p. 220; Verbeek, 2004, p. 87)

$$Est.Asy.Var(\mathbf{b}) = \frac{1}{n}\left(\frac{\mathbf{X}^T\mathbf{X}}{n}\right)^{-1}\left(\frac{1}{n}\sum_{i=1}^{n}\hat{\varepsilon}_i^2\mathbf{x}_i\mathbf{x}_i^T\right)\left(\frac{\mathbf{X}^T\mathbf{X}}{n}\right)^{-1}$$
$$= n(\mathbf{X}^T\mathbf{X})^{-1}\mathbf{S}_0(\mathbf{X}^T\mathbf{X})^{-1},$$

where $\hat{\varepsilon}_i$ is the ith least squares residual. As discussed in Greene (2003), it has been argued that in small samples the White's estimator tends to underestimate the true variance–covariance matrix, resulting in higher t-statistic ratios. In other words, using this estimator leads to liberal hypothesis tests involving the least square estimators. Davidson and MacKinnon (1993) offered two alternative versions of this estimator. Their first recommendation involves scaling up the White's estimator by a factor of $n/(n-k)$. Their second recommendation involves replacing $\hat{\varepsilon}_i^2$ with $\hat{\varepsilon}_i^2/m_{ii}$, where

$$m_{ii} = 1 - \mathbf{x}_i^T(\mathbf{X}^T\mathbf{X})^{-1}\mathbf{x}_i \quad \text{(Greene, 2003, p. 220)}.$$

We can compute Whites Estimator in SAS by using the Proc Model statements with the HCCME option in the Fit statement. Here, HCCME is the acronym for Heteroscedastic-Corrected Covariance Matrix. Using the credit card data, the following statements will be used and modified to generate standard errors using the different robust covariance matrices. See SAS/ETS User's Guide 9.1 and the Proc Panel Documentation pages 58–59 from SAS Institute, Inc. for more details on this.

```
proc model data=Expense noprint;
    parms Const C_Age C_OwnRent C_Income C_IncomeSq;
    AvgExp = Const + C_Age*Age + C_OwnRent*OwnRent +
    C_Income*Income + C_IncomeSq*Income*Income;
    fit AvgExp/HCCME=NO;
run;
```

The HCCME $= 0$ option calculates the standard errors based on Whites estimator. Here,

$$\hat{\mathbf{\Omega}}_0 = diag(\hat{\varepsilon}_1^2, \hat{\varepsilon}_2^2, \dots, \hat{\varepsilon}_n^2).$$

The HCCME $= 1$ option calculates the first alternative suggested by Davidson and MacKinnon. Here, the estimator is

$$\hat{\mathbf{\Omega}}_1 = \frac{n}{n-k} diag(\hat{\varepsilon}_1^2, \hat{\varepsilon}_2^2, \dots, \hat{\varepsilon}_n^2).$$

The HCCME $= 2$ option calculates the second alternative suggested by Davidson and MacKinnon. Here, the estimator is

$$\hat{\mathbf{\Omega}}_2 = diag\left(\hat{\varepsilon}_1^2\frac{1}{1-m_{11}}, \hat{\varepsilon}_2^2\frac{1}{1-m_{22}}, \dots, \hat{\varepsilon}_n^2\frac{1}{1-m_{nn}}\right)$$

where m_{ii} was defined earlier.

The HCCME $= 3$ option produces yet another modification of the White's estimator. Here, the estimator is

$$\hat{\mathbf{\Omega}}_3 = diag\left(\hat{\varepsilon}_1^2\frac{1}{(1-m_{11})^2}, \hat{\varepsilon}_2^2\frac{1}{(1-m_{22})^2}, \dots, \hat{\varepsilon}_n^2\frac{1}{(1-m_{nn})^2}\right).$$

Notice that in this case, the denominator of the second version of Davidson and MacKinnon's estimator has been adjusted to get a smaller estimator of the variances.

The following SAS statements can be used to calculate the parameter estimates and the covariance matrices. This code has been modified from a code set written by the SAS Institute (SAS/ETS Users Guide 9.1). The analysis results are given in Output 5.6.

```
proc model data=credit_card noprint;
    parms Const C_Age C_OwnRent C_Income C_IncomeSq;
    AvgExp = Const + C_Age*Age + C_OwnRent*OwnRent +
    C_Income*Income + C_IncomeSq*Income*Income;
    fit AvgExp/HCCME=NO outest=ols covout;
    fit AvgExp/HCCME=0 outest=H0 covout;
    fit AvgExp/HCCME=1 outest=H1 covout;
    fit AvgExp/HCCME=2 outest=H2 covout;
```

Obs	_NAME_	_TYPE_	C_Age	C_OwnRent	C_Income	C_IncomeSq
1	C_Age	OLS	30.412	-138.44	-116.39	8.575
2	C_OwnRent	OLS	-138.440	6876.11	-863.30	15.427
3	C_Income	OLS	-116.391	-863.30	6458.69	-574.810
4	C_IncomeSq	OLS	8.575	15.43	-574.81	55.791
5	C_Age	HCCME0	10.901	-104.40	96.54	-7.285
6	C_OwnRent	HCCME0	-104.400	8498.59	-4631.67	318.887
7	C_Income	HCCME0	96.543	-4631.67	7897.23	-612.393
8	C_IncomeSq	HCCME0	-7.285	318.89	-612.39	48.227
9	C_Age	HCCME1	11.714	-112.19	103.75	-7.829
10	C_OwnRent	HCCME1	-112.191	9132.81	-4977.32	342.684
11	C_Income	HCCME1	103.748	-4977.32	8486.57	-658.094
12	C_IncomeSq	HCCME1	-7.829	342.68	-658.09	51.826
13	C_Age	HCCME3	11.887	-115.10	103.87	-7.805
14	C_OwnRent	HCCME3	-115.099	9153.15	-4997.53	343.810
15	C_Income	HCCME3	103.871	-4997.53	8479.40	-657.620
16	C_IncomeSq	HCCME3	-7.805	343.81	-657.62	51.833
17	C_Age	HCCME3	12.993	-126.98	111.49	-8.332
18	C_OwnRent	HCCME3	-126.985	9863.32	-5392.00	370.544
19	C_Income	HCCME3	111.492	-5392.00	9116.74	-707.670
20	C_IncomeSq	HCCME3	-8.332	370.54	-707.67	55.896

OUTPUT 5.6. Comparing the HCCME estimators for the credit card data.

```
        fit AvgExp/HCCME=3 outest=H3 covout;
run;
data all;
        set ols H0 H1 H2 H3;
        if _name_=' ' then _name_='Parameter Estimates';
        if _n_ in (1,2,3,4,5,6) then _type_='OLS';
        else if _n_ in (7,8,9,10,11,12) then _type_='HCCME0';
        else if _n_ in (13,14,15,16,17,18) then _type_='HCCME1';
        else if _n_ in (19,20,21,22,23,24) then _type_='HCCME2';
        else _type_='HCCME3';
        drop _status_ _nused_ const;
        if _n_ in (1,2,7,8,13,14,19,20,25,26) then delete;
run;
proc print data=all;
run;
```

The parameter estimates are given in the first row of the output. This is followed by the covariance matrix using OLS estimation and the HCCME estimators. Overall, the two Davidson and MacKinnon estimators for the variance of the parameters are almost identical. As expected, the variance estimators of the parameters using White's estimation are smaller than the variance estimators calculated using the Davidson and MacKinnon's estimators. The OLS variance estimators for OwnRent and Income are the smallest when compared to the robust estimators. On the other hand, the robust variance estimators for Age are all significantly smaller than then OLS variance estimator. In general the three Davidson and MacKinnon estimators appear to be very similar to each other.

We now show how the computations are carried out in Proc IML. The following code will calculate the standard errors of the OLS estimates using the different HCCME options.

```
proc iml;
    use Expense;
    read all var {'age' 'ownrent' 'income' 'incomesq'} into X;
```

```
read all var {'avgexp'} into Y;
n=nrow(X);
X=J(n,1,1)||X;
k=ncol(X);
```

The next step is to calculate the ordinary least squares residuals.

```
C=inv(X'*X);
beta_hat=C*X'*y;
resid=y-X*beta_hat;
```

Once the ordinary least squares estimator is calculated, we can start calculating White's estimator. The variable *S0* is nothing more than the middle term of the formula for White's estimator.

```
S0=J(k,k,0);
do i=1 to n;
     S0=S0 + resid[i,]*resid[i,]*X[i,]'*X[i,];
end;
S0=S0/n;
White=n*C*S0*C;
```

Davidson and MacKinnon's two alternative versions (DM1, DM2) and the third version of the estimator can now be calculated.

```
DM1=n/(n-k) * White;
S0=J(k,k,0);
S0T=J(k,k,0);
do i=1 to n;
     m_ii=1-X[i,]*C*X[i,]';
     Temp_Ratio=resid[i,]*resid[i,]/m_ii;
     Temp_Ratio2= resid[i,]*resid[i,]/(m_ii*m_ii)
     S0=S0+Temp_Ratio*X[i,]'*X[i,];
     S0T=S0T+Temp_Ratio2*X[i,]'*X[i,];
end;
S0=S0/n;
S0T=S0T/n;
DM2=n*C*S0*C;
JK=n*C*S0T*C;
```

The calculated estimates can now be printed by using the following statements.

```
SE_White=SQRT(vecdiag(White));
SE_DM1=SQRT(vecdiag(DM1));
SE_DM2=SQRT(vecdiag(DM2));
SE_JK=SQRT(vecdiag(JK));
STATS=SE_White||SE_DM1||SE_DM2||SE_JK;
STATS=STATS';
print 'Whites Estimator, David and MacKinnons alternatives+
Jack Knife (third version)';
Print STATS (|Colname={Constant Age OwnRent Income
IncomeSq} rowname={White_Est DM1 DM2 JK} format=8.3|);
```

Output 5.7 contains the analysis results. For completeness, we have included the least squares estimators, their standard errors, *t* test statistic and the *p* values under the assumption of homoscedastic disturbances.

Least Squares Regression Results

STATS					
	CONSTANT	AGE	OWNRENT	INCOME	INCOMESQ
COEFFICIENT	-237.147	-3.082	27.941	234.347	-14.997
SE	199.352	5.515	82.922	80.366	7.469
T_RATIO	-1.190	-0.559	0.337	2.916	-2.008
WHITE_EST	212.991	3.302	92.188	88.866	6.945
DM1	220.795	3.423	95.566	92.123	7.199
DM2	221.089	3.448	95.672	92.084	7.200
JK	229.574	3.605	99.314	95.482	7.476

OUTPUT 5.7. Proc IML output of the robust estimators of the variance–covariance matrix of the credit card data.

5.6 WEIGHTED LEAST SQUARES AND FGLS ESTIMATION

In this section, we discuss estimating $\boldsymbol{\beta}$ using weighted least squares and FGLS. As shown earlier, the GLS estimator is given by $\hat{\boldsymbol{\beta}} = (\mathbf{X}^T \boldsymbol{\Omega}^{-1} \mathbf{X})^{-1} \mathbf{X}^T \boldsymbol{\Omega}^{-1} \mathbf{y}$. Now, let $Var(\varepsilon_i | \mathbf{x}_i) = \sigma_i^2 = \sigma^2 \omega_i$, where ω_i can be viewed as the weight associated with the ith residual. Therefore, $\boldsymbol{\Omega}$ is given by (Greene, 2003, p. 225; Meyers, 1990, p. 279) $\boldsymbol{\Omega} = diag(\omega_1, \omega_2, \ldots, \omega_n)$ and $\boldsymbol{\Omega}^{-1/2} = diag(1/\sqrt{\omega_1}, 1/\sqrt{\omega_2}, \ldots, 1/\sqrt{\omega_n})$. The easiest approach to conducting weighted least squares regression is to use this in the equation of the GLS estimator. Another approach (Greene, 2003, p. 225) is to premultiply both $\mathbf{y}$ and $\mathbf{X}$ by $\boldsymbol{\Omega}^{-1/2}$ thereby distributing the appropriate weights of the residuals across their corresponding observations.

The GLS estimator is then calculated by regressing the transformed response variable against the transformed explanatory variables. As given in Greene (2003, p. 226) and Verbeek (2004, p. 85), a common approach used to obtain the weights is to specify that the variance of the disturbances is proportional to one of the regressors.

We will illustrate weighted regression methods by using the credit card data. In the first illustration, we will assume that the variance of the disturbance is proportional to income, while in the second illustration, we will assume that it is proportional to the square of income. Performing weighted least squares regression in SAS is straightforward. The weights are calculated and stored in the data step statement and then used with the "Weights" option in Proc Reg. The following SAS statements can be used. The analysis results are provided in Output 5.8.

```
proc import out=CCExp
     datafile="C:\Temp\TableF91"
     dbms=Excel Replace;
     getnames=yes;
run;
data CCExp;
     set CCExp;
     Income_Sq=Income*Income;
     if AvgExp>0;
     wt1=1/Income;
     wt2=1/(Income_Sq);
run;
proc reg data=CCExp;
     model AvgExp=Age OwnRent Income Income_Sq;
     weight wt1;
run;
proc reg data=CCExp;
     model AvgExp=Age OwnRent Income Income_Sq;
     weight wt2;
run;
```

The REG Procedure
Model: MODEL1
Dependent Variable: AvgExp AvgExp

Number of Observations Read	72
Number of Observations Used	72

Weight: wt1

Analysis of Variance					
Source	DF	Sum of Squares	Mean Square	F Value	Pr > F
Model	4	438889	109722	5.73	0.0005
Error	67	1283774	19161		
Corrected Total	71	1722663			

Root MSE	138.42258	R-Square	0.2548
Dependent Mean	207.94463	Adj R-Sq	0.2103
Coeff Var	66.56704		

Parameter Estimates						
Variable	Label	DF	Parameter Estimate	Standard Error	t Value	Pr > |t|
Intercept	Intercept	1	-181.87064	165.51908	-1.10	0.2758
Age	Age	1	-2.93501	4.60333	-0.64	0.5259
OwnRent	OwnRent	1	50.49364	69.87914	0.72	0.4724
Income	Income	1	202.16940	76.78152	2.63	0.0105
Income_Sq		1	-12.11364	8.27314	-1.46	0.1478

Number of Observations Read	72
Number of Observations Used	72

Weight: wt2

Analysis of Variance					
Source	DF	Sum of Squares	Mean Square	F Value	Pr > F
Model	4	112636	28159	5.73	0.0005
Error	67	329223	4913.78353		
Corrected Total	71	441860			

Root MSE	70.09838	R-Square	0.2549
Dependent Mean	168.79218	Adj R-Sq	0.2104
Coeff Var	41.52940		

Parameter Estimates						
Variable	Label	DF	Parameter Estimate	Standard Error	t Value	Pr > |t|
Intercept	Intercept	1	-114.10887	139.68750	-0.82	0.4169
Age	Age	1	-2.69419	3.80731	-0.71	0.4816
OwnRent	OwnRent	1	60.44877	58.55089	1.03	0.3056
Income	Income	1	158.42698	76.39115	2.07	0.0419
Income_Sq		1	-7.24929	9.72434	-0.75	0.4586

OUTPUT 5.8. Weighted least squares regression for the credit card expenditure data.

Notice that the effect of income is significant in both outputs. As should be expected, the standard errors of the parameters using the square of income as weights are smaller than the standard errors when the weights were based on income. The signs of the parameters are the same across both analysis and the variable income is significant in both analyses as well. Comparing the magnitudes of the parameter estimates, we see that except for rent, the magnitude of the parameter values for the first regression (using income as weights) is higher than those from the second regression (using the square of income as weights). Note the dependent variable means are different from each other and from the dependent variable mean under classical OLS because here, the response variable is transformed by using the weights.

We now discuss the case when $\mathbf{\Omega}$ is assumed to be unknown. As shown previously, the unrestricted heteroscedastic regression model has too many parameters that need estimation given the limitations on the sample size. As discussed in Greene (2003, pp. 227–228), we can work around this issue by expressing $\sigma^2\mathbf{\Omega}$ as a function of only a few parameters. In the credit card data, we may focus our attention on income, and the square of income. For instance, if we let $z_i = income$, then we can express σ_i^2 as $\sigma_i^2 = \sigma^2 z_i^\alpha$. Of course, we could have more than one variable making the parameter $\boldsymbol{\alpha}$ a vector. The modified variance–covariance matrix is now denoted as $\mathbf{\Omega}(\boldsymbol{\alpha})$. Therefore, estimating $\mathbf{\Omega}$ is now restricted to estimating $\boldsymbol{\alpha}$. How do we calculate a consistent estimator of $\boldsymbol{\alpha}$? As discussed in both Green (2003) and Verbeek (2004, p. 86), there are two ways of doing this. The first method involves the two-step GLS technique and the second method involves maximum likelihood estimation (MLE). We restrict our discussion to the two-step FGLS estimator.

The FGLS estimator is straightforward once $\mathbf{\Omega}(\hat{\boldsymbol{\alpha}})$ is computed. Simply use this estimator in the formula for the GLS estimator to get the FGLS estimators. That is,

$$\hat{\hat{\boldsymbol{\beta}}} = (\mathbf{X}^T\mathbf{\Omega}^{-1}(\hat{\boldsymbol{\alpha}})\mathbf{X})^{-1}\mathbf{X}^T\mathbf{\Omega}^{-1}(\hat{\boldsymbol{\alpha}})\mathbf{y}.$$

The general procedure for calculating the two-step FGLS estimator is as follows:

1. Obtain estimates of σ_i^2 using OLS residuals. Note that the estimates are simply $\hat{\varepsilon}_i^2$, the squared OLS least square residuals. Next, consider the model $\hat{\varepsilon}_i^2 = \mathbf{z}_i^{\boldsymbol{\alpha}} + v_i$.
2. OLS can be used to estimate $\boldsymbol{\alpha}$ by regressing $log(\hat{\varepsilon}_i^2)$ on $\log(\mathbf{z}_i)$.

The computations can be easily carried out using Proc Reg. We give two examples using the credit card data. In the first example, we let $\mathbf{z}_i = (1, income, incomesq)$ and assume that $\sigma_i^2 = \sigma^2 e^{\mathbf{z}_i\boldsymbol{\alpha}}$. An estimate of $\boldsymbol{\alpha} = (\alpha_0, \alpha_1, \alpha_2)$ is obtained by running the following regression:

$$\log(\hat{\varepsilon}_i^2) = \alpha_0 + \alpha_1 income + \alpha_2 incomesq + v_i.$$

The steps in SAS are as follows:

1. First, run the OLS regression model to estimate ε_i. The OLS residuals are stored in the variable labeled *residual* in the SAS data set *resid1*. The Proc Reg output was suppressed by using the *noprint* option as we are only interested in generating the OLS residuals at this stage.

```
proc reg noprint data=CCExp;
    model AvgExp=Age OwnRent Income Income_Sq;
    output out=resid1 r=residual;
run;
```

2. We can now compute $\log(\hat{\varepsilon}_i^2)$ and regress it against income and the square of income. Note that in this stage, we are interested in the predicted values $\mathbf{z}_i^{\hat{\alpha}}$ and therefore suppress the regression output again. The following statements can be used to conduct this step.

```
data test;
    set resid1;
    log_e=log(residual*residual);
run;
proc reg noprint data=test;
    model log_e=income income_sq;
    output out=resid2 p=pred;
run;
```

The REG Procedure
Model: MODEL1
Dependent Variable: AvgExp AvgExp

Number of Observations Read	72
Number of Observations Used	72

Weight: wt3

Analysis of Variance					
Source	DF	Sum of Squares	Mean Square	F Value	Pr > F
Model	4	1123.73425	280.93356	69.69	<0.0001
Error	67	270.08589	4.03113		
Corrected Total	71	1393.82015			

Root MSE	2.00777	R-Square	0.8062
Dependent Mean	401.66162	Adj R-Sq	0.7947
Coeff Var	0.49987		

Parameter Estimates						
Variable	Label	DF	Parameter Estimate	Standard Error	t Value	Pr > |t|
Intercept	Intercept	1	−117.86745	101.38621	−1.16	0.2491
Age	Age	1	−1.23368	2.55120	−0.48	0.6303
OwnRent	OwnRent	1	50.94976	52.81429	0.96	0.3382
Income	Income	1	145.30445	46.36270	3.13	0.0026
Income_Sq		1	−7.93828	3.73672	−2.12	0.0373

OUTPUT 5.9. FGLS estimation using the credit card data.

3. With the first stage complete, we can get the FGLS estimates by using weighted least squares regression described earlier. Here, the weights are just the exponential of the predicted values from stage 2. That is, $w_i = \exp(\hat{\mathbf{z}}_i^{\alpha})$. The following statements can be used.

```
data test;
     set resid2;
     wt3=1/exp(pred);
run;
proc reg data=test;
     model AvgExp=Age OwnRent Income Income_Sq;
     weight wt3;
run;
```

The analysis results are given in Output 5.9. The standard errors of the estimates are now significantly lower than the standard errors of the estimates when income and square of income were used as weights. Also note that the root mean square has reduced substantially over what was previously observed. The signs of the coefficients are the same, however, now both income and the square of income are significant.

The reader is asked to verify that using $\sigma_i^2 = \sigma^2 z_i^{\alpha}$ with $z_i =\log(income)$, gives the FGLS estimators as shown in Output 5.10. The parameter estimates and their standard errors are similar to the ones obtained when income was used as a weight. Also note that the root mean square errors are similar and that only income is significant.

5.7 AUTOREGRESSIVE CONDITIONAL HETEROSCEDASTICITY

We now turn our attention to heteroscedasticity in the time series setting. As discussed by Enders (2004), in a typical econometric model, the variance of the disturbances is assumed to be stable (constant) over time. However, there are instances when economic time series data exhibit periods of high "volatility" followed by periods of low "volatility" or

The REG Procedure
Model: MODEL1
Dependent Variable: AvgExp AvgExp

Number of Observations Read	72
Number of Observations Used	72

Weight: wt4

Analysis of Variance					
Source	DF	Sum of Squares	Mean Square	F Value	Pr > F
Model	4	562700	140675	5.69	0.0005
Error	67	1655217	24705		
Corrected Total	71	2217918			

Root MSE	157.17741	R-Square	0.2537
Dependent Mean	216.57420	Adj R-Sq	0.2092
Coeff Var	72.57439		

Parameter Estimates						
Variable	Label	DF	Parameter Estimate	Standard Error	t Value	Pr > |t|
Intercept	Intercept	1	-193.27961	171.06009	-1.13	0.2626
Age	Age	1	-2.95778	4.76203	-0.62	0.5366
OwnRent	OwnRent	1	47.37065	72.12961	0.66	0.5136
Income	Income	1	208.84940	77.19611	2.71	0.0086
Income_Sq		1	-12.76626	8.08456	-1.58	0.1190

OUTPUT 5.10. FGLS estimation using $z_i = \log(income)$ for the credit card data.

"calmness." Greene (2003, p. 238) analyzes the well-known Bollerslev and Ghysel's data on the daily percentage nominal return for the Deuschmark/Pound exchange rate. We analyzed the data using SAS—Figure 5.4 shows that there are periodic spikes in the data on both the high and low sides. It is obvious that the variability appears to be unstable or shifting over time. In particular, large shocks appear to follow each other and vice versa, small shocks appear to follow each other. The variance of the disturbance at a given time period is therefore assumed to depend on the variance of the disturbance in the previous time periods. Therefore, the homoscedastic variance assumption in this case is violated. The disturbance terms in the linear models must therefore take into account the dependence of its variance on past disturbances. This is the basic principle behind Engle's (1982) autoregressive, conditionally heteroscedastic models (ARCH). He proposed a methodology where the variance of the disturbances (ε_t) are allowed to depend on its history. That is, the variance of the series itself is an autoregressive time series.

Bollerslev (1986) extended the ARCH process by allowing an autoregressive moving average process for the error variance. Their resulting formulation is referred to as the generalized autoregressive conditional heteroscedastic model or GARCH. Both the ARCH and the GARCH models forecast the variance of the disturbance at time t. The ARCH models uses the weighted averages of the past values of the squared disturbances, while the GARCH model uses the weighted average of the past values of both the squared disturbances and the variances.

5.7.1 The Arch Model

The simplest form of Engle's ARCH model is the ARCH(1) model. The main idea behind the model is that the conditional variance of the disturbance at time t depends on the squared disturbance term at time $t-1$. To see this, first, consider the basic model given by $y_t = \sigma_t z_t$, where $z_t \sim i.i.d.N(0,1)$. In an ARCH(1) model, $\sigma_t^2 = \alpha_0 + \alpha_1 y_{t-1}^2$ where $\alpha_0 > 0$ and $\alpha_1 \geq 0$. It can be shown that

$$E(y_t|y_{t-1}) = 0 \quad \text{and} \quad \text{Var}(y_t|y_{t-1}) = \alpha_0 + \alpha_1 y_{t-1}^2.$$

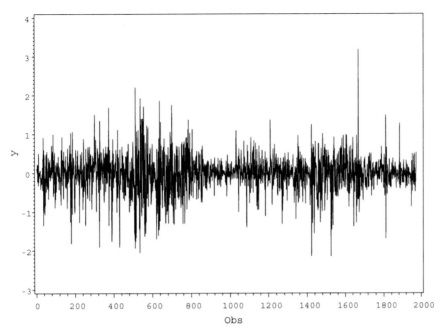

FIGURE 5.4. Time series plot for the nominal returns.

Therefore, the conditional variance of the disturbance at time t depends on the past values of the squared disturbances. The unconditional variance on the other hand is constant. To see this note that $Var(y_t) = E(y_t^2) = \alpha_0 + \alpha_1 E(y_{t-1}^2)$. Simplifying this gives $Var(y_t) = \alpha_0/(1 - \alpha_1)$, $\alpha_0 > 0, 0 \leq \alpha_1 < 1$.

The ARCH(1) process can easily be extended to linear models with explanatory variables. First, consider the linear model $y_t = \mathbf{x}_t^T \boldsymbol{\beta} + \varepsilon_t$ with $\varepsilon_t = \sigma_t z_t$, where $z_t \sim i.i.d.N(0, 1)$ and $\sigma_t^2 = \alpha_0 + \alpha_1 \varepsilon_{t-1}^2$. It can be easily shown that the conditional variance $Var(y_t | \mathbf{x}_t) = \alpha_0 + \alpha_1 \varepsilon_{t-1}^2$ while the unconditional variance $Var(y_t) = Var(\varepsilon_t) = \alpha_0/(1 - \boldsymbol{\alpha}_1)$ with $0 \leq \alpha_1 < 1$ (Greene, 2003, pp. 238–239)

5.7.2   ARCH(q) and the Generalized ARCH Models

Extending the simple ARCH(1) model to the more general case we get the ARCH(q) process given by $y_t = \sigma_t z_t$, where $z_t \sim i.i.d.N(0, 1)$ and

$$\sigma_t^2 = \alpha_0 + \sum_{i=1}^{q} \alpha_i y_{t-i}^2.$$

Note that the unconditional variance is now given by

$$Var(\varepsilon_t) = \alpha_0 / \left(1 - \sum_{i=1}^{q} \alpha_i \right) \text{ with } 0 \leq \sum_{i=1}^{q} \alpha_i < 1.$$

The ARCH(q) process can easily be extended to the linear regression setting in a similar fashion.

5.7.3   The GARCH Model

Bollerslev (1986) extended the ARCH models where the variance of the disturbance at time t depends on its own lag as well as the lag of the squared disturbances. In the GARCH(1,1) model,

$$\sigma_t^2 = \alpha_0 + \alpha_1 \varepsilon_{t-1}^2 + \beta_1 \sigma_{t-1}^2.$$

The basic principle is to make the forecast of the variance at time t more accurate. In a GARCH(p,q) model

$$\sigma_t^2 = \alpha_0 + \sum_{i=1}^{p} \delta_i \sigma_{t-i}^2 + \sum_{j=1}^{q} \alpha_j \varepsilon_{t-j}^2.$$

It can be shown that the unconditional variance of the disturbance at time t is

$$\sigma_\varepsilon^2 = \frac{\alpha_0}{1 - \sum_{i=1}^{q} \alpha_i - \sum_{j=1}^{p} \beta_j}$$

where $\alpha_0 \geq 0$, $\alpha_i \geq 0$, $\beta_i \geq 0$, and $0 \leq 1 - \sum_{i=1}^{q} \alpha_i - \sum_{j=1}^{p} \beta_j < 1$.

The GARCH(p,q) models can easily be extended to the linear regression setting as well.

5.7.4   Testing for ARCH Effects

The Lagrange Multiplier test (LM) can be used to test for ARCH(q) effects. The hypothesis tested is

$$H_0 : \alpha_1 = \alpha_2 = \ldots = \alpha_q = 0$$
$$H_1 : \text{at least one } \alpha_i \neq 0, i = 1, \ldots q.$$

The steps are as follows:

1. Estimate $y_t = \mathbf{x}_t^T \boldsymbol{\beta} + \varepsilon_t$ using OLS and calculate $\hat{\varepsilon}_{t-i}^2$ for $i = 1, \ldots, q$.
2. Conduct a regression of $\hat{\varepsilon}_t^2$ on a constant and $\hat{\varepsilon}_{t-1}^2, \ldots, \hat{\varepsilon}_{t-q}^2$ and calculate the coefficient of determination, R^2.
3. Calculate the test statistic TR^2, where T is the number of observations. Under the null hypothesis, $TR^2 \sim \chi_q^2$. We reject the null hypothesis of no ARCH effects if the calculated value of the test statistic exceeds the tabled value from the chi-squared distribution.

The LM test can also be used for testing GARCH effects. In a test for a GARCH(p,q) model, however, the hypothesis tested is the null of a ARCH(q) process versus a ARCH($p+q$) process (Baltagi, 2008, p. 370). Here, the LM test is based on the regression of $\hat{\varepsilon}_t^2$ on $p + q$ lagged values $\hat{\varepsilon}_{t-1}^2, \ldots, \hat{\varepsilon}_{t-p-q}^2$. The test statistic is the same as before.

Maximum likelihood estimation can be used to estimate the parameters of both the ARCH and GARCH models. Details can be found in Greene (2003) page 239 (ARCH) and pages 242–243 (GARCH).

We will now illustrate the estimation of these models in SAS by using the Bollerslev and Ghysels nominal exchange rate data. For illustration purposes, we will use a GARCH(1,1) model.

The Proc Autoreg module in SAS can be used to fit this model. Recall that this procedure should be used to perform regression analysis when the underlying assumption of heteroscedasticity and autocorrelation are violated. It can also be used to perform ARCH and GARCH calculations. First, we will use the procedure to test for heteroscedasticity in the data by using the "archtest" option. The following commands can be used. Note that a temporary SAS data set named "garch" was created prior to invoking this procedure. The analysis results are given in Output 5.11.

```
proc autoreg data=garch;
  model y=/archtest;
run;
```

The first table of the output gives the OLS estimates. The values for SSE and MSE are for the error and mean sums of squares. The MSE is really the unconditional variance of the series. The Durbin–Watson statistic is used to test for serial correlation and

The AUTOREG Procedure

Dependent Variable	y
	y

Ordinary Least Squares Estimates			
SSE	436.289188	DFE	1973
MSE	0.22113	Root MSE	0.47024
SBC	2629.78062	AIC	2624.1928
Regress R-Square	0.0000	Total R-Square	0.0000
Durbin-Watson	1.9805		

Q and LM Tests for ARCH Disturbances				
Order	Q	Pr > Q	LM	Pr > LM
1	96.4249	<0.0001	96.3422	<0.0001
2	157.1627	<0.0001	129.5878	<0.0001
3	196.7515	<0.0001	142.6618	<0.0001
4	227.4684	<0.0001	150.3655	<0.0001
5	297.7401	<0.0001	183.3808	<0.0001
6	314.1284	<0.0001	183.3929	<0.0001
7	328.6768	<0.0001	183.8867	<0.0001
8	347.5464	<0.0001	186.8223	<0.0001
9	364.7738	<0.0001	188.8952	<0.0001
10	392.9791	<0.0001	194.1606	<0.0001
11	397.5269	<0.0001	194.9219	<0.0001
12	404.9266	<0.0001	195.1401	<0.0001

Variable	DF	Estimate	Standard Error	t Value	Approx Pr > |t|
Intercept	1	-0.0164	0.0106	-1.55	0.1208

OUTPUT 5.11. Testing for the heteroscedasticity in the nominal exchange data.

will be discussed in detail in Chapter 6. DFE is simply the degrees of freedom and is the total number of observations -1. The values of AIC and BIC are information criteria values that are used to assess model fit. Smaller values of the statistics are desirable. The Durbin–Watson statistics will be discussed in Chapter 6.

The output also contains the Q and LM tests. Both statistics test for heteroscedasticity in the time series. The Q statistic proposed by McLeod and Li (1983) (see the Proc Autoreg reference guide from SAS Institute) checks for changing variability over time. The test is highly significant across the 12 lag windows. The LM statistic was discussed earlier. It is also highly significant across all 12 lag windows indicating that a higher order ARCH process needs to be used to model the data.

As discussed earlier, the GARCH process introduces the lagged values of the variances also and thus introduces a "longer memory" (Proc Autoreg reference guide, SAS Institute, Inc.). Therefore, we start our initial model at the GARCH(1,1) process. The following statements can be used. Note that the option "garch" can be changed to allow for an ARCH process. For example, using the option "garch=(q=1)" will request an ARCH(1) process for the dataset. The analysis results are given in Output 5.12.

```
proc autoreg data=garch;
       model y=/Garch=(p=1,q=1);
run;
```

The output indicates that there is strong evidence of GARCH effects (p value < 0.0001). The unconditional variance for the GARCH model is 0.2587 compared to 0.2211 for the OLS model. The normality test is highly significant

```
The AUTOREG Procedure
```

Dependent Variable	y
	y

Ordinary Least Squares Estimates			
SSE	436.289188	DFE	1973
MSE	0.22113	Root MSE	0.47024
SBC	2629.78062	AIC	2624.1928
Regress R-Square	0.0000	Total R-Square	0.0000
Durbin-Watson	1.9805		

Variable	DF	Estimate	Standard Error	t Value	Approx Pr > |t|
Intercept	1	-0.0164	0.0106	-1.55	0.1208

```
Algorithm converged.
```

GARCH Estimates			
SSE	436.495992	Observations	1974
MSE	0.22112	Uncond Var	0.25876804
Log Likelihood	-1106.6908	Total R-Square	.
SBC	2243.73289	AIC	2221.38163
Normality Test	1081.7663	Pr > ChiSq	<0.0001

Variable	DF	Estimate	Standard Error	t Value	Approx Pr > |t|
Intercept	1	-0.006191	0.008426	-0.73	0.4625
ARCH0	1	0.0108	0.001327	8.15	<0.0001
ARCH1	1	0.1524	0.0139	10.97	<0.0001
GARCH1	1	0.8058	0.0166	48.61	<0.0001

OUTPUT 5.12. GARCH$(1, 1)$ model for the nominal exchange rate data.

(p value < 0.0001), which indicates that the residuals from the GARCH model are not normally distributed—a clear contradiction to the normality assumption. ARCH0 gives the estimate of α_0, ARCH1 gives the estimate of α_1, and GARCH1 gives the estimate of β_1.

6

AUTOCORRELATION

6.1 INTRODUCTION

In Chapters 4 and 5 we discussed estimation methods for $\boldsymbol{\beta}$ under departures from the exogeneity and homoscedasticity assumptions. This chapter extends the discussion to the case when the assumption of independent disturbances is violated. That is, we will relax the assumption that the disturbance related to an observation is independent of the disturbance related to another observation. We call this situation *serial correlation* or *autocorrelation*. Simply put, in autocorrelation $Cov(\varepsilon_t, \varepsilon_s) \neq 0$ for $t \neq s$ where t and s are two time periods. Autocorrelation most often occurs in time series data where the observation at a given point in time is dependent on the observations from the previous time periods

The texts by Greene (2003, Chapter 12), Meyers (1990, Chapter 7), and Verbeek (2004, Chapter 4) offer a good discussion on autocorrelation models. Brocklebank and Dickey (2003) offer a thorough treatment of how SAS can be used to fit autocorrelation models.

Autocorrelation in regression models often occurs when models are misspecified or when variables are mistakenly omitted from the model. In the omitted variable case, unobserved or omitted variables that are correlated over time are now absorbed in the error term, causing autocorrelation. As an example, consider the gasoline consumption data in Greene (2003). Gasoline consumption along with measurements on other variables was observed from 1960 to 1995. Note that this data was analyzed in Chapter 2. The full equation for this model is (Greene, 2003, p. 136)

$$\ln(G_t/Pop_t) = \beta_1 + \beta_2\ln(Pg_t) + \beta_3\ln(I_t/Pop_t) + \beta_4\ln(Pnc_t) + \beta_5\ln(Puc_t)$$
$$+ \beta_6\ln(PpT_t) + \beta_7\ln(PN_t) + \beta_8\ln(PD_t) + \beta_9\ln(PS_t) + \beta_{10}t + \varepsilon_t.$$

Assume that we fit the model

$$\ln(G_t/Pop_t) = \beta_1 + \beta_2\ln(Pg_t) + \beta_3\ln(I_t/Pop_t) + \beta_4\ln(Pnc_t) + \beta_5\ln(Puc_t) + \varepsilon_t.$$

The residuals from the fitted model and the full model are shown in Figures 6.1 and 6.2. Note that the residuals from both models show that autocorrelation should be investigated. However, the fitted model shows a higher degree of autocorrelation.

Applied Econometrics Using the SAS® System, by Vivek B. Ajmani
Copyright © 2009 John Wiley & Sons, Inc.

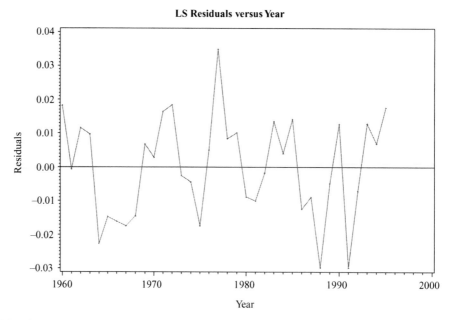

FIGURE 6.1. Time series plot of the residuals from the full model in the gasoline consumption data.

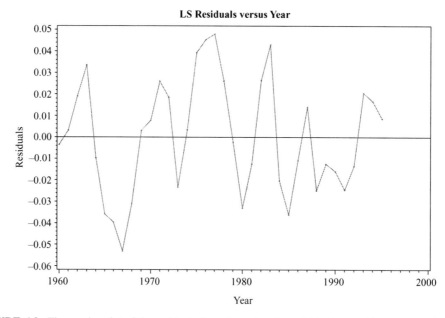

FIGURE 6.2. Time series plot of the residuals from the reduced model in the gasoline consumption data.

Although, time series plots of the residuals (as the ones above) can be used to detect autocorrelation issues quickly, we will discuss more formal procedures to detect the presence of autocorrelation in the data.

6.2 PROBLEMS ASSOCIATED WITH OLS ESTIMATION UNDER AUTOCORRELATION

We start our discussion of estimation under autocorrelation by considering the simple autoregressive first-order autocorrelation model (AR1) with exogenous explanatory variables. That is, a model of the form $\mathbf{y}_i = \mathbf{x}_i^T \boldsymbol{\beta} + \varepsilon_i i = 1, 2, \ldots$, with $\varepsilon_i = \rho \varepsilon_{i-1} + u_i$

where $|\rho|<1$ is required to ensure $Var(\varepsilon_t) < \infty$. Assume that $E(u_i) = 0$, $Var(u_i) = \sigma^2$, and $Cov(u_t, u_s) = 0$ for $t \neq s$. That is, the error term u_i has mean zero, has constant variance, and exhibits no serial correlation. It can be easily shown that $E(\varepsilon_i) = 0$ and that the variance $Var(\varepsilon_i) = \sigma_u^2/(1 - \rho^2)$ where $|\rho|<1$. Also note that the covariance between ε_i and ε_{i-1} denoted by $Cov(\varepsilon_i, \varepsilon_{i-1})$ is given by

$$Cov(\varepsilon_i, \varepsilon_{i-1}) = \rho \frac{\sigma_u^2}{1 - \rho^2}$$

and that the covariance between two disturbances, which are two periods apart, $Cov(\varepsilon_i, \varepsilon_{i-2})$ is given by

$$Cov(\varepsilon_i, \varepsilon_{i-2}) = \rho^2 \frac{\sigma_u^2}{1 - \rho^2}.$$

Extending this to two disturbances, which are j time periods apart, we get $Cov(\varepsilon_i, \varepsilon_{i-j})$ given by

$$Cov(\varepsilon_i, \varepsilon_{i-j}) = \frac{\sigma_u^2}{1 - \rho^2}\rho^j.$$

Autocorrelation therefore implies that the errors are heteroscedastic (Greene, 2003, p. 258; Meyers, 1990, p. 289). As shown in Chapter 5, OLS estimators, although unbiased will be inefficient and will have incorrect standard errors.

$$\mathbf{y}_i = \mathbf{x}_i^T \boldsymbol{\beta} + \varepsilon_i, \quad i = 1, 2, \ldots,$$
$$\varepsilon_i = \rho\varepsilon_{i-1} + u_i,$$

6.3   ESTIMATION UNDER THE ASSUMPTION OF SERIAL CORRELATION

Estimation techniques under the assumption of serial correlation parallel the estimation methods for heteroscedasticity that was discussed in Chapter 5. That is, we need to first estimate the variance–covariance matrix. Using the variance and covariance of the disturbances given in the previous section, we can easily construct the covariance matrix $\sigma^2\boldsymbol{\Omega}$ as (Greene, 2003, p. 259)

$$\sigma^2\boldsymbol{\Omega} = \frac{\sigma_u^2}{1-\rho^2}\begin{bmatrix} 1 & \rho & \rho^2 & \rho^3 & \cdots & \rho^{T-1} \\ \rho & 1 & \rho & \rho^2 & \cdots & \rho^{T-2} \\ \rho^2 & \rho & 1 & \rho & \cdots & \rho^{T-3} \\ \vdots & \vdots & \vdots & \ddots & \vdots & \vdots \\ \vdots & \vdots & \vdots & \vdots & \ddots & \rho \\ \rho^{T-1} & \rho^{T-2} & \rho^{T-3} & \cdots & \rho & 1 \end{bmatrix}.$$

If ρ is known, then using the discussion from Chapter 5, one can get a GLS estimator of $\boldsymbol{\beta}$ as

$$\mathbf{b}_{GLS} = (\mathbf{X}^T\boldsymbol{\Omega}^{-1}\mathbf{X})^{-1}\mathbf{X}^T\boldsymbol{\Omega}^{-1}\mathbf{y}.$$

It can be shown that the GLS estimator is an unbiased, consistent estimator for $\boldsymbol{\beta}$ with variance-covariance matrix given by

$$Est.Var[\mathbf{b}_{GLS}] = \sigma_\varepsilon^2[\mathbf{X}^T\boldsymbol{\Omega}^{-1}\mathbf{X}]^{-1},$$

where

$$\sigma_\varepsilon^2 = \frac{(\mathbf{y}-\mathbf{X}\mathbf{b}_{GLS})^T \boldsymbol{\Omega}^{-1}(\mathbf{y}-\mathbf{X}\mathbf{b}_{GLS})}{T} \quad \text{(Greene, 2003, p. 271)}.$$

These computations can be carried out easily by simply transforming $\mathbf{y}$ and $\mathbf{X}$ and running OLS on the transformed values. The transformations for the first-order autocorrelation process are

$$y_1^* = \sqrt{1-\rho^2}\, y_1 \quad \text{and} \quad \mathbf{x}_1^* = \sqrt{1-\rho^2}\, \mathbf{x}_1 \quad \text{for } t = 1,$$
$$y_t^* = y_t - \rho y_{t-1} \quad \text{and} \quad \mathbf{x}_t^* = \mathbf{x}_t - \rho \mathbf{x}_{t-1} \quad \text{for } t = 2, \ldots, T \text{ (Greene, 2003, p. 272)}.$$

These transformations are called the Prais–Winsten transformations. The traditional approach (Cochrane and Orcutt, 1949) used the same transformation but dropped the first observation for computational ease. As discussed in Verbeek (2004, p. 100), deleting the observation leads to an approximate GLS estimator that is not as efficient as the GLS estimator obtained by including all the observations.

The GLS estimator can then be calculated as follows:

$$\mathbf{b}_{GLS} = (X_*^T X_*)^{-1} X_*^T y_*.$$

Greene (2003) extends the process to the second-order autocorrelation process. As can be seen from the transformation formulas the author provides, the process becomes very complex as the order of the autoregressive process increases.

The estimation techniques discussed so far are based on the assumption that ρ is known. However, in reality, ρ is unknown and has to be estimated. Estimation of the least squares parameters is straightforward under the assumption that ρ is unknown. In the case of the first-order autocorrelation model, the steps are as follows:

1. Estimate the model $\mathbf{y}_i = \mathbf{x}_i^T \boldsymbol{\beta} + \varepsilon_i$, $i = 1, 2, \ldots$ using OLS and save the residuals.
2. Using the residuals, fit the model $\hat{\varepsilon}_i = \rho \hat{\varepsilon}_{i-1} + u_i$ and get an estimate of ρ.
3. Use the residuals from Step 2 to estimate σ_u^2. We can use the estimated values of ρ and σ_u^2 to construct $\hat{\boldsymbol{\Omega}}$ This can then be used to calculate a FGLS estimator of $\boldsymbol{\beta}$. Another alternative is to simply use the estimated value of ρ to transform both $\mathbf{y}$ and $\mathbf{X}$ using the Prais–Winsten transformation and then conduct OLS with the transformed values (Greene, 2003, p. 273).

The Proc Autoreg procedure in SAS will allow us to estimate the least squares parameters by FGLS, MLE, or the iterated FGLS method. We will discuss ways of conducting the analysis in SAS a bit later. For now, let us discuss methods of finding out whether the regression model suffers from autocorrelation.

6.4   DETECTING AUTOCORRELATION

The Durbin–Watson test is perhaps the most commonly used test for detecting autocorrelation. It is based on the statistic (Meyers, 1990, p. 289)

$$DW = \frac{\sum_{i=2}^{T} (\hat{\varepsilon}_i - \hat{\varepsilon}_{i-1})^2}{\sum_{i=1}^{T} \hat{\varepsilon}_i^2}$$

and tests the hypothesis

$$H_0: \quad \rho = 0,$$
$$H_1: \quad \rho \neq 0.$$

Proc Autoreg can be used to conduct the Durbin–Watson test. On the other hand, even though Proc Reg does give the value of the test statistic and the estimated first-order correlation, it does not output the p value associated with the test.

The Lagrange multiplier test suggested by Breusch and Godfrey (1978) is an alternative to the Durbin–Watson test. The test statistic is $LM = TR^2$, where R^2 is the R-squared value obtained by regressing the OLS residuals against p-lagged residuals $\hat{\varepsilon}_{t-1}, \ldots, \hat{\varepsilon}_{t-p}$ along with the original variables in **X** where the residuals can be obtained by OLS. The test statistic has a chi-squared distribution with p degrees of freedom.

Two other alternatives to the Durbin–Watson test are the Box and Pierce's test (B&P) and Ljung's modification of the B&P test (Greene 2003, p. 269). The B&Q test is based on the test statistic

$$Q = T \sum_{j=1}^{p} r_j^2,$$

where

$$r_j = \frac{\sum\limits_{t=j+1}^{T} e_t e_{t-j}}{\sum\limits_{t=1}^{T} e_t^2}.$$

That is, r_j measures the autocorrelation between e_t and e_{t-j}. This test statistic has a limiting chi-squared distribution with p degrees of freedom. Here, p refers to the number of lags used.

As stated in Greene (2003, p. 269) the Ljung's refinement of the B&P test is given by

$$Q' = T(T+2) \sum_{j=1}^{p} \frac{r_j^2}{T-j},$$

where r_j is defined as before. This test statistic has a chi-squared distribution with p degrees of freedom also.

We will now illustrate the computations involved by analyzing the gas consumption data in Greene (2003). We restrict our attention to the reduced model

$$\ln(G_t/Pop_t) = \beta_1 + \beta_2 \ln(Pg_t) + \beta_3 \ln(I_t/Pop_t) + \beta_4 \ln(Pnc_t) + \beta_5 \ln(Puc_t) + \varepsilon_t$$

to force some autocorrelation behavior. The following SAS statements can be used to fit the OLS model to the data and also compute the Durbin–Watson and Breusch–Godfrey statistic along with their p values. The "dwprob" and "godfrey" options are used to generate the statistics. The analysis results are given in Output 6.1.

```
proc autoreg data=gasoline;
    model Ln_G_Pop=Ln_Pg Ln_Income Ln_Pnc Ln_Puc/dwprob
    godfrey;
run;
```

Both tests indicate a presence of autocorrelation in the data. The p value for the Durbin–Watson test indicates the presence of a positive autocorrelation. The LM test indicates that the significance extends to the higher order AR(4). Adding the option "dw = 5" to the model statement yields the Durbin–Watson statistics for the first five autocorrelations. The analysis results are given in Output 6.2. Note that there is strong evidence of a positive first-order correlation.

We can use the correlations calculated by Proc Autoreg to conduct the B&P test and the Ljung test. First, we calculate the first five autocorrelations by using the following statements (see Output 6.3):

```
proc autoreg data=gasoline;
    model Ln_G_Pop=Ln_Pg Ln_Income Ln_Pnc Ln_Puc/nlag=5;
run;
```

The AUTOREG Procedure

Dependent Variable	Ln_G_Pop

Ordinary Least Squares Estimates			
SSE	0.02487344	DFE	31
MSE	0.0008024	Root MSE	0.02833
SBC	−141.90789	AIC	−149.82548
Regress R-Square	0.9691	Total R-Square	0.9691
Durbin–Watson	0.8909	Pr < DW	<0.0001
Pr > DW	1.0000		

Note: Pr < DW is the *p* value for testing positive autocorrelation, and Pr > DW is the *p* value for testing negative autocorrelation.

Godfrey's Serial Correlation Test		
Alternative	LM	Pr > LM
AR(1)	11.2170	0.0008
AR(2)	17.1932	0.0002
AR(3)	17.2414	0.0006
AR(4)	17.6825	0.0014

Variable	DF	Estimate	Standard Error	t Value	Approx Pr > |t|
Intercept	1	−7.7892	0.3593	−21.68	<0.0001
Ln_pg	1	−0.0979	0.0283	−3.46	0.0016
Ln_Income	1	2.1175	0.0988	21.44	<0.0001
Ln_Pnc	1	0.1224	0.1121	1.09	0.2830
Ln_Puc	1	−0.1022	0.0693	−1.48	0.1502

OUTPUT 6.1. Using Proc Autoreg to detect autocorrelation in the gasoline data.

The AUTOREG Procedure

Dependent Variable	Ln_G_Pop

Ordinary Least Squares Estimates			
SSE	0.02487344	DFE	31
MSE	0.0008024	Root MSE	0.02833
SBC	−141.90789	AIC	−149.82548
Regress R-Square	0.9691	Total R-Square	0.9691

Durbin–Watson Statistics			
Order	DW	Pr < DW	Pr > DW
1	0.8909	<0.0001	1.0000
2	2.0276	0.4049	0.5951
3	2.3773	0.8792	0.1208
4	2.1961	0.8211	0.1789
5	1.9231	0.6318	0.3682

Note: Pr < DW is the *p* value for testing positive autocorrelation, and Pr > DW is the *p* value for testing negative autocorrelation.

Variable	DF	Estimate	Standard Error	t Value	Approx Pr > |t|
Intercept	1	−7.7892	0.3593	−21.68	<0.0001
Ln_pg	1	−0.0979	0.0283	−3.46	0.0016
Ln_Income	1	2.1175	0.0988	21.44	<0.0001
Ln_Pnc	1	0.1224	0.1121	1.09	0.2830
Ln_Puc	1	−0.1022	0.0693	−1.48	0.1502

OUTPUT 6.2. Durbin–Watson statistics for the first five autocorrelations in the gasoline data.

The AUTOREG Procedure

Dependent Variable	Ln_G_Pop

Ordinary Least Squares Estimates			
SSE	0.02487344	DFE	31
MSE	0.0008024	Root MSE	0.02833
SBC	−141.90789	AIC	−149.82548
Regress R-Square	0.9691	Total R-Square	0.9691
Durbin–Watson	0.8909		

Variable	DF	Estimate	Standard Error	t Value	Approx Pr > |t|
Intercept	1	−7.7892	0.3593	−21.68	<0.0001
Ln_pg	1	−0.0979	0.0283	−3.46	0.0016
Ln_Income	1	2.1175	0.0988	21.44	<0.0001
Ln_Pnc	1	0.1224	0.1121	1.09	0.2830
Ln_Puc	1	−0.1022	0.0693	−1.48	0.1502

Estimates of Autocorrelations			
Lag	Covariance	Correlation	−1 9 8 7 6 5 4 3 2 1 0 1 2 3 4 5 6 7 8 9 1
0	0.000691	1.000000	| |********************|
1	0.000382	0.552840	| |*********** |
2	−0.00001	−0.021286	| | |
3	−0.00015	−0.211897	| ****| |
4	−0.00010	−0.147596	| ***| |
5	−0.00002	−0.025090	| *| |

Preliminary MSE	0.000365

Estimates of Autoregressive Parameters			
Lag	Coefficient	Standard Error	t Value
1	−0.883472	0.195936	−4.51
2	0.623294	0.260538	2.39
3	−0.217351	0.284615	−0.76
4	0.117484	0.260538	0.45
5	−0.042809	0.195936	−0.22

Yule–Walker Estimates			
SSE	0.01258864	DFE	26
MSE	0.0004842	Root MSE	0.02200
SBC	−147.56116	AIC	−163.39635
Regress R-Square	0.9568	Total R-Square	0.9844
Durbin–Watson	1.8728		

Variable	DF	Estimate	Standard Error	t Value	Approx Pr > |t|
Intercept	1	−7.5542	0.4583	−16.48	<0.0001
Ln_pg	1	−0.0706	0.0337	−2.09	0.0461
Ln_Income	1	2.0520	0.1259	16.30	<0.0001
Ln_Pnc	1	0.1344	0.1336	1.01	0.3239
Ln_Puc	1	−0.1257	0.0833	−1.51	0.1434

OUTPUT 6.3. Proc Autoreg output showing the first five autocorrelations from the gasoline data.

The AUTOREG Procedure

Dependent Variable	Ln_G_Pop

Ordinary Least Squares Estimates			
SSE	0.02487344	DFE	31
MSE	0.0008024	Root MSE	0.02833
SBC	−141.90789	AIC	−149.82548
Regress R-Square	0.9691	Total R-Square	0.9691
Durbin–Watson	0.8909	Pr < DW	<0.0001
Pr > DW	1.0000		

Note: Pr < DW is the *p* value for testing positive autocorrelation, and Pr > DW is the *p* value for testing negative autocorrelation.

Variable	DF	Estimate	Standard Error	t Value	Approx Pr > |t|
Intercept	1	−7.7892	0.3593	−21.68	<0.0001
Ln_pg	1	−0.0979	0.0283	−3.46	0.0016
Ln_Income	1	2.1175	0.0988	21.44	<0.0001
Ln_Pnc	1	0.1224	0.1121	1.09	0.2830
Ln_Puc	1	−0.1022	0.0693	−1.48	0.1502

Estimates of Autocorrelations			
Lag	Covariance	Correlation	-1 9 8 7 6 5 4 3 2 1 0 1 2 3 4 5 6 7 8 9 1
0	0.000691	1.000000	| |********************|
1	0.000382	0.552840	| |********** |

Preliminary MSE	0.000480

Estimates of Autoregressive Parameters			
Lag	Coefficient	Standard Error	t Value
1	−0.552840	0.152137	−3.63

Yule–Walker Estimates			
SSE	0.01604701	DFE	30
MSE	0.0005349	Root MSE	0.02313
SBC	−153.73763	AIC	−163.23874
Regress R-Square	0.9271	Total R-Square	0.9801
Durbin–Watson	1.3707	Pr < DW	0.0077
Pr > DW	0.9923		

Note: Pr < DW is the *p* value for testing positive autocorrelation, and Pr > DW is the *p* value for testing negative autocorrelation.

Variable	DF	Estimate	Standard Error	t Value	Approx Pr > |t|
Intercept	1	−7.1940	0.5152	−13.96	<0.0001
Ln_pg	1	−0.1239	0.0366	−3.39	0.0020
Ln_Income	1	1.9534	0.1418	13.77	<0.0001
Ln_Pnc	1	0.1221	0.1317	0.93	0.3611
Ln_Puc	1	−0.0548	0.0784	−0.70	0.4900

OUTPUT 6.4. AR(1) model for the gasoline consumption data.

The B&P test statistic value is

$$Q = 36 \times \lfloor 0.5528^2 + 0.0213^2 + 0.212^2 + 0.1476^2 + 0.0251^2 \rfloor = 13.44,$$

while the Ljung test statistic value is

$$Q' = 36 \times 38 \times \left[\frac{0.5528^2}{35} + \frac{0.0213^2}{34} + \frac{0.212^2}{33} + \frac{0.1476^2}{32} + \frac{0.0251^2}{31} \right] = 14.78.$$

Both test statistic values exceed the chi-squared critical value of $\chi^2_{0.05,5} = 11.07$ leading to the rejection of the null hypothesis of no autocorrelation. Therefore, the OLS estimates are not efficient and we need to get estimates that are adjusted for the autocorrelations.

6.5 USING SAS TO FIT THE AR MODELS

Having detected the presence of autocorrelation, we must now estimate the parameters by using either GLS or FGLS. As mentioned in the earlier sections, in reality ρ is assumed to be unknown and therefore has to be estimated leading to the FGLS estimator. We will use Proc Autoreg to fit the AR models. In this instance, we use it to conduct OLS regression where we suspect that the disturbances are correlated. Proc Reg can also be used to conduct the analysis. However, the data will have to be first transformed by using the Prais–Winsten (or the Cochrane and Orcutt) methods. The following statements can be used to fit the AR (1) model to the gasoline consumption data set. The "nlag=1" option requests the AR(1) model while the options "dw" and "dwprob" are used for the Durbin–Watson test statistic and p values. The analysis results are given in Output 6.4.

The AUTOREG Procedure

Dependent Variable Ln_G_Pop

Ordinary Least Squares Estimates			
SSE	0.02487344	DFE	31
MSE	0.0008024	Root MSE	0.02833
SBC	–141.90789	AIC	–149.82548
Regress R-Square	0.9691	Total R-Square	0.9691

Durbin–Watson Statistics			
Order	DW	Pr < DW	Pr > DW
1	0.8909	<0.0001	1.0000
2	2.0276	0.4049	0.5951

Note: Pr<DW is the p value for testing positive autocorrelation, and Pr>DW is the p value for testing negative autocorrelation.

Variable	DF	Estimate	Standard Error	t Value	Approx Pr > |t|
Intercept	1	–7.7892	0.3593	–21.68	<0.0001
Ln_pg	1	–0.0979	0.0283	–3.46	0.0016
Ln_Income	1	2.1175	0.0988	21.44	<0.0001
Ln_Pnc	1	0.1224	0.1121	1.09	0.2830
Ln_Puc	1	–0.1022	0.0693	–1.48	0.1502

Estimates of Autocorrelations			
Lag	Covariance	Correlation	-1 9 8 7 6 5 4 3 2 1 0 1 2 3 4 5 6 7 8 9 1
0	0.000691	1.000000	| |********************|
1	0.000382	0.552840	| |********** |
2	–0.00001	–0.021286	| | |

Preliminary MSE 0.000373

OUTPUT 6.5. AR(2) model for the gasoline consumption data.

Estimates of Autoregressive Parameters			
Lag	Coefficient	Standard Error	t Value
1	−0.813124	0.163827	−4.96
2	0.470813	0.163827	2.87

Yule–Walker Estimates			
SSE	0.01289972	DFE	29
MSE	0.0004448	Root MSE	0.02109
SBC	−157.51232	AIC	−168.59695
Regress R-Square	0.9610	Total R-Square	0.9840

Durbin-Watson Statistics			
Order	DW	Pr < DW	Pr > DW
1	1.8505	0.2150	0.7850
2	2.2454	0.6761	0.3239

Note: Pr<DW is the *p* value for testing positive autocorrelation, and Pr>DW is the *p* value for testing negative autocorrelation.

Variable	DF	Estimate	Standard Error	t Value	Approx Pr > |t|
Intercept	1	−7.7456	0.4280	−18.10	<0.0001
Ln_pg	1	−0.0743	0.0318	−2.34	0.0265
Ln_Income	1	2.1044	0.1175	17.91	<0.0001
Ln_Pnc	1	0.1870	0.1302	1.44	0.1616
Ln_Puc	1	−0.1616	0.0816	−1.98	0.0572

OUTPUT 6.5. (*Continued*).

```
proc autoreg data=gasoline;
    model Ln_G_Pop=Ln_Pg Ln_Income Ln_Pnc Ln_Puc/nlag=1 dw=1
    dwprob;
run;
```

The first part of the output gives the least squares estimator values along with the Durbin–Watson statistic. Notice again that the null hypothesis of no autocorrelation is rejected. The next part of the output gives the value of ρ. The value reported is reversed in sign and is 0.553. The FGLS estimates are then reported, assuming the AR(1) model. Notice that the Durbin–Watson statistic is still significant, indicating that the AR(1) may be inadequate for the data and that a higher order autocorrelation model may be more appropriate. The parameter estimates table indicates that both the price of gasoline and the income are significant in explaining the variation in gasoline consumption. These variables also show up as significant in the OLS model. Furthermore, the signs of the coefficients between these two models are the same. The option "nlag = 2" is used to fit an AR(2) model (see Output 6.5).

Notice that the Durbin–Watson statistic is now insignificant. We can therefore conclude that the AR(2) model is more appropriate than the AR(1) model and that the data set used did suffer from second-order autocorrelation. The values of the two autoregressive parameters θ_1 and θ_2 are (using the opposite signs) 0.813 and −0.471, respectively. However, the variable ln_Pvc, which is the log of the price of used cars, is significant at the 10% level in the AR(2) model, whereas it was not significant in the AR(1) model. The magnitudes of the coefficients for fuel price in the AR(2) model is significantly lower than in the AR(1) model. The magnitude of the price of used cars in the AR(2) model is significantly larger than in the AR(1) model.

Proc Autoreg can be used to generate estimates based on maximum likelihood estimation by using the "method = ml" option in the model statement. Results of this analysis are provided in Outputs 6.6 and 6.7. Notice that the parameter estimates are quite different from both OLS and FGLS estimators.

The model statement option "method = ityw" will result in the iterated FGLS analysis. The results of the analysis are provided in Outputs 6.8 and 6.9. Notice that the results are very similar to the results produced using maximum likelihood estimation, thus confirming that in general the iterated estimation technique converges to the maximum likelihood estimates (Greene, 2003).

It may be of interest to compare the predicted values from Proc Reg (OLS estimation) and Proc Autoreg. We will compare the OLS predicted values with the predicted values from the AR(2) model. The following statements can be used. The analysis results are given in Figure 6.3.

```
proc reg data=gasoline noprint;
     model Ln_G_Pop=Ln_Pg Ln_Income Ln_Pnc Ln_Puc;
     output out=a r=r_g;
run;
proc autoreg data=gasoline noprint;
```

The AUTOREG Procedure

Dependent Variable	Ln_G_Pop

Ordinary Least Squares Estimates			
SSE	0.02487344	DFE	31
MSE	0.0008024	Root MSE	0.02833
SBC	−141.90789	AIC	−149.82548
Regress R-Square	0.9691	Total R-Square	0.9691
Durbin–Watson	0.8909	Pr < DW	<0.0001
Pr > DW	1.0000		

Note: Pr<DW is the *p* value for testing positive autocorrelation, and Pr>DW is the *p* value for testing negative autocorrelation.

Variable	DF	Estimate	Standard Error	t Value	Approx Pr > |t|
Intercept	1	−7.7892	0.3593	−21.68	<0.0001
Ln_pg	1	−0.0979	0.0283	−3.46	0.0016
Ln_Income	1	2.1175	0.0988	21.44	<0.0001
Ln_Pnc	1	0.1224	0.1121	1.09	0.2830
Ln_Puc	1	−0.1022	0.0693	−1.48	0.1502

Estimates of Autocorrelations			
Lag	Covariance	Correlation	−1 9 8 7 6 5 4 3 2 1 0 1 2 3 4 5 6 7 8 9 1
0	0.000691	1.000000	| |********************|
1	0.000382	0.552840	| |*********** |

Preliminary MSE	0.000480

Estimates of Autoregressive Parameters			
Lag	Coefficient	Standard Error	t Value
1	−0.552840	0.152137	−3.63

Algorithm converged.

OUTPUT 6.6. MLE estimates of the AR(1) model for the gasoline consumption data.

Maximum Likelihood Estimates			
SSE	0.01262941	DFE	30
MSE	0.0004210	Root MSE	0.02052
SBC	−160.53233	AIC	−170.03344
Regress R-Square	0.7830	Total R-Square	0.9843
Durbin–Watson	1.4455	Pr < DW	0.0193
Pr > DW	0.9807		

Note: Pr<DW is the *p* value for testing positive autocorrelation, and Pr>DW is the *p* value for testing negative autocorrelation.

Variable	DF	Estimate	Standard Error	t Value	Approx Pr > |t|
Intercept	1	−5.1710	0.7617	−6.79	<0.0001
Ln_pg	1	−0.1939	0.0409	−4.74	<0.0001
Ln_Income	1	1.3896	0.2041	6.81	<0.0001
Ln_Pnc	1	0.2509	0.1480	1.70	0.1003
Ln_Puc	1	−0.004280	0.0698	−0.06	0.9515
AR1	1	−0.9425	0.0915	−10.30	<0.0001

Autoregressive parameters assumed given					
Variable	DF	Estimate	Standard Error	t Value	Approx Pr > |t|
Intercept	1	−5.1710	0.6808	−7.60	<0.0001
Ln_pg	1	−0.1939	0.0380	−5.11	<0.0001
Ln_Income	1	1.3896	0.1868	7.44	<0.0001
Ln_Pnc	1	0.2509	0.1353	1.86	0.0734
Ln_Puc	1	−0.004280	0.0697	−0.06	0.9515

OUTPUT 6.6. (*Continued*).

```
    model Ln_G_Pop=Ln_Pg Ln_Income Ln_Pnc Ln_Puc/nlag=2;
    output out=b r=ra_g;
run;
data c;
    merge a b;
run;
    proc gplot data=c;
    plot r_g*year=1 ra_g*year=2/overlay href=0 haxis=1960 to
    1995 by 5;
run;
```

It does appear that magnitudes of the residuals from OLS (solid line) have been reduced by using the AR(2) model (dotted line) confirming that the AR(2) model is more appropriate for the gasoline consumption data set than the OLS model.

In the discussion so far, we used an AR(2) model because the data set suffered from second-order autocorrelation. As it turns out, there is not much gain (if any) in using a higher autocorrelation model. The following statements in SAS can be used to compare the residuals from various models. The "nlag=" option with values 1 through 5 is used to fit models ranging from the AR(1) to the AR(5) models. The residuals from each model are stored and compared against each other by using the "overlay" option of Proc Gplot (Freund and Littell, 2000, p. 93). The analysis results are given in Figure 6.4.

```
proc autoreg data=gasoline noprint;
    model Ln_G_Pop=Ln_Pg Ln_Income Ln_Pnc Ln_Puc/nlag=1;
    output out=a r=r1;
```

```
run;
proc autoreg data=gasoline noprint;
     model Ln_G_Pop=Ln_Pg Ln_Income Ln_Pnc Ln_Puc/nlag=2;
     output out=b r=r2;
run;
proc autoreg data=gasoline noprint;
     model Ln_G_Pop=Ln_Pg Ln_Income Ln_Pnc Ln_Puc/nlag=3;
     output out=c r=r3;
run;
proc autoreg data=gasoline noprint;
     model Ln_G_Pop=Ln_Pg Ln_Income Ln_Pnc Ln_Puc/nlag=4;
     output out=d r=r4;
run;
proc autoreg data=gasoline noprint;
     model Ln_G_Pop=Ln_Pg Ln_Income Ln_Pnc Ln_Puc/nlag=5;
     output out=e r=r5;
```

The AUTOREG Procedure

Dependent Variable	Ln_G_Pop

Ordinary Least Squares Estimates			
SSE	0.02487344	DFE	31
MSE	0.0008024	Root MSE	0.02833
SBC	−141.90789	AIC	−149.82548
Regress R-Square	0.9691	Total R-Square	0.9691

Durbin–Watson Statistics			
Order	DW	Pr < DW	Pr > DW
1	0.8909	<0.0001	1.0000
2	2.0276	0.4049	0.5951

Note: Pr<DW is the p value for testing positive autocorrelation, and Pr>DW is the p value for testing negative autocorrelation.

Variable	DF	Estimate	Standard Error	t Value	Approx Pr > |t|
Intercept	1	−7.7892	0.3593	−21.68	<0.0001
Ln_pg	1	−0.0979	0.0283	−3.46	0.0016
Ln_Income	1	2.1175	0.0988	21.44	<0.0001
Ln_Pnc	1	0.1224	0.1121	1.09	0.2830
Ln_Puc	1	−0.1022	0.0693	−1.48	0.1502

Estimates of Autocorrelations			
Lag	Covariance	Correlation	-1 9 8 7 6 5 4 3 2 1 0 1 2 3 4 5 6 7 8 9 1
0	0.000691	1.000000	| |******************* |
1	0.000382	0.552840	| |********** |
2	−0.00001	−0.021286	| | |

Preliminary MSE	0.000373

OUTPUT 6.7. MLE estimates of the AR(2) model for the gasoline consumption data.

Estimates of Autoregressive Parameters			
Lag	Coefficient	Standard Error	t Value
1	−0.813124	0.163827	−4.96
2	0.470813	0.163827	2.87

Algorithm converged.

Maximum Likelihood Estimates			
SSE	0.0126731	DFE	29
MSE	0.0004370	Root MSE	0.02090
SBC	−157.92648	AIC	−169.01111
Regress R-Square	0.9683	Total R-Square	0.9843

Durbin–Watson Statistics			
Order	DW	Pr < DW	Pr > DW
1	1.9062	0.2627	0.7373
2	2.1296	0.5341	0.4659

Note: Pr<DW is the *p* value for testing positive autocorrelation, and Pr>DW is the *p* value for testing negative autocorrelation.

Variable	DF	Estimate	Standard Error	t Value	Approx Pr > |t|
Intercept	1	−7.8531	0.3973	−19.77	<0.0001
Ln_pg	1	−0.0673	0.0301	−2.24	0.0332
Ln_Income	1	2.1337	0.1091	19.56	<0.0001
Ln_Pnc	1	0.2078	0.1253	1.66	0.1081
Ln_Puc	1	−0.1850	0.0797	−2.32	0.0276
AR1	1	−0.8220	0.1554	−5.29	<0.0001
AR2	1	0.5643	0.1567	3.60	0.0012

Autoregressive parameters assumed given.					
Variable	DF	Estimate	Standard Error	t Value	Approx Pr > |t|
Intercept	1	−7.8531	0.3905	−20.11	<0.0001
Ln_pg	1	−0.0673	0.0295	−2.28	0.0299
Ln_Income	1	2.1337	0.1072	19.91	<0.0001
Ln_Pnc	1	0.2078	0.1247	1.67	0.1065
Ln_Puc	1	−0.1850	0.0786	−2.35	0.0256

OUTPUT 6.7. (*Continued*).

```
run;
data f;
    merge a b c d e;
run;
proc gplot data=f;
    title 'Comparing the residuals from the AR(1)--AR(5) models';
    plot r1*year=1 r2*year=2 r3*year=3 r4*year=4
    r5*year=5/overlay href=0
    haxis=1960 to 1995 by 5;
run;
```

The AUTOREG Procedure

Dependent Variable	Ln_G_Pop

Ordinary Least Squares Estimates			
SSE	0.02487344	DFE	31
MSE	0.0008024	Root MSE	0.02833
SBC	-141.90789	AIC	-149.82548
Regress R-Square	0.9691	Total R-Square	0.9691
Durbin–Watson	0.8909	Pr < DW	<0.0001
Pr > DW	1.0000		

Note: Pr<DW is the *p* value for testing positive autocorrelation, and Pr>DW is the *p* value for testing negative autocorrelation.

Variable	DF	Estimate	Standard Error	t Value	Approx Pr > |t|
Intercept	1	-7.7892	0.3593	-21.68	<0.0001
Ln_pg	1	-0.0979	0.0283	-3.46	0.0016
Ln_Income	1	2.1175	0.0988	21.44	<0.0001
Ln_Pnc	1	0.1224	0.1121	1.09	0.2830
Ln_Puc	1	-0.1022	0.0693	-1.48	0.1502

Estimates of Autocorrelations			
Lag	Covariance	Correlation	-1 9 8 7 6 5 4 3 2 1 0 1 2 3 4 5 6 7 8 9 1
0	0.000691	1.000000	| |********************|
1	0.000382	0.552840	| |********** |

Preliminary MSE	0.000480

Estimates of Autoregressive Parameters			
Lag	Coefficient	Standard Error	t Value
1	-0.552840	0.152137	-3.63

Algorithm converged.

Yule–Walker Estimates			
SSE	0.01271242	DFE	30
MSE	0.0004237	Root MSE	0.02059
SBC	-160.48746	AIC	-169.98857
Regress R-Square	0.7887	Total R-Square	0.9842
Durbin–Watson	1.4458	Pr < DW	0.0191
Pr > DW	0.9809		

Note: Pr<DW is the *p* value for testing positive autocorrelation, and Pr>DW is the *p* value for testing negative autocorrelation.

Variable	DF	Estimate	Standard Error	t Value	Approx Pr > |t|
Intercept	1	-5.2466	0.6761	-7.76	<0.0001
Ln_pg	1	-0.1915	0.0381	-5.02	<0.0001
Ln_Income	1	1.4108	0.1858	7.59	<0.0001
Ln_Pnc	1	0.2482	0.1349	1.84	0.0757
Ln_Puc	1	-0.004509	0.0703	-0.06	0.9493

OUTPUT 6.8. Iterated FGLS estimates of AR(1) model for the gasoline data.

The AUTOREG Procedure

Dependent Variable	Ln_G_Pop

Ordinary Least Squares Estimates			
SSE	0.02487344	DFE	31
MSE	0.0008024	Root MSE	0.02833
SBC	–141.90789	AIC	–149.82548
Regress R-Square	0.9691	Total R-Square	0.9691

Durbin–Watson Statistics			
Order	DW	Pr < DW	Pr > DW
1	0.8909	<0.0001	1.0000
2	2.0276	0.4049	0.5951

Note: Pr<DW is the *p* value for testing positive autocorrelation, and Pr>DW is the *p* value for testing negative autocorrelation.

Variable	DF	Estimate	Standard Error	t Value	Approx Pr > |t|
Intercept	1	–7.7892	0.3593	–21.68	<0.0001
Ln_pg	1	–0.0979	0.0283	–3.46	0.0016
Ln_Income	1	2.1175	0.0988	21.44	<0.0001
Ln_Pnc	1	0.1224	0.1121	1.09	0.2830
Ln_Puc	1	–0.1022	0.0693	–1.48	0.1502

Estimates of Autocorrelations			
Lag	Covariance	Correlation	-1 9 8 7 6 5 4 3 2 1 0 1 2 3 4 5 6 7 8 9 1
0	0.000691	1.000000	| |********************|
1	0.000382	0.552840	| |********** |
2	–0.00001	–0.021286	| | |

Preliminary MSE	0.000373

Estimates of Autoregressive Parameters			
Lag	Coefficient	Standard Error	t Value
1	–0.813124	0.163827	–4.96
2	0.470813	0.163827	2.87

Algorithm converged.

Yule–Walker Estimates			
SSE	0.01265611	DFE	29
MSE	0.0004364	Root MSE	0.02089
SBC	–157.9093	AIC	–168.99393
Regress R-Square	0.9690	Total R-Square	0.9843

Durbin–Watson Statistics			
Order	DW	Pr < DW	Pr > DW
1	1.9257	0.2822	0.7178
2	2.1120	0.5122	0.4878

Note: Pr<DW is the *p* value for testing positive autocorrelation, and Pr>DW is the *p* value for testing negative autocorrelation.

Variable	DF	Estimate	Standard Error	t Value	Approx Pr > |t|
Intercept	1	–7.8665	0.3862	–20.37	<0.0001
Ln_pg	1	–0.0661	0.0292	–2.27	0.0311
Ln_Income	1	2.1374	0.1060	20.17	<0.0001
Ln_Pnc	1	0.2118	0.1241	1.71	0.0987
Ln_Puc	1	–0.1891	0.0783	–2.42	0.0223

OUTPUT 6.9. Iterated FGLS estimates of AR(2) model for the gasoline data.

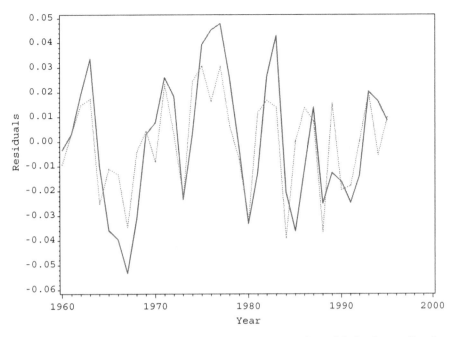

FIGURE 6.3. Comparing the residuals of the OLS versus AR(2) models for the gasoline data.

The residuals from the AR(1) process are indicated by the line that goes beyond all other lines across the range of year. The residuals from the AR(2) through the AR(5) models are highly coincidental and cannot be distinguished. Notice that the residuals appear to be more stable for the AR(2) through AR(5) models. Therefore, there is no significant improvement beyond the AR(2) model.

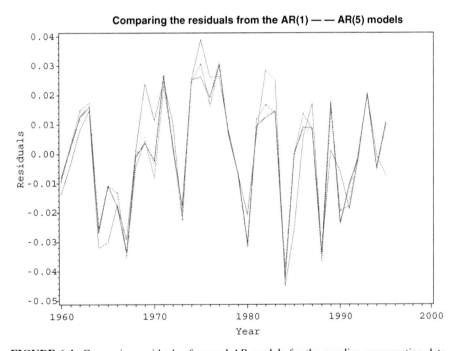

FIGURE 6.4. Comparing residuals of several AR models for the gasoline consumption data.

7

PANEL DATA ANALYSIS

7.1 WHAT IS PANEL DATA?

The discussion in the previous chapters focused on analytical methods for cross-section data and (to some extent) time series data. This chapter deals with the analysis of data from panel studies. Data from such studies consist of repeated measurements on cross sections over a period of time. In other words, in panel data there are repeated observations on the same subject over a time period as in longitudinal studies where, for instance, subjects are surveyed or followed over time. Here the term subjects will be used to refer to people, countries, companies, and so on.

As an example of panel data, consider the Cost for US Airlines example from Greene (2003). The data set has 90 observations for six firms from 1970 to 1984. This data set has been analyzed in Chapter 3, and it will be used to illustrate the different analytical techniques in this chapter. We have a cross section if we take data from a single year. The selected cross section simply gives a snapshot of the six airlines for the selected year. We have panel data if we use data for the six airlines from every year in the time period 1970–1984.

As another example of panel data, consider the case where a financial planning company collects data on profits generated by its financial advisors. For instance, the profit of the financial advisors may be monitored for several years. Various factors such as the regional location of the advisor, their age, and the wealth profile of the advisor's clients may be used to explain the differences (if any) between the advisor's profit over time. The collected data can be viewed as panel data since we have a cross section of advisors who are followed for several years.

Notice that both examples may include explanatory variables that are either observed (controllable) or unobserved (uncontrollable). For instance, in the second example above, the data on the "salesmanship ability" of the financial advisor may not be available. We can therefore, in principle, partition the set of explanatory variables into two sets—one, consisting of observed variables and the other consisting of uncontrollable or unobserved variables. The set of explanatory variables can also be comprised of variables that are time dependent as well as variables that are time independent. For example, gender and marital status can be treated as time independent whereas experience in the workforce and the age of the subject are time dependent.

7.1.1 Advantages of Panel Data

There are two main reasons for using panel data methods.

1. *Increased Sample Sizes.* A natural consequence of using panel data is that the available sample size is increased. In the Cost of US Airlines data, we had six airlines for which data were collected over a 15-year time period. A single cross section

Applied Econometrics Using the SAS® System, by Vivek B. Ajmani would have resulted in six data points, whereas in the panel data format we have 90 data points. In general, if there are n subjects and T time periods, then there will potentially be a total of nT observations for analysis.

2. *The Ability to Control for Unobserved Heterogeneity.* The increased sample size is a natural and trivial consequence of using panel data methods. The main reason the preferring a panel approach to estimation is that one can control the unobserved heterogeneity among the individual subjects or unit effects. As will be discussed in the next section, a complete formulation of the panel data model includes both observed and unobserved explanatory variables. As mentioned earlier, the unobserved explanatory variables may include variables that we cannot control or that are just not observable. Omitting these variables from the model will lead to omitted variable bias (Chapter 4). The panel data approach allows us to include the unobserved heterogeneity effects in the model. If the unobserved subject-specific effects are treated as constants, then we have a fixed effects model and if they are treated as random, then we have a random effects model.

7.2 PANEL DATA MODELS

Using the notation given in Greene (2003) page 285, we can write the basic panel data model as $y_{it} = \mathbf{x}_{it}^T \boldsymbol{\beta} + \mathbf{z}_i^T \boldsymbol{\alpha} + \varepsilon_{it}$, where $i = 1, \ldots, n$ and $t = 1, \ldots, T$. Here, n is the number of subjects and T is the number of time periods. The number of time periods may be different for the different subjects in the study, leading to unbalanced panel data and this may arise if some subjects drop out of the study prior to completion of the study. Although there are well-established analysis methods available for unbalanced panel data, we will focus our attention on analysis of balanced panel data only. The term y_{it} is the observation collected on the ith subject at time period t. The term $\mathbf{x}_{it}$ is a vector of k observed explanatory variables. The time-independent term $\mathbf{z}_i^T \boldsymbol{\alpha}$ captures the unobserved heterogeneity of the subjects and is assumed to contain a constant term. Some authors make a distinction between the constant term and the unobserved heterogeneity term by expressing the panel data model as $y_{it} = \mathbf{x}_{it}^T \boldsymbol{\beta} + (\alpha_0 + \mathbf{z}_i^T \boldsymbol{\alpha}) + \varepsilon_{it}$, where $i = 1, \ldots, n$ and $t = 1, \ldots, T$. Both formulations lead to the same exact results and we use the first notation for simplicity and convenience.

It is trivial to see that if the variables in $\mathbf{z}_i$ is observed for all subjects, then the panel data model can be estimated by using ordinary least squares. In this case, the variables in $\mathbf{x}$ and $\mathbf{z}$ can be combined for the analysis. Here, we may assume that the set of controllable variables $\mathbf{x}$ is exhaustive implying that $\mathbf{z}$ just has a constant term.

Since $\mathbf{z}_i^T \boldsymbol{\alpha}$ is assumed to be unobserved, it is convenient to write $\alpha_i = \mathbf{z}_i^T \boldsymbol{\alpha}$ and re-express the model as

$$y_{it} = \mathbf{x}_{it}^T \boldsymbol{\beta} + \alpha_i + \varepsilon_{it}, \quad i = 1, \ldots, n; \ t = 1, \ldots, T.$$

The treatment of the heterogeneity effects determines the type of model that is used to analyze panel data. The various models that are considered when analyzing panel data are as follows:

a. *Pooled Regression:* It is trivial to show that if $\mathbf{z}_i$ contains only a constant term, that is, if $\alpha_1 = \alpha_2 = \ldots = \alpha_n = \alpha$, then the general model can be written as $y_{it} = \alpha + \mathbf{x}_{it}^T \boldsymbol{\beta} + \varepsilon_{it}$, where $i = 1, \ldots, n$ and $t = 1, \ldots, T$, and the parameters can be estimated via OLS.

b. *Fixed Effects:* In the fixed effects model, we relax the assumption that $\alpha_1 = \alpha_2 = \ldots = \alpha_n = \alpha$ and write the model as $y_{it} = \alpha_i + \mathbf{x}_{it}^T \boldsymbol{\beta} + \varepsilon_{it}$. Here, α_i can be viewed as the subject-specific intercept terms. This representation results in a common coefficients vector but different intercept terms—the intercept terms being the subject-specific constant terms α_i. In the simple linear regression case, this will result in different regression lines for the different subjects where the lines are parallel to each other (same slope) but have different intercepts.

c. *Random Effects:* In the fixed effects analysis, it is assumed that the selected subjects represent the entire population of subjects who are available for the study. On the other hand, if the subjects were selected from a much larger population, then it may be reasonable to assume that the differences among the subjects are randomly distributed across the population. The random effects model can be easily formed by assuming that $\mathbf{z}_i^T \boldsymbol{\alpha} = \alpha_i = \alpha + u_i$, where $E(u_i) = 0$ and $Var(u_i) = \sigma_u^2$. That is, the unobserved effect is partitioned into a component that is fixed or common to all subjects and a disturbance that is subject specific. The general linear model can now be expressed as $y_{it} = \mathbf{x}_{it}^T \boldsymbol{\beta} + \alpha + u_i + \varepsilon_{it}$. A key assumption in the random effects model is that the unobserved subject-specific heterogeneity, $\mathbf{z}_i$, is distributed independently of $\mathbf{x}_{it}$. It is easy to see why violations of this assumption will lead to endogeneity of the observed explanatory variables leading to biased and inconsistent random effects estimates.

To illustrate the computations for the various panel data models, we will make use of the cost function of US Airlines data from Greene (2003), which was analyzed in the earlier chapters. The data in this example consists of repeated measurements from 1970 to 1984 on six airlines. Therefore, this can be viewed as a panel data with six subjects and 15 time periods. The following model will be estimated from the data (Greene, 2003, p. 286):

$$\ln(\cos t_{it}) = \beta_1 + \beta_2 \ln(output_{it}) + \beta_3 \ln(Fuel\_Price_{it}) + \beta_4 Load\_Factor_{it} + \varepsilon_{it}$$

where $i = 1, \ldots, 6$ and $t = 1, \ldots, 15$. As described by the author, the variable output gives the "revenue passenger miles," which is the number of revenue paying passengers times the number of miles flown by the airline in the given time period. The variable load factor measures the percentage of available seating capacity that is filled with passengers.

7.3 THE POOLED REGRESSION MODEL

In the pooled regression model, we assume that the individuals effects are fixed and more importantly common across all subjects, such that $z_i^T \alpha_i = \alpha_i = \alpha$, $\forall i = 1, \ldots, n$. The model parameters can therefore be estimated using OLS. The following SAS statements can be used to fit a pooled regression model to the data. Note that we are assuming that a temporary SAS data set named airline was created in a data step module. The analysis results are given in Output 7.1.

```
proc reg data=airline;
     model LnC=LnQ LnPF LF;
run;
```

The REG Procedure
Model: MODEL1
Dependent Variable: LnC

Number of Observations Read	90
Number of Observations Used	90

Analysis of Variance					
Source	DF	Sum of Squares	Mean Square	F Value	Pr > F
Model	3	112.70545	37.56848	2419.34	<0.0001
Error	86	1.33544	0.01553		
Corrected Total	89	114.04089			

Root MSE	0.12461	R-Square	0.9883
Dependent Mean	13.36561	Adj R-Sq	0.9879
Coeff Var	0.93234		

Parameter Estimates						
Variable	Label	DF	Parameter Estimate	Standard Error	t Value	Pr > |t|
Intercept	Intercept	1	9.51692	0.22924	41.51	<0.0001
LnQ		1	0.88274	0.01325	66.60	<0.0001
LnPF		1	0.45398	0.02030	22.36	<0.0001
LF	LF	1	−1.62751	0.34530	−4.71	<0.0001

OUTPUT 7.1. The pooled regression model for the airlines data using OLS.

The pooled regression model is given by

$$\ln(cos\,t_{it}) = 9.52 + 0.88\ln(output_{it}) + 0.45\ln(Fuel\_Price_{it}) - 1.63 Load\_Factor_{it}.$$

The coefficient of determination is $R^2 = 0.988$, and an estimate of the root mean square error is, $\hat{\sigma}^2 = 0.1246$. The coefficients for the explanatory variables are all highly significant. The signs of the coefficients in the model make intuitive sense. We should expect the cost of the airline to increase with increases in output and fuel prices and decreases in load factor.

7.4 THE FIXED EFFECTS MODEL

As seen earlier, in the fixed effects model, we assume that the individual effects are constant but are not common across the subjects. That is, $z_i^T \alpha_i = \alpha_i \neq \alpha$, $\forall i = 1, \ldots, n$. Therefore, each α_i will have to be estimated along with β. There are three main methods used for estimating fixed effects models: the least squares dummy variable approach, the within-group effects approach, and the between-group effects approach.

The least squares dummy variables model (LSDV) incorporates the individual subject unobserved effects via dummy variables into the model, whereas the within-group effects method does not, since by construction, the unobserved effects are "swept" from the model. Both these strategies produce identical slopes for the nondummy independent variables. The between-group effects model also does not bring the dummy variables into the model and produces different parameter estimates from the LSDV and the within-group since the model relates the subject means of the dependent variable to the subject means of the explanatory variables plus an overall subject fixed effect that is a constant. A major issue with the between-groups estimation is that the analysis is based on a total of n (the number of subjects) observations only, which becomes restrictive if the model of interest has a large number of explanatory variables. This is because we need the number of observations in the data set to be at least as large as the number of model parameters to be estimated.

A disadvantage of the LSDV approach is that it becomes restrictive in the presence of a large number of subjects in the panel data. As mentioned earlier, this approach involves calculating estimates of the dummy variable parameters along with the estimates of the coefficient vector of the explanatory variables. The number of parameters to be estimated therefore increases as the number of subjects in the panel data increases. As discussed in Baltagi (2005, p. 13), if the number of time periods (T) is fixed and if the number of subjects increases ($n \to \infty$), then only the fixed effect parameters of the explanatory variables is consistent.

Baltagi (2005) lists three disadvantages of the within-group model. First, the dummy variables have to be estimated separately if the researcher is interested in the dummy variable parameters. This is because, by construction, the within-group model "sweeps" the dummy variables from the model. An advantage here is that the parameter estimates will be consistent even if the unobserved subject-specific heterogeneity is correlated to the observed explanatory variables. The dummy variable estimators can be computed by using the formula $a_i = \bar{y}_{i.} - b^T \bar{x}_{i.}$ (Greene, 2003, p. 288). Notice that this formula is just the least squares formula to compute the subject-specific intercepts (Chapter 1). Here, b is the LSDV (or within-group) estimator.

Second, since the within-group model does not incorporate the dummy variables, the degree of freedom for the error term gets large. This, in turn, results in a smaller root mean square error of the regression model. As discussed in Baltagi (2005, p. 14), the variance–covariance obtained from this model will have to be adjusted by a factor equal to the ratio of the error degrees of freedom of the within-group and the LSDV models to get the correct variance–covariance matrix. That is, the variance–covariance matrix is multiplied by $(nT - k)/(nT - n - k)$.

Finally, since the within-group model does not contain an intercept, the coefficient of determination (R^2) is incorrect. Meyers (1990, p. 39) gives the coefficient of determination for the nonintercept in the simple linear regression model as

$$R_{(0)}^2 = \frac{\sum_{i=1}^{n} \hat{y}_i^2}{\sum_{i=1}^{n} y_i^2}.$$

Notice that this is different from the formulation of R^2 in the case of the simple linear model with an intercept term. The argument can easily be extended to the multiple regression case.

We now begin our discussion of estimation of the fixed effects model parameters by using the LSDV approach. Consider the model

$$y_i = X_i\beta + i\alpha_i + \varepsilon_i$$

where $\mathbf{y}_i$ and $\mathbf{X}_i$ are the T observations on the response and the explanatory variables, $\mathbf{i}$ is a $T \times 1$ column of ones, and ε_i is the $T \times 1$ disturbance vector. If we stack the data for all subjects, we can write this model in matrix form as

$$\mathbf{y} = \begin{bmatrix} \mathbf{X} & \mathbf{d}_1 & \mathbf{d}_2 & \ldots & \mathbf{d}_n \end{bmatrix} \begin{bmatrix} \boldsymbol{\beta} \\ \boldsymbol{\alpha} \end{bmatrix} + \boldsymbol{\varepsilon}.$$

Notice that by construction,

$$[\mathbf{d}_1, \ldots, \mathbf{d}_n] = \begin{bmatrix} \mathbf{i} & \mathbf{0} & \ldots & \mathbf{0} \\ \mathbf{0} & \mathbf{i} & \ldots & \mathbf{0} \\ \vdots & \vdots & \ddots & \vdots \\ \mathbf{0} & \mathbf{0} & \ldots & \mathbf{i} \end{bmatrix}.$$

Here $\mathbf{d}_i$ is a dummy variable vector for the ith subject, that is, $\mathbf{d}_i$ is a $nT \times 1$ column vector where the elements are equal to 1 for the ith subject and 0 otherwise. At first glance this method of estimation appears to be analogous to dummy variables regression. However, we do not run into the "dummy-variable trap" here because we did not assume the presence of a constant in $\mathbf{x}_{it}$. Therefore, this estimation technique allows us to get clean estimates of all the model parameters (Greene, 2003, p. 287).

If we let the $nT \times n$ matrix be $\mathbf{D} = \begin{bmatrix} \mathbf{d}_1 & \mathbf{d}_2 & \ldots & \mathbf{d}_n \end{bmatrix}$, we can rewrite the fixed effects model as the least squares dummy variable (LSDV) model $\mathbf{y} = \mathbf{X}\boldsymbol{\beta} + \mathbf{D}\boldsymbol{\alpha} + \boldsymbol{\varepsilon}$. Using the Frisch–Waugh theorem (Chapter 1), it can be shown that the least squares estimator of $\boldsymbol{\beta}$ is given by $\mathbf{b} = [\mathbf{X}^T\mathbf{M}_D\mathbf{X}]^{-1}\mathbf{X}^T\mathbf{M}_D\mathbf{y}$, where $\mathbf{M}_D = \mathbf{I} - \mathbf{D}(\mathbf{D}^T\mathbf{D})^{-1}\mathbf{D}^T$ and is idempotent. If we let the vector $\mathbf{a}$ denote an estimator of $\boldsymbol{\alpha}$, then it can be shown (using Frisch–Waugh theorem again) that $\mathbf{a} = [\mathbf{D}^T\mathbf{D}]^{-1}\mathbf{D}^T(\mathbf{y} - \mathbf{X}\mathbf{b})$, which implies that for the ith subject, $a_i = \bar{y}_i - \mathbf{b}^T\bar{\mathbf{x}}_i$. Again, notice that this is the formula for calculating the intercept in a multiple linear regression model. It can also be shown that the asymptotic covariance matrix of $\mathbf{b}$ is $Est.Asy.Var(\mathbf{b}) = s^2(\mathbf{X}^T\mathbf{M}_D\mathbf{X})^{-1}$ with

$$s^2 = \frac{(\mathbf{y} - \mathbf{M}_D\mathbf{X}_b)^T(\mathbf{y} - \mathbf{M}_D\mathbf{X}_b)}{nT - n - k}.$$

The asymptotic variance of a_i is given by

$$Asy.Var(a_i) = \frac{\sigma^2}{T} + \bar{\mathbf{x}}_{i.}^T(Asy.Var(\mathbf{b}))\bar{\mathbf{x}}_{i.}$$

See Greene (2003, p. 288) for more details.

The differences across subjects can be tested by using a F test which tests the hypothesis that the constant terms are all equal. That is, $H_0 : \alpha_i = \alpha, \forall i = 1, \ldots, n$. The F statistic used for this test is given by

$$F(n-1, nT-n-k) = \frac{(R^2_{LSDV} - R^2_{pooled})/(n-1)}{(1 - R^2_{LSDV})/(nT-n-k)},$$

where LSDV indicates the dummy variable model and pooled indicates the pooled model. Notice that this test is identical to the F test that was discussed in Chapter 3 to compare a restricted model with an unrestricted model. Here, the pooled model is the restricted model as it restricts the fixed heterogeneity terms to be constant across all the subjects (Greene, 2003, p. 289).

We will use the airline cost data to illustrate the computations involved in the LSDV model. For simplicity, we will estimate only the parameters and their standard errors. First, we will analyze the data by using Proc IML and then by using the Proc Panel procedure. The following Proc IML statements can be used. We assume that a temporary SAS dataset names airline was created in the data step module. The analysis results are given in Output 7.2.

```
proc iml;
   * Read the data into matrices and create some constants.;
      use airline;
      read all var{'lq','lpf','lf'} into X;
      read all var{'lc'} into y;
      T=15;N=6;k=ncol(X);
   * Create the MD matrix.;
      i=J(T,1,1);
```

I	Nobs	Variable	MEAN
1	15	LNQ	0. 31927
		LNPF	12. 73180
		LF	0. 59719
2	15	LNQ	−0. 03303
		LNPF	12. 75171
		LF	0. 54709
3	15	LNQ	−0. 91226
		LNPF	12. 78972
		LF	0. 58454
4	15	LNQ	−1. 63517
		LNPF	12. 77803
		LF	0. 54768
5	15	LNQ	−2. 28568
		LNPF	12. 79210
		LF	0. 56649
6	15	LNQ	−2. 49898
		LNPF	12. 77880
		LF	0. 51978
All	90	LNQ	−1. 17431
		LNPF	12. 77036
		LF	0. 56046

The LSDV estimates are

TABLE1		
	BETA_LSDV	**SE**
LNQ	0.9193	0.0299
LNPF	0.4175	0.0152
LF	-1.0704	0.2017

TABLE2		
	ALPHA	**SE**
ALPHA1	9.7059	0.1931
ALPHA2	9.6647	0.1990
ALPHA3	9.4970	0.2250
ALPHA4	9.8905	0.2418
ALPHA5	9.7300	0.2609
ALPHA6	9.7930	0.2637

OUTPUT 7.2. LSDV estimates for the airlines data using Proc IML.

```
    NT=nrow(X);
    D=block(i,i,i,i,i,i);
    I=I(NT);
    MD=I-D*inv(D`*D)*D`;
* Calculate the LSDV estimates and their standard errors.;
    b_LSDV=inv(X`*MD*X)*X`*MD*y;
    a=inv(D`*D)*D`*(y-X*b_LSDV);
    sigma2=(MD*y-MD*X*b_LSDV)`*(MD*y-MD*X*b_LSDV)/(NT-N-K);
    Var_B=sqrt(vecdiag(sigma2*inv(X`*MD*X)));
    summary var {lq lpf lf} class {i} stat{mean}
    opt{save};
    X_Mean=LQ||LPF||LF;
```

```
       Var_A=Vecdiag(SQRT(sigma2/T +
       X_mean*sigma2*inv(X`*MD*X)*X_Mean`));
   * Print the results.;
       print 'The LSDV estimates are';
       Table1=b_LSDV||Var_B;
       Table2=a||Var_A;
       Print Table1 (|Colname={Beta_LSDV SE} rowname={LNQ LNPF LF}
format=8.4|);
       Print Table2 (|Colname={ALPHA SE} rowname={Alpha1 Alpha2 Alpha3
   Alpha4 Alpha5 Alpha6} format=8.4|); run;
```

It is trivial to show that an alternative way of estimating the parameters of the LSDV model is by using OLS to estimate the parameter vector δ in the model $\mathbf{y} = \mathbf{X}_d \delta + \varepsilon$, where $\mathbf{X}_d = [\, \mathbf{X} \; \mathbf{d}_1 \; \mathbf{d}_2 \; \ldots \; \mathbf{d}_n \,]$ and $\delta = [\, \beta \;\; \alpha \,]^T$. In the cost of airlines example, $\mathbf{X}_d$ is a 90×9 matrix, and δ is a 9×1 vector of unknown coefficients. The OLS estimator of δ is given by $\hat{\delta} = (\mathbf{X}_d^T \mathbf{X}_d)^{-1} \mathbf{X}_d^T \mathbf{y}$. The asymptotic variance of δ is given by $s^2 (\mathbf{X}_d^T \mathbf{X}_d)^{-1}$, where

$$ s^2 = \frac{(\mathbf{y} - \mathbf{X}_d \hat{\delta})^T (\mathbf{y} - \mathbf{X}_d \hat{\delta})}{nT - n - k}. $$

The following Proc IML statements can be used to estimate the parameters using this alternative formulation.

```
proc iml;
* Read the data into matrices and create some constants.;
    use airline;
    read all var{'lq','lpf','lf'} into X;
    read all var{'lc'} into y;
    T=15;N=6;k=ncol(X);
* Create the Xd matrix.;
    i=J(T,1,1);
    NT=nrow(X);
    D=block(i,i,i,i,i,i);
    I=I(NT);
    X=X||D;
* Calculate the LSDV estimator and its standard error.;
    Delta_LSDV=inv(X`*X)*X`*y;
    sigma2=(y-X*Delta_LSDV)`*(y-X*Delta_LSDV)/(NT-N-K);
    Var_Delta=sqrt(vecdiag(sigma2*inv(X`*X)));
* Print out the results.;
    Print Table1 (|Colname={LSDV_Estimates SE} rowname={Intercept
    LNQ LNPF LF Alpha1 Alpha2 Alpha3 Alpha4 Alpha5 Alpha6}
    format=8.4|); run;
```

The analysis results are given in Output 7.3. Notice that the results for the coefficients and their standard errors are identical to the results given in Output 7.2.

We can analyze the data using the Proc Panel or the Proc TSCSREG procedure. Proc Panel is an enhancement over Proc TSCSREG procedure and it can be used to analyze simple panel data models (fixed effects, random effects, one-way and two-way models) as well as more complex models in the panel data setting (heteroscedasticity violations, autocorrelation violations, dynamic panel data models). Readers are encouraged to refer to the Proc Panel Procedure reference guide from SAS Institute, Inc. for more details on this procedure. Although the procedure offers a wide range of options for analyzing panel data models, we will use the minimal required to illustrate the methods discussed in this chapter.

The following statements can be used for the LSDV model. The option "fixone" specifies that the first variable in the id statement will be treated as fixed. The analysis results are given in Output 7.4.

TABLE1		
	LSDV_ESTIMATES	**SE**
LNQ	0.9193	0.0299
LNPF	0.4175	0.0152
LF	-1.0704	0.2017
ALPHA1	9.7059	0.1931
ALPHA2	9.6647	0.1990
ALPHA3	9.4970	0.2250
ALPHA4	9.8905	0.2418
ALPHA5	9.7300	0.2609
ALPHA6	9.7930	0.2637

OUTPUT 7.3. The LSDV model of the airlines data using OLS calculations.

The PANEL Procedure
Fixed One Way Estimates

Dependent Variable: LnC

Model Description	
Estimation Method	FixOne
Number of Cross Sections	6
Time Series Length	15

Fit Statistics			
SSE	0.2926	**DFE**	81
MSE	0.0036	**Root MSE**	0.0601
R-Square	0.9974		

F Test for No Fixed Effects			
Num DF	**Den DF**	**F Value**	**Pr > F**
5	81	57.73	<0.0001

Parameter Estimates						
Variable	**DF**	**Estimate**	**Standard Error**	**t Value**	**Pr > |t|**	**Label**
CS1	1	-0.08706	0.0842	-1.03	0.3042	Cross Sectional Effect 1
CS2	1	-0.1283	0.0757	-1.69	0.0941	Cross Sectional Effect 2
CS3	1	-0.29598	0.0500	-5.92	<0.0001	Cross Sectional Effect 3
CS4	1	0.097494	0.0330	2.95	0.0041	Cross Sectional Effect 4
CS5	1	-0.06301	0.0239	-2.64	0.0100	Cross Sectional Effect 5
Intercept	1	9.793004	0.2637	37.14	<0.0001	Intercept
LnQ	1	0.919285	0.0299	30.76	<0.0001	
LnPF	1	0.417492	0.0152	27.47	<0.0001	
LF	1	-1.0704	0.2017	-5.31	<0.0001	LF

OUTPUT 7.4. LSDV estimates for the airlines data using Proc panel.

```
proc panel data=airline;
     id i t;
     model LnC=LnQ LnPF LF/fixone;
run;
```

The estimates for the group effects are easy to calculate by adding the variables CS_i to the intercept. The group effects are $a_1 = 9.706$, $a_2 = 9.665$, $a_3 = 9.497$, $a_4 = 9.890$, and $a_5 = 9.73$. The value for the intercept is the group effects value for the sixth firm. That is, $a_6 = 9.79$. Using Proc IML to fit the LSDV model allows us to get the actual group effects estimates and their standard errors. The fitted LSDV models for the six firms are given by

$$\text{Airline 1}: \quad \ln C = 9.706 + 0.9192 \quad \ln Q + 0.4174 \quad \ln PF - 1.070LF,$$

$$\text{Airline 2}: \quad \ln C = 9.665 + 0.9192 \quad \ln Q + 0.4174 \quad \ln PF - 1.070LF,$$

$$\text{Airline 3}: \quad \ln C = 9.497 + 0.9192 \quad \ln Q + 0.4174 \quad \ln PF - 1.070LF,$$

$$\text{Airline 4}: \quad \ln C = 9.890 + 0.9192 \quad \ln Q + 0.4174 \quad \ln PF - 1.070LF$$

$$\text{Airline 5}: \quad \ln C = 9.729 + 0.9192 \quad \ln Q + 0.4174 \quad \ln PF - 1.070LF$$

$$\text{Airline 6}: \quad \ln C = 9.793 + 0.9192 \quad \ln Q + 0.4174 \quad \ln PF - 1.070LF$$

Note that the equations only differ in the constant term and therefore represents a parametric shift in the regression lines. Comparing the LSDV output to the pooled output, we find that the root mean square for the LSDV is significantly smaller than the root mean square for the pooled model. This should not be surprising since the LSDV model essentially blocks out the subject effects and therefore gives a more precise estimate of the root mean square error. Also note that the error degrees of freedom for the LSDV model take into account the inclusion of the subject terms in the model. The coefficient of determination for the LSDV model is slightly higher than the coefficient of determination for the pooled model. The signs of the parameter estimates are the same between the two models. The magnitude of the coefficient for LF from the LSDV model is significantly lower than that from the pooled model.

Proc GLM can also be used to fit a fixed effects model to the airlines cost data set. The following statements can be used.

```
proc glm data=airline;
     class i;
     model LC=i LQ LPF LF/solution;
run;
```

The class statement with input "i" instructs the program to treat the airlines id as a classification variable and to treat the explanatory variables as covariates. The solution option for the model statement requests the parameter estimates for the terms in the model. Output 7.5 contains the analysis results. A description of Proc GLM was given in Chapter 3. Notice that the calculated estimates are identical to the ones calculated by using Proc Panel. The X^TX matrix was found to be singular simply because the procedure creates a column of ones in the X matrix. Proc GLM gives the F test for differences between the individual fixed effects for the airlines. The p values from both Type 1 and Type 3 sums of squares indicate high significance, implying that the null hypothesis of equality of the individual effects is to be rejected. The Type 1 sums of squares also referred to as the sequential sums of squares measures how much the residual sums of squares is reduced by adding a particular variable to the model containing all the variables before it. As an example, the Type 1 sums of squares for the airline effect tell us by how much the residual sums of squares for a model with just a constant term is reduced by adding the airlines effects to the model. On the other hand, the Type 3 sums of squares tell us by how much the residual sums of squares is reduced if the particular variable is added to a model containing all other variables. Both sums of squares measure the importance of the variable in question. We now move on to estimation using the within-group and the between-group methods. The functional form of the within-group model is

$$y_{it} - \bar{y}_{i.} = \left(x_{it} - \bar{x}_{i.}\right)^T \beta + \varepsilon_{it} - \bar{\varepsilon}_{i.}$$

whereas that of the between-group model is

$$\bar{y}_{i.} = \alpha + \bar{x}_i^T \beta + \bar{\varepsilon}_{i.}$$

The GLM Procedure

Class Level Information		
Class	**Levels**	**Values**
I	6	1 2 3 4 5 6

Number of Observations Read	90
Number of Observations Used	90

The GLM Procedure

Dependent Variable: LnC

Source	DF	Sum of Squares	Mean Square	F Value	Pr > F
Model	8	113.7482727	14.2185341	3935.80	<0.0001
Error	81	0.2926222	0.0036126		
Corrected Total	89	114.0408949			

R-Square	Coeff Var	Root MSE	LnC Mean
0.997434	0.449699	0.060105	13.36561

Source	DF	Type I SS	Mean Square	F Value	Pr > F
I	5	74.67988205	14.93597641	4134.39	<0.0001
LnQ	1	36.33305337	36.33305337	10057.3	<0.0001
LnPF	1	2.63358517	2.63358517	729.00	<0.0001
LF	1	0.10175213	0.10175213	28.17	<0.0001

Source	DF	Type III SS	Mean Square	F Value	Pr > F
I	5	1.04281997	0.20856399	57.73	<0.0001
LnQ	1	3.41718518	3.41718518	945.90	<0.0001
LnPF	1	2.72571947	2.72571947	754.50	<0.0001
LF	1	0.10175213	0.10175213	28.17	<0.0001

Parameter		Estimate		Standard Error	t Value	Pr > |t|
Intercept		9.793003883	B	0.26366188	37.14	<0.0001
I	1	-0.087061966	B	0.08419945	-1.03	0.3042
I	2	-0.128297833	B	0.07572803	-1.69	0.0941
I	3	-0.295983079	B	0.05002302	-5.92	<.0001
I	4	0.097494011	B	0.03300923	2.95	0.0041
I	5	-0.063006988	B	0.02389185	-2.64	0.0100
I	6	0.000000000	B	.	.	.
LnQ		0.919284650		0.02989007	30.76	<0.0001
LnPF		0.417491776		0.01519912	27.47	<0.0001
LF		-1.070395844		0.20168974	-5.31	<0.0001

Note: The X'X matrix has been found to be singular, and a generalized inverse was used to solve the normal equations. Terms whose estimates are followed by the letter 'B' are not uniquely estimable.

OUTPUT 7.5. LSDV estimates for the airlines data using Proc GLM.

The number of observations and the error degrees of freedom for these two representations are $(nT, nT - k)$ and $(n, n - k)$, respectively.

The following SAS code can be used to estimate the within-group effects model. The reader is asked to verify that the parameter estimates is the same as the estimates from the LSDV analysis. Note that the root mean square of the within-group model is larger than that of the LSDV model and that the coefficient of determination is slightly different. As discussed earlier, this value is incorrect, given that we do not have an intercept term in the model. The correct standard errors of the coefficients can be obtained by using the adjustment factor given in Baltagi (2005). The temporary data set "airline" was created prior to sorting and includes all the required transformed variables.

```
/*Sort the data by airline to facilitate correct calculations of group means*/
proc sort data=airline;
     by i;
run;
/*Calculate the group means*/
proc univariate data=airline noprint;
     var LnC LnQ LnPF LF;
     by i;
     output out=junk mean=meanc meanq meanpf meanlf;
run;
/*Merge the summary statistics to the original dataset and calculate the group deviations*/
data test;
     merge airline(in=a) junk(in=b);
     by i;
if a and b;
     lnc=lnc-meanc;
     lnq=lnq-meanq;
     lnpf=lnpf-meanpf;
     lf=lf-meanlf;
run;
/*Conduct the OLS regression*/
proc reg data=test;
     model lnc=lnq lnpf lf/noint;
run;
```

The between-group analysis can be conducted using Proc Panel with the "btwng" option in the model statement. The following statements can be used:

```
proc panel data=airline;
     id i t;
     model LnC=LnQ LnPF LF/btwng;
run;
```

The reader is asked to verify that the parameter estimates and their standard errors are given by Intercept/Constant 85.809 (56.483), LnQ 0.784 (0.109), LnPF -5.524 (4.479), and LF -1.751 (2.743). Note that only the coefficient for output LnQ is significant. The sign on the coefficient for fuel price $LnPF$ is now reversed.

7.4.1   Fixed Time and Group Effects

The general panel data model can easily be adjusted to incorporate a term for the time effect if it is of interest to determine whether the time periods are significantly different from each other. As shown in Greene (2003, p. 291, the LSDV model with a time-specific effect is given by

$$y_{it} = \mathbf{x}_{it}^T \boldsymbol{\beta} + \alpha_i + \gamma_t + \varepsilon_{it}, \quad i = 1, \dots, n; \quad t = 1, \dots, T,$$

where γ_t is the tth fixed time effect.

We will analyze the airlines data by incorporating the time effect into the fixed effects model. Note that we are still working with a one-way fixed effects model. The analysis is conducted by using Proc Panel with the "fixonetime" option. The following SAS statements in Proc Panel can be used to fit this model.

```
proc panel data=airline;
    id i t;
    model LnC=LnQ LnPF LF/fixonetime;
run;
```

The analysis results are given in Output 7.6. Notice that the *p* values associated with the *F* test for fixed time effects are not significant.

The fixed time effects analysis can also be done in Proc GLM by using the following statements. Output 7.7 contains the analysis results.

The PANEL Procedure
Fixed One Way Estimates Time-Wise

Dependent Variable: LnC

Model Description	
Estimation Method	FixOneTm
Number of Cross Sections	6
Time Series Length	15

Fit Statistics			
SSE	1.0882	DFE	72
MSE	0.0151	Root MSE	0.1229
R-Square	0.9905		

F Test for No Fixed Effects			
Num DF	Den DF	F Value	Pr > F
14	72	1.17	0.3178

Parameter Estimates						
Variable	DF	Estimate	Standard Error	t Value	Pr > |t|	Label
TS1	1	-2.04096	0.7347	-2.78	0.0070	Time Series Effect 1
TS2	1	-1.95873	0.7228	-2.71	0.0084	Time Series Effect 2
TS3	1	-1.88104	0.7204	-2.61	0.0110	Time Series Effect 3
TS4	1	-1.79601	0.6988	-2.57	0.0122	Time Series Effect 4
TS5	1	-1.33694	0.5060	-2.64	0.0101	Time Series Effect 5
TS6	1	-1.12515	0.4086	-2.75	0.0075	Time Series Effect 6
TS7	1	-1.03342	0.3764	-2.75	0.0076	Time Series Effect 7
TS8	1	-0.88274	0.3260	-2.71	0.0085	Time Series Effect 8
TS9	1	-0.7072	0.2947	-2.40	0.0190	Time Series Effect 9
TS10	1	-0.42296	0.1668	-2.54	0.0134	Time Series Effect 10
TS11	1	-0.07144	0.0718	-1.00	0.3228	Time Series Effect 11
TS12	1	0.114572	0.0984	1.16	0.2482	Time Series Effect 12
TS13	1	0.07979	0.0844	0.95	0.3477	Time Series Effect 13
TS14	1	0.015463	0.0726	0.21	0.8320	Time Series Effect 14
Intercept	1	22.53678	4.9405	4.56	<0.0001	Intercept
LnQ	1	0.867727	0.0154	56.32	<0.0001	
LnPF	1	-0.48448	0.3641	-1.33	0.1875	
LF	1	-1.9544	0.4424	-4.42	<0.0001	LF

OUTPUT 7.6. Fixed time effects analysis for the airlines data.

The GLM Procedure

Class Level Information		
Class	Levels	Values
T	15	1 2 3 4 5 6 7 8 9 10 11 12 13 14 15

Number of Observations Read	90
Number of Observations Used	90

The GLM Procedure

Dependent Variable: LnC

Source	DF	Sum of Squares	Mean Square	F Value	Pr > F
Model	17	112.9527040	6.6442767	439.62	<0.0001
Error	72	1.0881909	0.0151138		
Corrected Total	89	114.0408949			

R-Square	Coeff Var	Root MSE	LnC Mean
0.990458	0.919809	0.122938	13.36561

Source	DF	Type I SS	Mean Square	F Value	Pr > F
T	14	37.30676742	2.66476910	176.31	<0.0001
LnQ	1	75.30317703	75.30317703	4982.42	<0.0001
LnPF	1	0.04776504	0.04776504	3.16	0.0797
LF	1	0.29499451	0.29499451	19.52	<0.0001

Source	DF	Type III SS	Mean Square	F Value	Pr > F
T	14	0.24725125	0.01766080	1.17	0.3178
LnQ	1	47.93302463	47.93302463	3171.48	<0.0001
LnPF	1	0.02675904	0.02675904	1.77	0.1875
LF	1	0.29499451	0.29499451	19.52	<0.0001

Parameter		Estimate		Standard Error	t Value	Pr > |t|
Intercept		22.53678445	B	4.94053826	4.56	<.0001
T	1	-2.04096367	B	0.73469041	-2.78	0.0070
T	2	-1.95872954	B	0.72275187	-2.71	0.0084
T	3	-1.88103769	B	0.72036547	-2.61	0.0110
T	4	-1.79600992	B	0.69882566	-2.57	0.0122
T	5	-1.33693575	B	0.50604558	-2.64	0.0101
T	6	-1.12514656	B	0.40862234	-2.75	0.0075
T	7	-1.03341601	B	0.37641681	-2.75	0.0076
T	8	-0.88273866	B	0.32601349	-2.71	0.0085
T	9	-0.70719587	B	0.29470154	-2.40	0.0190
T	10	-0.42296351	B	0.16678941	-2.54	0.0134
T	11	-0.07143815	B	0.07176388	-1.00	0.3228
T	12	0.11457178	B	0.09841217	1.16	0.2482

OUTPUT 7.7. Fixed time effects analysis for the airlines data using Proc GLM.

Parameter		Estimate		Standard Error	t Value	Pr > |t|
T	13	0.07978953	B	0.08441708	0.95	0.3477
T	14	0.01546270	B	0.07263977	0.21	0.8320
T	15	0.00000000	B			
LnQ		0.86772671		0.01540820	56.32	<0.0001
LnPF		-0.48448499		0.36410896	-1.33	0.1875
LF		-1.95440278		0.44237789	-4.42	<0.0001

Note: The X'X matrix has been found to be singular, and a generalized inverse was used to solve the normal equations. Terms whose estimates are followed by the letter 'B' are not uniquely estimable.

OUTPUT 7.7. *(Continued)*

```
proc glm data=airline;
    class t;
    model LC=t LQ LPF LF/solution;
run;
```

Notice that the output from Proc Panel matches the Proc GLM output from the Type 3 sums of squares table. Also note that as before, the individual time estimates can be calculated by adding the variables *TS* (*T* for Proc GLM) to the intercept. The following Proc Panel statements can be used when treating both time and firm effects as fixed. Output 7.8 contains the analysis results.

```
proc panel data=airline;
    id i t;
    model LnC=LnQ LnPF LF/fixtwo;
run;
```

The two-way fixed effects model can be easily estimated using Proc GLM. The following statements can be used. The analysis results are given in Output 7.9.

```
proc glm data=airline;
    class i t;
    model LC=i t LQ LPF LF/solution;
run;
```

The LSDV model in both the one-way and the two-way effects cases can be easily written down and the equations for the specific airline–time combination can be easily extracted by using the dummy variables. We avoid specific details on the analysis results as the output can be interpreted in a similar fashion to the analysis outputs given earlier in this chapter.

7.5   RANDOM EFFECTS MODELS

As mentioned in the previous sections, the fixed effects model is appropriate when differences between the subjects may be viewed as parametric shifts in the regression model. Furthermore, the interpretations resulting from the fixed effects analysis is only applicable to the subjects who were selected for the study. On the other hand, in a random effects model, the subjects in the study are assumed to be selected from a much large population of available subjects. Therefore, the interpretations from the random are effects analysis applicable to the larger population. We also assumed that the unobserved subject-specific heterogeneity is uncorrelated to the observed explanatory variables. In the fixed effects model, violations of this assumption is not really an issue since the analysis "sweeps" the unobserved heterogeneity component from the model.

To motivate our discussion on analysis techniques for a random effects model, consider the general random effects model given in Section 7.2.

$$y_{it} = \mathbf{x}_{it}^T \boldsymbol{\beta} + \alpha + u_i + \varepsilon_{it}$$

The PANEL Procedure
Fixed Two Way Estimates

Dependent Variable: LnC

Model Description	
Estimation Method	FixTwo
Number of Cross Sections	6
Time Series Length	15

Fit Statistics			
SSE	0.1768	DFE	67
MSE	0.0026	Root MSE	0.0514
R-Square	0.9984		

F Test for No Fixed Effects			
Num DF	Den DF	F Value	Pr > F
19	67	23.10	<0.0001

Parameter Estimates						
Variable	DF	Estimate	Standard Error	t Value	Pr > |t|	Label
CS1	1	0.174282	0.0861	2.02	0.0470	Cross Sectional Effect 1
CS2	1	0.111451	0.0780	1.43	0.1575	Cross Sectional Effect 2
CS3	1	-0.14351	0.0519	-2.77	0.0073	Cross Sectional Effect 3
CS4	1	0.180209	0.0321	5.61	<0.0001	Cross Sectional Effect 4
CS5	1	-0.04669	0.0225	-2.08	0.0415	Cross Sectional Effect 5
TS1	1	-0.69314	0.3378	-2.05	0.0441	Time Series Effect 1
TS2	1	-0.63843	0.3321	-1.92	0.0588	Time Series Effect 2
TS3	1	-0.5958	0.3294	-1.81	0.0750	Time Series Effect 3
TS4	1	-0.54215	0.3189	-1.70	0.0938	Time Series Effect 4
TS5	1	-0.47304	0.2319	-2.04	0.0454	Time Series Effect 5
TS6	1	-0.4272	0.1884	-2.27	0.0266	Time Series Effect 6
TS7	1	-0.39598	0.1733	-2.28	0.0255	Time Series Effect 7
TS8	1	-0.33985	0.1501	-2.26	0.0268	Time Series Effect 8
TS9	1	-0.27189	0.1348	-2.02	0.0477	Time Series Effect 9
TS10	1	-0.22739	0.0763	-2.98	0.0040	Time Series Effect 10
TS11	1	-0.1118	0.0319	-3.50	0.0008	Time Series Effect 11
TS12	1	-0.03364	0.0429	-0.78	0.4357	Time Series Effect 12
TS13	1	-0.01773	0.0363	-0.49	0.6263	Time Series Effect 13

OUTPUT 7.8. Fixed time and firm effects for the airlines data.

with k regressors and $E(\mathbf{z}_i^T \boldsymbol{\alpha}) = \alpha + u_i$, where $\boldsymbol{\mu}$ is a constant and can be viewed as a common fixed effect and u_i, the disturbance, is the random subject-specific effect. We make the following assumptions: $u_i \sim i.i.d(0, \sigma_u^2)$, $\varepsilon_{it} \sim i.i.d.(0, \sigma_\varepsilon^2$, $E(u_i|\mathbf{x}_{it}) = 0$, $E(\varepsilon_{it}|\mathbf{x}_{it}) = 0$, and $Cov(u_i, \varepsilon_{it}) = 0$. Notice that the exogeneity assumption of the regressors with respect to u_i arises from the original assumption that $\mathbf{z}_i$ is independent of $\mathbf{x}_i$. An additional assumption is that $E(\varepsilon_{it}\varepsilon_{js}|\mathbf{X}) = 0$ if $t \neq s$ or $i \neq j$, and $E(u_i u_j|\mathbf{X}) = 0$ if $i \neq j$. That is, we assume that the disturbances are uncorrelated among themselves across time and across subjects (Greene, 2003, p. 294; Verbeek, 2004, p. 348).

Parameter Estimates						
Variable	DF	Estimate	Standard Error	t Value	Pr > |t|	Label
TS14	1	−0.01865	0.0305	−0.61	0.5432	Time Series Effect 14
Intercept	1	12.94003	2.2182	5.83	<0.0001	Intercept
LnQ	1	0.817249	0.0319	25.66	<0.0001	
LnPF	1	0.168611	0.1635	1.03	0.3061	
LF	1	−0.88281	0.2617	−3.37	0.0012	LF

OUTPUT 7.8. *(Continued)*

If we denote the covariance structure for the ith subject as $\mathbf{\Sigma}$, then it is easy to prove that $\mathbf{\Sigma} = \sigma_u^2 \mathbf{i}_T \mathbf{i}_T^T + \sigma_\varepsilon^2 \mathbf{I}_T$ where $\mathbf{i}_T$ is a vector of 1's. That is, the diagonal elements of the covariance matrix are all equal to $\sigma_u^2 + \sigma_\varepsilon^2$ while the off-diagonal elements are equal to σ_u^2. Combining the covariance matrices across the n subjects and taking into consideration the assumptions for a random effects model stated earlier, the disturbance covariance matrix for the entire set of nT observations can be written as $\mathbf{\Omega} = \mathbf{I}_n \otimes \mathbf{\Sigma}$.

7.5.1   Generalized Least Squares Estimation

As shown in Chapter 5, the generalized least squares estimator can easily be calculated by first premultiplying $\mathbf{y}_i$ and $\mathbf{X}_i$ by $\mathbf{\Sigma}^{-1/2}$. If we let $\mathbf{y}^*$ and $\mathbf{X}^*$ represent the stacked transformed data across all n subjects, then the GLS estimator is obtained regressing $\mathbf{y}^*$ against $\mathbf{X}^*$.

In reality, the variance components are unknown and FGLS estimation has to be used. As discussed in Greene (2003, pp. 296–297), a commonly used approach to estimating the variance components is to use standard OLS and LSDV to estimate σ^2 (pooled) and σ_ε^2 (LSDV), respectively. As discussed by the author, the OLS estimator can be assumed to provide an estimate for $\sigma_u^2 + \sigma_\varepsilon^2$ while the LSDV estimator provides an estimator for σ_ε^2. Therefore, the difference between these two can be used to estimate σ_u^2. That is, $\hat{\sigma}_u^2 = s_{pooled}^2 - s_{LSDV}^2$. An alternate method is to use the expression for the expected mean square of the random effect and then solve for $\hat{\sigma}_u^2$ using the mean squares. These expressions are available in Proc GLM. However, all complexities are avoided by simply using Proc Panel.

We will discuss estimation under the assumption of a random effects models subsequently. For now, we discuss ways of determining whether a fixed or random effects model should be used for the panel data.

7.5.2   Testing for Random Effects

The Breusch and Pagan (1980) Lagrange Multiplier (LM) test and the Hausman Specification tests are the two most commonly used tests for determining whether a random effect or a fixed effect should be used for the data. The LM test tests the hypothesis that $\sigma_u^2 = 0$ versus $\sigma_u^2 > 0$. If the null hypothesis is not rejected, then $\mathbf{\Sigma}$ is diagonal, which may imply that a random effects model should not be used for the panel data. The LM test statistic is given by (Greene, 2003, p. 299, The Proc Panel Procedure, p. 60, SAS Institute, Inc.)

$$LM = \frac{nT}{2(T-1)} \left[\frac{\sum_{i=1}^{n} (T\bar{\hat{\varepsilon}}_{i.})^2}{\sum_{i=1}^{n}\sum_{t=1}^{T} \hat{\varepsilon}_{it}^2} - 1 \right]^2$$

and is distributed as a chi-squared distribution under the null hypothesis.

We will use the airlines cost equation example to illustrate the computations of the LM test. The residuals from the pooled model are first saved. We make use of Proc GLM to conduct this portion of the analysis. The following SAS statements can be used. Notice that we are suppressing the output since we simply want to save the residuals at this stage.

The GLM Procedure

Class Level Information		
Class	Levels	Values
I	6	1 2 3 4 5 6
T	15	1 2 3 4 5 6 7 8 9 10 11 12 13 14 15

Number of Observations Read	90
Number of Observations Used	90

The GLM Procedure

Dependent Variable: LnC

Source	DF	Sum of Squares	Mean Square	F Value	Pr > F
Model	22	113.8640466	5.1756385	1960.82	<0.0001
Error	67	0.1768483	0.0026395		
Corrected Total	89	114.0408949			

R-Square	Coeff Var	Root MSE	LnC Mean
0.998449	0.384392	0.051376	13.36561

Source	DF	Type I SS	Mean Square	F Value	Pr > F
I	5	74.67988205	14.93597641	5658.58	<0.0001
T	14	37.30676742	2.66476910	1009.56	<0.0001
LnQ	1	1.84507227	1.84507227	699.02	<0.0001
LnPF	1	0.00229645	0.00229645	0.87	0.3543
LF	1	0.03002842	0.03002842	11.38	0.0012

Source	DF	Type III SS	Mean Square	F Value	Pr > F
I	5	0.91134261	0.18226852	69.05	<0.0001
T	14	0.11577389	0.00826956	3.13	0.0009
LnQ	1	1.73776357	1.73776357	658.36	<0.0001
LnPF	1	0.00280788	0.00280788	1.06	0.3061
LF	1	0.03002842	0.03002842	11.38	0.0012

Parameter		Estimate		Standard Error	t Value	Pr > |t|
Intercept		12.94003049	B	2.21823061	5.83	<0.0001
I	1	0.17428210	B	0.08611999	2.02	0.0470
I	2	0.11145059	B	0.07795501	1.43	0.1575
I	3	−0.14351138	B	0.05189334	−2.77	0.0073
I	4	0.18020869	B	0.03214429	5.61	<0.0001
I	5	−0.04669433	B	0.02246877	−2.08	0.0415
I	6	0.00000000	B	.	.	.
T	1	−0.69313650	B	0.33783841	−2.05	0.0441
T	2	−0.63843490	B	0.33208013	−1.92	0.0588
T	3	−0.59580170	B	0.32944723	−1.81	0.0750
T	4	−0.54215223	B	0.31891384	−1.70	0.0938

OUTPUT 7.9. Fixed time and firm effects for the airlines data using Proc GLM.

Parameter	Estimate		Standard Error	t Value	Pr > |t|
T 5	−0.47304191	B	0.23194587	−2.04	0.0454
T 6	−0.42720347	B	0.18843991	−2.27	0.0266
T 7	−0.39597739	B	0.17329687	−2.28	0.0255
T 8	−0.33984567	B	0.15010620	−2.26	0.0268
T 9	−0.27189295	B	0.13481748	−2.02	0.0477
T 10	−0.22738537	B	0.07634948	−2.98	0.0040
T 11	−0.11180326	B	0.03190050	−3.50	0.0008
T 12	−0.03364114	B	0.04290077	−0.78	0.4357
T 13	−0.01773478	B	0.03625539	−0.49	0.6263
T 14	−0.01864518	B	0.03050793	−0.61	0.5432
T 15	0.00000000	B	.	.	.
LnQ	0.81724884		0.03185093	25.66	<0.0001
LnPF	0.16861074		0.16347803	1.03	0.3061
LF	−0.88281211		0.26173699	−3.37	0.0012

Note: The X'X matrix has been found to be singular, and a generalized inverse was used to solve the normal equations. Terms whose estimates are followed by the letter 'B' are not uniquely estimable.

OUTPUT 7.9. *(Continued)*

```
proc glm data=airline noprint;
    model LnC=LnQ LnPF LF/solution;
    output out=resid residual=res;
run;
```

Proc Univariate is now used to calculate the means of the OLS residuals for each firm. The analysis results are given in Output 7.10.

```
proc univariate data=resid noprint;
    var res;
    by i;
    output out=junk mean=mean;
run;
```

The sums of squares of the OLS residuals is 1.3354. Substituting all the values into the LM formula, we get

$$LM = \frac{6 \times 15}{2(15-1)} \left[\frac{15^2 \times (0.06887^2 + 0.01388^2 + 0.19422^2 + 0.15273^2 + 0.02158^2 + 0.00809^2)}{1.3354} - 1 \right]^2 = 334.85.$$

Obs	I	mean
1	1	0.06887
2	2	−0.01388
3	3	−0.19422
4	4	0.15273
5	5	−0.02158
6	6	0.00809

OUTPUT 7.10. Mean of residuals for each of the airlines.

The tabled value from the chi-squared table is $\chi^2_{1,0.05} = 3.84$, and so we reject the null hypothesis that $\sigma^2_u = 0$ and claim that there is evidence that the random effects model is more appropriate for the airlines data. As discussed in Greene (2003, p. 299) "a fixed effects model may produce the same results," that is, lead to the same conclusion. He suggests erring on the side of caution and concludes that the pooled model is inappropriate for the airlines data set rather than stating that the random effects model is more appropriate. In the next section, we will discuss the Hausman's specification test (the preferred approach) to determine if the fixed effects or the random effects model is more appropriate for the panel data.

7.5.3   Hausman's Test

Hausman's (1978) test can be used to determine whether the fixed effects or the random effects model is more appropriate for the panel data. The procedure tests the null hypothesis of no correlation between the unobserved subject-specific effects and the observed explanatory variables versus the alternative hypothesis that the unobserved subject-specific effects are correlated to the observed explanatory variables. The test is based on the covariance matrix of the difference vector $\mathbf{b}_{FE} - \mathbf{b}_{RE}$, where $\mathbf{b}_{FE}$ is the fixed effects estimator and $\mathbf{b}_{RE}$ is the random effects estimator. Under the null hypothesis of no correlation, both estimators are consistent estimators for $\boldsymbol{\beta}$. However, under the alternative hypothesis, only $\mathbf{b}_{FE}$ is consistent for $\boldsymbol{\beta}$. A significant difference between the two estimators will lead to the rejection of the null hypothesis (Greene, 2003, pp. 301–302).

The Hausman's test statistic is given by the following:

$$W = (\mathbf{b}_{FE} - \mathbf{b}_{RE})^T \Phi^{-1} (\mathbf{b}_{FE} - \mathbf{b}_{RE})$$

where $\Phi = Var(\mathbf{b}_{FE} - \mathbf{b}_{RE})$. Under the null hypothesis of no correlation, the test statistic is distributed as a chi-squared random variable with k degrees of freedom, where k is the number of observed explanatory variables.

To illustrate Hausmans's test, we will again make use of the airlines data from Greene (2003). To compute the test statistic, we need to first generate the covariance matrices for both the fixed and random effects models. The following statements using Proc Panel can be used to store the covariance matrices for both models. Note that we will use the "ranone" option again subsequently when estimating a random effects model. The two covariance matrices are given in Output 7.11.

```
proc panel data=airline outest=out1 covout noprint;
title 'This is the Fixed Effects Analysis';
      id i t;
      model LnC=LnQ LnPF LF/fixone;
run;
proc panel data=airline outest=out2 covout noprint;
title 'This is the Random Effects Model';
      id i t;
      model LnC=LnQ LnPF LF/ranone;
run;
```

This is the Fixed Effects Results

Obs	LnQ	LnPF	LF
1	0.000893416	-0.000317817	-0.001884
2	-0.000317817	0.000231013	-0.000769
3	-0.001884262	-0.000768569	0.040679

This is the Random Effects Results

Obs	LnQ	LnPF	LF
1	0.000676608	-0.000235445	-0.001554
2	-0.000235445	0.000198785	-0.000879
3	-0.001554439	-0.000878566	0.039785

OUTPUT 7.11. Covariance matrices of the estimates for the fixed and random effects model of the airline data.

The PANEL Procedure
Fuller and Battese Variance Components (RanOne)

Dependent Variable: LnC

Model Description	
Estimation Method	RanOne
Number of Cross Sections	6
Time Series Length	15

Fit Statistics			
SSE	0.3090	DFE	86
MSE	0.0036	Root MSE	0.0599
R-Square	0.9923		

Variance Component Estimates	
Variance Component for Cross Sections	0.018198
Variance Component for Error	0.003613

Hausman Test for Random Effects		
DF	m Value	Pr > m
3	0.92	0.8209

Breusch Pagan Test for Random Effects (One Way)		
DF	m Value	Pr > m
1	334.85	<0.0001

Parameter Estimates						
Variable	DF	Estimate	Standard Error	t Value	Pr > |t|	Label
Intercept	1	9.637	0.2132	45.21	<0.0001	Intercept
LnQ	1	0.908024	0.0260	34.91	<0.0001	
LnPF	1	0.422199	0.0141	29.95	<0.0001	
LF	1	−1.06469	0.1995	−5.34	<0.0001	LF

OUTPUT 7.12. Using proc panel to generate Hausman and Breusch–Pagan tests for the random effects model.

Using the covariance matrices and the fixed and random effects coefficients that were calculated using Proc Panel, we can substitute all the values in the formula for Hausman's test in Proc IML to get a test statistic value of 4.16. The chi-squared tabled value with a type 1 error rate of 0.05 and 3 degrees of freedom is 7.814. The coding in Proc IML is straightforward and is therefore not included here. The results of the analysis indicate that we cannot reject the null hypothesis that the unobserved heterogeneity subject-specific effects are uncorrelated with the observed explanatory variables. Therefore, both the fixed effects estimator and the random effects estimator are consistent estimators of β. On the other hand, the *LM* test rejected the null hypothesis $H_0 : \sigma_u^2 = 0$, thus indicating that a random effects model was more appropriate than the pooled regression model. As discussed by Greene (2003) in Example 13.5 on page 302, based on the results of both the *LM* test and the Hausman test, we would conclude that the random effects model is more appropriate for the airlines data. The Hausman's test is given by default in the output of Proc Panel when we specify a random effects model. The "LM" option can be used in the model statement to get the Breusch–Pagan Lagrange Multiplier test. Output 7.12 contains the Hausman and Breusch–Pagan Tests from the Proc Panel procedure. Notice that the test statistic value for Hausman's test is different from the one given above. The value reported uses the covariance matrix of the parameters including the parameter for firm. We used Proc IML with the modified covariance matrices and found a value very close to the one in the SAS output.

```
proc panel data=airline;
     id i t;
     model LnC=LnQ LnPF LF/ranone bp;
run;
```

7.5.4 Random Effects Model Estimation

In Section 7.5.2, we used Proc Panel to estimate the fixed and random effects model. The outputs provided us with estimates of the two variance components. For instance, from Output 7.12, we see that an estimate for σ_ε^2 is 0.003613 while the estimate for σ_u^2 is 0.018198. Therefore, we can use these estimators to construct Σ to perform FGLS using Proc IML. In the previous section, we used Proc Panel to perform the calculations. The analysis results are given in Output 7.12. We will now briefly discuss the output results.

The PANEL Procedure
Fuller and Battese Variance Components (RanTwo)

Dependent Variable: LnC

Model Description	
Estimation Method	RanTwo
Number of Cross Sections	6
Time Series Length	15

Fit Statistics			
SSE	0.2322	DFE	86
MSE	0.0027	Root MSE	0.0520
R-Square	0.9829		

Variance Component Estimates	
Variance Component for Cross Sections	0.017439
Variance Component for Time Series	0.001081
Variance Component for Error	0.00264

Hausman Test for Random Effects		
DF	m Value	Pr > m
3	6.93	0.0741

Parameter Estimates						
Variable	DF	Estimate	Standard Error	t Value	Pr > |t|	Label
Intercept	1	9.362676	0.2440	38.38	<0.0001	Intercept
LnQ	1	0.866448	0.0255	33.98	<0.0001	
LnPF	1	0.436163	0.0172	25.41	<0.0001	
LF	1	−0.98053	0.2235	−4.39	<0.0001	LF

OUTPUT 7.13. Random effects model assuming both firms and time are random.

The first table gives the estimation method and the number of cross sections and time periods. The second table gives some basic statistics including the coefficient of determination and the root mean square error. Notice that the error degree of freedom is $90 - 4 = 86$ as there are four estimated parameters. The next table gives the variance components for both the cross section and the LSDV model. The last table gives the parameter estimates and the associated p values. Notice that the signs for the coefficients match the signs of the coefficients from the fixed effects model. Also note that the magnitudes of the coefficient values are similar. All explanatory variables are highly significant in the model.

The following Proc Panel code analyzes the data assuming that both the firm and the time effects are random. The analysis results are given in Output 7.13. Note that now the output contains three variance components: one for the LSDV model, one for the cross sections, and one for the time effect. The rest of the output can be interpreted as before.

```
proc panel data=airline;
     id i t;
     model LnC=LnQ LnPF LF/rantwo;
run;
```

8

SYSTEMS OF REGRESSION EQUATIONS

8.1 INTRODUCTION

The previous chapters discussed estimation of single linear equation models. In practice, it is not uncommon to encounter models that are characterized by several linear or nonlinear equations where the disturbance vectors from the equations are involved in cross-equation correlations. As an example, consider the well-known Grunfeld's (1958) investment model given by

$$I_{it} = \beta_0 + \beta_1 F_{it} + \beta_2 C_{it} + \varepsilon_{it}, \quad i = 1, \ldots, n; \ t = 1, \ldots, T,$$

where I_{it} is the investment for firm i in time period t, F_{it} is the market value of the firm, and C_{it} is the value of capital stock. The original data set was comprised of 10 large US manufacturing firms, which were followed from 1935 to 1954. As discussed by Greene (2003, p. 339), the disturbance vectors in each equation are characterized by shocks that may be common to all the firms. For instance, the general health of the economy may have an impact on the investment behavior of each firm. On the other hand, certain industries exhibit a cyclical nature where they are heavily dependent upon the economy whereas other industries are not cyclical and are not impacted by the economy. Therefore, another component of the disturbance term may be shocks that are specific to the industry the company belongs to.

A naïve approach to analysis may treat the system of equations as unrelated or independent. However, analysis of the residuals from the system of equations may reveal a covariance structure that consists of cross-correlations between the equations. Estimation of the parameters must take the inter-equation cross-correlation into account. Zellner (1962) introduced the seemingly unrelated regression (SUR) models that takes into account the cross-equation correlation when analyzing systems of regression equations. This chapter deals with using SAS to analyze SUR models.

The seemingly unrelated regressions (SUR) model is characterized by a system of n equations and is given by (Greene, 2003, p. 340)

$$\mathbf{y}_1 = \mathbf{X}_1 \boldsymbol{\beta}_1 + \boldsymbol{\varepsilon}_1,$$
$$\mathbf{y}_2 = \mathbf{X}_2 \boldsymbol{\beta}_2 + \boldsymbol{\varepsilon}_2,$$
$$\vdots$$
$$\mathbf{y}_n = \mathbf{X}_n \boldsymbol{\beta}_n + \boldsymbol{\varepsilon}_n.$$

In this formulation, $\mathbf{y}_i$ is the $T \times 1$ dependent variable, $\mathbf{X}_i$ is the $T \times K_i$ matrix of regressors, $\boldsymbol{\beta}_i$ is the $K_i \times 1$ parameter vector, and $\boldsymbol{\varepsilon}_i$ is the $T \times 1$ vector of disturbances. This setup results in a total of nT observations. Note that in order to estimate the parameters of the system, we require each equation in the system to satisfy the constraint that $T > K_i$.

Applied Econometrics Using the SAS® System, by Vivek B. Ajmani
Copyright © 2009 John Wiley & Sons, Inc.

An implied assumption in the above setup is that the conditional mean is not fixed across the groups as the regressors are not restricted to be the same across all the equations. Therefore, we may also view this system within a panel data framework with n cross-sections and T time periods.

Each equation in the SUR is assumed to satisfy all the assumptions of the classical OLS model. However, the disturbance terms of the equations are assumed to be correlated, that is

$$E(\varepsilon_i \varepsilon_j^T | \mathbf{X}_1, \mathbf{X}_2, \dots, \mathbf{X}_n) = \sigma_{ij}\mathbf{I}_T.$$

This implies that the covariance of the disturbance of the ith and jth equations are correlated and is constant across all the observations. Writing the full disturbance vector as $\varepsilon = \begin{bmatrix} \varepsilon_1^T & \varepsilon_2^T & \dots & \varepsilon_n^T \end{bmatrix}^T$, these assumptions imply that $E(\varepsilon|\mathbf{X}_1, \mathbf{X}_2, \dots, \mathbf{X}_n) = \mathbf{0}$ and $E(\varepsilon\varepsilon^T|\mathbf{X}_1, \mathbf{X}_2, \dots, \mathbf{X}_n) = \mathbf{\Omega}$. If we let $\mathbf{\Sigma} = \lfloor \sigma_{ij} \rfloor$, then we can write $E(\varepsilon\varepsilon^T|\mathbf{X}_1, \mathbf{X}_2, \dots, \mathbf{X}_n) = \mathbf{\Sigma} \otimes \mathbf{I}_T$ (Greene, 2003, p. 341).

8.2   ESTIMATION USING GENERALIZED LEAST SQUARES

The potential presence of inter-equation cross-correlations renders the OLS equation by equation estimation inefficient. A more efficient approach is to use the generalized least squares (GLS) approach as described below (Greene, 2003, pp. 342–343).

If we stack the n equations, we get

$$\begin{bmatrix} \mathbf{y}_1 \\ \mathbf{y}_2 \\ \vdots \\ \mathbf{y}_M \end{bmatrix} = \begin{bmatrix} \mathbf{X}_1 & \mathbf{0} & \dots & \mathbf{0} \\ \mathbf{0} & \mathbf{X}_2 & \dots & \mathbf{0} \\ \vdots & \vdots & \vdots & \vdots \\ \mathbf{0} & \mathbf{0} & \dots & \mathbf{X}_M \end{bmatrix} \begin{bmatrix} \boldsymbol{\beta}_1 \\ \boldsymbol{\beta}_2 \\ \vdots \\ \boldsymbol{\beta}_M \end{bmatrix} + \begin{bmatrix} \varepsilon_1 \\ \varepsilon_2 \\ \vdots \\ \varepsilon_M \end{bmatrix} = \mathbf{X}\boldsymbol{\beta} + \varepsilon,$$

where $E(\varepsilon|\mathbf{X}) = \mathbf{0}$ and $E(\varepsilon\varepsilon^T|\mathbf{X}) = \mathbf{\Omega} = \mathbf{\Sigma} \otimes \mathbf{I}_T$. The GLS estimator of $\boldsymbol{\beta}$ is therefore given by

$$\hat{\boldsymbol{\beta}}_{GLS} = [\mathbf{X}^T(\mathbf{\Sigma}^{-1} \otimes \mathbf{I})\mathbf{X}]^{-1}\mathbf{X}^T(\mathbf{\Sigma}^{-1} \otimes \mathbf{I})\mathbf{y}$$

with asymptotic covariance matrix $[\mathbf{X}^T(\mathbf{\Sigma}^{-1} \otimes \mathbf{I})\mathbf{X}]^{-1}$. As is always the case, $\mathbf{\Omega}$ is assumed to be unknown and therefore the FGLS method has to be used to estimate $\boldsymbol{\beta}$.

8.3   SPECIAL CASES OF THE SEEMINGLY UNRELATED REGRESSION MODEL

1. GLS is the same as equation-by-equation OLS if the system equations are uncorrelated. This is easy to show by realizing that if $\sigma_{ij} = 0$ for $i \neq j$, then $\mathbf{\Sigma}$ is diagonal and that the variance terms of each equation simply drop out of the GLS estimator giving

$$\hat{\boldsymbol{\beta}}_{GLS} = \begin{bmatrix} (\mathbf{X}_1^T\mathbf{X}_1)^{-1}\mathbf{X}_1^T\mathbf{y}_1 \\ (\mathbf{X}_2^T\mathbf{X}_2)^{-1}\mathbf{X}_2^T\mathbf{y}_2 \\ \vdots \\ (\mathbf{X}_M^T\mathbf{X}_M)^{-1}\mathbf{X}_M^T\mathbf{y}_M \end{bmatrix},$$

   which is the equation-by-equation OLS (Greene, 2003, p. 343–344).

2. If the equations have identical explanatory variables, then GLS is equation-by-equation OLS.

This is easy to show by first assuming that $\mathbf{X}_1 = \mathbf{X}_2 = \ldots = \mathbf{X}_n = \mathbf{X}_c$ and realizing that $\mathbf{X} = \mathbf{X}_c \otimes \mathbf{I}_T$. Using this in the GLS estimator and making use of basic properties of Kronecker products we can show that

$$\hat{\boldsymbol{\beta}}_{GLS} = [\mathbf{X}^T(\boldsymbol{\Sigma}^{-1} \otimes \mathbf{I})\mathbf{X}]^{-1}\mathbf{X}^T(\boldsymbol{\Sigma}^{-1} \otimes \mathbf{I})\mathbf{y}$$
$$= (\mathbf{X}_c^T\mathbf{X}_c)^{-1}\mathbf{X}_c^T\mathbf{y},$$

which is the equation-by-equation OLS estimator.

8.4   FEASIBLE GENERALIZED LEAST SQUARES

The discussion so far assumed that $\boldsymbol{\Sigma}$ is known. In practice, $\boldsymbol{\Sigma}$ is almost always unknown and therefore has to be estimated. FGLS estimators (see Chapter 5) can be used to estimate $\boldsymbol{\beta}$ in this case. The analysis proceeds in two steps.

First, the OLS residuals for each equation are calculated using $\hat{\boldsymbol{\varepsilon}}_i = \mathbf{y}_i - \mathbf{X}_i\hat{\boldsymbol{\beta}}_i$, where $\hat{\boldsymbol{\varepsilon}}_i$ is the residual vector for the ith equation. The elements of $\boldsymbol{\Sigma}$ can then be constructed using

$$\hat{\sigma}_{ij} = \frac{1}{T}\mathbf{e}_i^T\mathbf{e}_j.$$

giving the FGLS estimator

$$\hat{\boldsymbol{\beta}}_{FGLS} = \left[\mathbf{X}^T(\hat{\boldsymbol{\Sigma}}^{-1} \otimes \mathbf{I})\mathbf{X}\right]^{-1}\mathbf{X}^T(\hat{\boldsymbol{\Sigma}}^{-1} \otimes \mathbf{I})\mathbf{y}$$

with asymptotic covariance matrix $[\mathbf{X}^T(\hat{\boldsymbol{\Sigma}}^{-1} \otimes \mathbf{I})\mathbf{X}]^{-1}$ (Greene, 2003, p. 344).

To illustrate the techniques involved in the estimation of $\boldsymbol{\beta}$, we will make use of the Grunfeld's Investment data from Greene (2003). Greene's version of the Grunfeld data set consists of a subset of five firms Grunfeld's model observed over 20 years.

In the analysis that follows, the coefficients are unrestricted and are allowed to vary across firms. The downloaded data set has a pooled data structure with common names for the model variables across all the firms. The input data set for analysis must therefore be adjusted to get firm specific names for the explanatory variables. The following statements can be used to create distinct variable names.

```
data GM CH GE WE US;
     set SUR;
     if firm=1 then output GM;
     else if firm=2 then output CH;
     else if firm=3 then output GE;
     else if firm=4 then output WE;
     else output US;
run;
data GM;
     set GM;
     rename i=i_gm f=f_gm c=c_gm;
run;
data CH;
     set CH;
     rename i=i_ch f=f_ch c=c_ch;
run;
data GE;
     set GE;
     rename i=i_ge f=f_ge c=c_ge;
run;
data WE;
     set WE;
     rename i=i_we f=f_we c=c_we;
run;
```

```
data US;
     set US;
     rename i=i_us f=f_us c=c_us;
run;
data grunfeld;
     merge gm ch ge we us;
     by year;
run;
```

We start the analysis of the Grunfeld data set by estimating the parameters of the model using pooled OLS. No change to the original data set is required, and the following statements can be used. The analysis results are given in Output 8.1.

```
proc reg data=SUR;
     model I=F C;
run;
```

The results indicate that both the firm's market value and the value of the firm's capital are highly significant in explaining the variability in investment. The positive coefficients indicate that the firm's investment will be higher if it's market value and the value of its capital is high.

The Grunfeld SAS data set consisting of separate variable names for each firm can be analyzed by using Proc Syslin. The following statements can be used. Output 8.2 contains the analysis results. This procedure will give the OLS estimates for each equation followed by the cross-equation covariance and correlation matrices. These are then followed by the FGLS estimates for each equation.

```
proc syslin data=grunfeld SUR;
     gm:model i_gm = f_gm c_gm;
     ch:model i_ch = f_ch c_ch;
```

The REG Procedure
Model: MODEL1
Dependent Variable: I I

Number of Observations Read	100
Number of Observations Used	100

Analysis of Variance					
Source	DF	Sum of Squares	Mean Square	F Value	Pr > F
Model	2	5532554	2766277	170.81	<0.0001
Error	97	1570884	16195		
Corrected Total	99	7103438			

Root MSE	127.25831	R-Square	0.7789
Dependent Mean	248.95700	Adj R-Sq	0.7743
Coeff Var	51.11658		

Parameter Estimates						
Variable	Label	DF	Parameter Estimate	Standard Error	t Value	Pr > |t|
Intercept	Intercept	1	-48.02974	21.48017	-2.24	0.0276
F	F	1	0.10509	0.01138	9.24	<0.0001
C	C	1	0.30537	0.04351	7.02	<0.0001

OUTPUT 8.1. Pooled OLS regression for the Grunfeld data.

```
The SYSLIN Procedure
Ordinary Least Squares Estimation
```

Model	GM
Dependent Variable	i_gm
Label	I

Analysis of Variance					
Source	DF	Sum of Squares	Mean Square	F Value	Pr > F
Model	2	1677687	838843.3	99.58	<0.0001
Error	17	143205.9	8423.875		
Corrected Total	19	1820893			

Root MSE	91.78167	R-Square	0.92135
Dependent Mean	608.02000	Adj R-Sq	0.91210
Coeff Var	15.09517		

Parameter Estimates						
Variable	DF	Parameter Estimate	Standard Error	t Value	Pr > |t|	Variable Label
Intercept	1	-149.782	105.8421	-1.42	0.1751	Intercept
f_gm	1	0.119281	0.025834	4.62	0.0002	F
c_gm	1	0.371445	0.037073	10.02	<0.0001	C

Model	CH
Dependent Variable	i_ch
Label	I

Analysis of Variance					
Source	DF	Sum of Squares	Mean Square	F Value	Pr > F
Model	2	31686.54	15843.27	89.86	<0.0001
Error	17	2997.444	176.3203		
Corrected Total	19	34683.99			

Root MSE	13.27856	R-Square	0.91358
Dependent Mean	86.12350	Adj R-Sq	0.90341
Coeff Var	15.41805		

Parameter Estimates						
Variable	DF	Parameter Estimate	Standard Error	t Value	Pr > |t|	Variable Label
Intercept	1	-6.18996	13.50648	-0.46	0.6525	Intercept
f_ch	1	0.077948	0.019973	3.90	0.0011	F
c_ch	1	0.315718	0.028813	10.96	<0.0001	C

OUTPUT 8.2. Grunfeld data analysis results using Proc Syslin SUR.

The SYSLIN Procedure
Ordinary Least Squares Estimation

Model	GE
Dependent Variable	i_ge
Label	I

Analysis of Variance					
Source	DF	Sum of Squares	Mean Square	F Value	Pr > F
Model	2	31632.03	15816.02	20.34	<0.0001
Error	17	13216.59	777.4463		
Corrected Total	19	44848.62			

Root MSE	27.88272	R-Square	0.70531
Dependent Mean	102.29000	Adj R-Sq	0.67064
Coeff Var	27.25850		

Parameter Estimates						
Variable	DF	Parameter Estimate	Standard Error	t Value	Pr > |t|	Variable Label
Intercept	1	-9.95631	31.37425	-0.32	0.7548	Intercept
f_ge	1	0.026551	0.015566	1.71	0.1063	F
c_ge	1	0.151694	0.025704	5.90	<0.0001	C

Model	WE
Dependent Variable	i_we
Label	I

Analysis of Variance					
Source	DF	Sum of Squares	Mean Square	F Value	Pr > F
Model	2	5165.553	2582.776	24.76	<0.0001
Error	17	1773.234	104.3079		
Corrected Total	19	6938.787			

Root MSE	10.21312	R-Square	0.74445
Dependent Mean	42.89150	Adj R-Sq	0.71438
Coeff Var	23.81153		

Parameter Estimates						
Variable	DF	Parameter Estimate	Standard Error	t Value	Pr > |t|	Variable Label
Intercept	1	-0.50939	8.015289	-0.06	0.9501	Intercept
f_we	1	0.052894	0.015707	3.37	0.0037	F
c_we	1	0.092406	0.056099	1.65	0.1179	C

OUTPUT 8.2. (*Continued*)

Model	US
Dependent Variable	i_us
Label	I

Analysis of Variance					
Source	DF	Sum of Squares	Mean Square	F Value	Pr > F
Model	2	139978.1	69989.04	6.69	0.0072
Error	17	177928.3	10466.37		
Corrected Total	19	317906.4			

Root MSE	102.30529	R-Square	0.44031
Dependent Mean	405.46000	Adj R-Sq	0.37447
Coeff Var	25.23191		

Parameter Estimates						
Variable	DF	Parameter Estimate	Standard Error	t Value	Pr > |t|	Variable Label
Intercept	1	−30.3685	157.0477	−0.19	0.8490	Intercept
f_us	1	0.156571	0.078886	1.98	0.0635	F
c_us	1	0.423866	0.155216	2.73	0.0142	C

The SYSLIN Procedure
Seemingly Unrelated Regression Estimation

Cross Model Covariance					
	GM	CH	GE	WE	US
GM	8423.88	−332.655	714.74	148.443	−2614.2
CH	−332.65	176.320	−25.15	15.655	491.9
GE	714.74	−25.148	777.45	207.587	1064.6
WE	148.44	15.655	207.59	104.308	642.6
US	−2614.19	491.857	1064.65	642.571	10466.4

Cross Model Correlation					
	GM	CH	GE	WE	US
GM	1.00000	−0.27295	0.27929	0.15836	−0.27841
CH	−0.27295	1.00000	−0.06792	0.11544	0.36207
GE	0.27929	−0.06792	1.00000	0.72896	0.37323
WE	0.15836	0.11544	0.72896	1.00000	0.61499
US	−0.27841	0.36207	0.37323	0.61499	1.00000

Cross Model Inverse Correlation					
	GM	CH	GE	WE	US
GM	1.41160	0.14649	−0.32667	−0.46056	0.74512
CH	0.14649	1.23373	0.27615	−0.08670	−0.45566
GE	−0.32667	0.27615	2.33055	−1.65117	−0.04531
WE	−0.46056	−0.08670	−1.65117	3.16367	−1.42618
US	0.74512	−0.45566	−0.04531	−1.42618	2.26642

OUTPUT 8.2. (*Continued*).

The SYSLIN Procedure
Seemingly Unrelated Regression Estimation

	GM	CH	GE	WE	US
	\multicolumn{5}{c}{Cross Model Inverse Covariance}				
GM	0.000168	0.000120	-0.000128	-0.000491	0.000079
CH	0.000120	0.006997	0.000746	-0.000639	-0.000335
GE	-0.000128	0.000746	0.002998	-0.005798	-0.000016
WE	-0.000491	-0.000639	-0.005798	0.030330	-0.001365
US	0.000079	-0.000335	-0.000016	-0.001365	0.000217

System Weighted MSE	0.9401
Degrees of freedom	85
System Weighted R-Square	0.8707

Model	GM
Dependent Variable	i_gm
Label	I

Parameter Estimates

Variable	DF	Parameter Estimate	Standard Error	t Value	Pr > |t|	Variable Label
Intercept	1	-162.364	97.03216	-1.67	0.1126	Intercept
f_gm	1	0.120493	0.023460	5.14	<0.0001	F
c_gm	1	0.382746	0.035542	10.77	<0.0001	C

Model	CH
Dependent Variable	i_ch
Label	I

Parameter Estimates

Variable	DF	Parameter Estimate	Standard Error	t Value	Pr > |t|	Variable Label
Intercept	1	0.504304	12.48742	0.04	0.9683	Intercept
f_ch	1	0.069546	0.018328	3.79	0.0014	F
c_ch	1	0.308545	0.028053	11.00	<0.0001	C

Model	GE
Dependent Variable	i_ge
Label	I

Parameter Estimates

Variable	DF	Parameter Estimate	Standard Error	t Value	Pr > |t|	Variable Label
Intercept	1	-22.4389	27.67879	-0.81	0.4287	Intercept
f_ge	1	0.037291	0.013301	2.80	0.0122	F
c_ge	1	0.130783	0.023916	5.47	<0.0001	C

Model	WE
Dependent Variable	i_we
Label	I

OUTPUT 8.2. (*Continued*).

Parameter Estimates						
Variable	DF	Parameter Estimate	Standard Error	t Value	Pr > |t|	Variable Label
Intercept	1	1.088877	6.788627	0.16	0.8745	Intercept
f_we	1	0.057009	0.012324	4.63	0.0002	F
c_we	1	0.041506	0.044689	0.93	0.3660	C

Model	US
Dependent Variable	i_us
Label	I

Parameter Estimates						
Variable	DF	Parameter Estimate	Standard Error	t Value	Pr > |t|	Variable Label
Intercept	1	85.42325	121.3481	0.70	0.4910	Intercept
f_us	1	0.101478	0.059421	1.71	0.1059	F
c_us	1	0.399991	0.138613	2.89	0.0103	C

OUTPUT 8.2. (*Continued*).

```
        ge:model i_ge = f_ge c_ge;
        we:model i_we = f_we c_we;
        us:model i_us = f_us c_us;
run;
```

The first part of the output gives the OLS equation-by-equation estimates of the parameters for each of the five firms. Notice that the F test for the global hypothesis H_0: $\beta_1 = \beta_2 = \beta_3 = 0$ versus H_1: At least one $\beta \neq 0$ is highly significant for each of the five firms. The root MSE value is the highest for US and lowest for WE. The R^2 values for GM and CH indicates a good fit, while the R^2 values GE and WE indicate a moderate fit. The R^2 value for US indicates a poor fit. Both explanatory variables are highly significant for firms GM and CH. Market value is not significant in the model for GE, WE, and US at the 5% significance level.

The OLS equation-by-equation output is followed by the cross-equation covariance and correlation matrices along with their inverses. The diagonal elements of the cross-model covariance matrix are the variances of the residuals for the five firms. For example, the variance of the residuals for GM is 8423.88. Taking the square root of this will yield 91.78, which is the root MSE of the OLS model for this firm. The off-diagonal elements of this matrix display the values of the covariances between the OLS residuals of each of the five firms. The cross-equation covariance is calculated by first calculating the residuals $(\mathbf{e}_i, i = 1, \ldots, M = 5)$ from the FGLS procedure for each firm and each time period. Let

$$\mathbf{E} = \begin{bmatrix} \mathbf{e}_1 & \mathbf{e}_2 & \mathbf{e}_3 & \mathbf{e}_4 & \mathbf{e}_5 \end{bmatrix}.$$

The covariance matrix is given by

$$\frac{1}{T}\mathbf{E}^T\mathbf{E} \text{ (Green, 2003, p. 322)},$$

where T in Grunfeld's model example is 20.

The cross-equation correlation is calculated by using the formula

$$\rho(x, y) = Cov(x, y)/(\sigma_x \sigma_y).$$

This results in all diagonal elements being equal to 1. To see how the off-diagonal elements are calculated, consider calculating the correlation between the residuals of firms GM and CH. Using this formula, one gets

$$\rho(GM, CH) = -332.655 \Big/ \left(\sqrt{8423.88}\sqrt{176.32} \right) = -0.273.$$

The cross-equation correlation matrix indicates high correlation between the residuals of GE and WE, and WE and US. The inverse of both the cross-equation covariance and correlation matrices wraps up this portion of the output.

The last part of the output consists of the FGLS estimates of the parameters of the two explanatory variables for each of the five firms. The output displays both the system weighted MSE and the system weighted R^2. Note that the degrees of freedom associated with the system weighted MSE is $17 \times 5 = 85$. The analysis results indicate that both market value and the value of stock of plant and equipment are highly significant for firms GM, CH, and GE. The value of stock of plant and equipments is not significant for firm WE. The market value is not significant for firm US.

9

SIMULTANEOUS EQUATIONS

9.1 INTRODUCTION

The previous chapters focused on single equations and on systems of single equation models that were characterized by dependent variables (endogenous) on the left-hand side and the explanatory variables (exogenous or endogenous) on the right-hand side of the equations. For example, Chapter 4 dealt with instrumental variables, where the endogenous variables were on the right-hand side. This chapter extends the concept of systems of linear equations where endogenous variables were determined one at a time (sequentially) to the case when they are determined simultaneously.

We begin our discussion of simultaneous equation models by considering the following wage–price equations

$$p_t = \beta_1 w_t + \beta_2 m_t + \varepsilon_{1t}$$
$$w_t = \alpha_1 p_t + \alpha_2 u_t + \varepsilon_{2t},$$

where p_t is the price inflation at time t, w_t is wage inflation at time t, m_t is the money supply at time t, u_1 is unemployment rate time t, ε_{1t} and ε_{2t} are the error terms with means 0 and constant variances σ_1^2 and σ_2^2, respectively, and $\gamma = (\beta_1, \beta_2, \alpha_1, \alpha_2)$ are the model parameters that need to be estimated. We also assume that the disturbance terms are uncorrelated. These equations are referred to as structural equations. In the wage–price inflation equation, we have two structural equations and four unknown parameters.

The first equation describes the relation of price inflation to wage inflation and money supply. As wages increase so do prices as demand for good and services tend to increase as well and this puts pressure on them. Furthermore, the increase in wage is typically passed on to the consumer resulting in price inflation. There is also a positive relationship between money supply and price inflation. The second equation describes the behavior of wage inflation vis-a-vis price inflation and the unemployment rate. As prices increase, workers tend to demand higher wages but the demand is offset by the unemployment rate since a higher unemployment rate tends to decrease the rate of wage increases as the demand of goods and services decreases and thus there is no pressure on price.

Notice that both equations are required to determine the price and wage inflations. The variables p and w are therefore endogenous variables. The unemployment rate and money supply are determined outside of the system of equations and therefore exogenous.

9.2 PROBLEMS WITH OLS ESTIMATION

In this section, we will show why ordinary least squares estimation for simultaneous equations is inappropriate. Recall that a critical assumption of OLS is that the explanatory variables are exogenous. This assumption is violated when an endogenous

Applied Econometrics Using the SAS® System, by Vivek B. Ajmani variable becomes an explanatory variable in another equation of the system. OLS estimates in this case will lead to biased and inconsistent estimatiors. We will illustrate this by considering the first equation of the wage–price inflation system of equations given above.

Consider the second equation in the wage–price inflation system. The expectation $E(w_t \varepsilon_{1t})$ is given by

$$E(w_t \varepsilon_{1t}) = \alpha_1 E(p_t \varepsilon_{1t}) + \alpha_2 E(u_t \varepsilon_{1t}) + E(\varepsilon_{2t} \varepsilon_{1t}).$$

Using the assumptions stated earlier, it can be shown that this simplifies to

$$E(w_t \varepsilon_{1t}) = \alpha_1 E(p_t \varepsilon_{1t})$$
$$E(p_t \varepsilon_{1t}) = \frac{1}{\alpha_1} E(w_t \varepsilon_{1t}).$$

Taking the expectation $E(p_t \varepsilon_{1t})$ and substituting the above in it, we get

$$\frac{1}{\alpha_1} E(w_t \varepsilon_{1t}) = \beta_1 E(w_t \varepsilon_{1t}) + \alpha_2 E(m_t \varepsilon_{1t}) + E(\varepsilon_{1t}^2).$$

Using the earlier assumptions, we can show that

$$E(w_t \varepsilon_{1t}) = \frac{\alpha_1}{1 - \alpha_1 \beta_1} \sigma^2 \neq 0.$$

Therefore, w_t is endogenous with respect to ε_{1t} and the OLS assumptions are violated. Using OLS, therefore, will lead to biased and inconsistent estimators of the parameters in the first equation.

The nature of the bias can be shown by writing down the OLS estimate expression for β_1

$$\hat{\beta 1} = \frac{\sum_{t=1}^{T} (w_t - \bar{w})(p_t - \bar{p})}{\sum_{t=1}^{T} (w_t - \bar{w})^2}.$$

Simplifying this, we get

$$\hat{\beta_1} = \beta_1 + \beta_2 \frac{\sum_{t=1}^{T} (w_t - \bar{w})(m_t - \bar{m})}{\sum_{t=1}^{T} (w_t - \bar{w})^2} + \frac{\sum_{t=1}^{T} (w_t - \bar{w})(\varepsilon_t - \bar{\varepsilon})}{\sum_{t=1}^{T} (w_t - \bar{w})^2}.$$

Asymptotically, the second term is simply the covariance between wage and money flow and is assumed to be zero based on the wage–price inflation structural equations. Asymptotically, the last expression gives the covariance between wage and the disturbance of the first equation and is nonzero as shown earlier. The denominator term is the asymptotic variance of wage. The OLS estimate is therefore biased and inconsistent with the direction of the bias depending on $Cov(w_t, \varepsilon_t)$.

Ashenfelter et al. (2003, pp. 222–223) use the simple Keynesian model

$$c_t = \beta_0 + \beta_1 y_t + \varepsilon_t$$
$$y_t = c_t + i_t$$

to show the nature of the bias of the OLS estimator for β_1. Here, c_t is the consumption at time t, y_t is the income at time t, i_t is investment at time t, ε_t is the disturbance with zero mean and variance σ_ε^2, and $\boldsymbol{\beta} = (\beta_0, \beta_1)$ are parameters to be estimated. The authors show that the OLS estimator of β_1 is biased upwards and is given by

$$bias\left(\hat{\beta}_1\right) = \frac{\sigma_\varepsilon^2}{(1-\beta_1)\sigma_y^2},$$

where

$$\sigma_y^2 = Var(y_t).$$

9.3   STRUCTURAL AND REDUCED FORM EQUATIONS

The wage–inflation system of equations given in Section 9.1 contains two structural equations in four unknown parameters. The basic goal of simultaneous equation modeling is to provide estimates of the parameters in the structural equations. A natural path to estimating these parameters can be seen by first expressing the endogenous variables in the system of equations as functions of the exogenous variables in the system. The resulting equations are called reduced form equations.

For the wage–price inflation structural equations, the reduced form equations can be obtained by substituting the equation for wage in the equation for price and vice-versa, the equation of price in the equation for wage. To see how the reduced form equation for price is constructed, consider

$$p_t = \beta_1(\alpha_1 p_t + \alpha_2 u_t + \varepsilon_{2t}) + \beta_2 m_t + \varepsilon_{1t}$$
$$p_t = \alpha_1\beta_1 p_t + \alpha_2\beta_1 u_t + \beta_1\varepsilon_{2t} + \beta_2 m_t + \varepsilon_{1t}$$
$$p_t - \alpha_1\beta_1 p_t = \alpha_2\beta_1 u_t + \beta_2 m_t + \beta_1\varepsilon_{2t} + \varepsilon_{1t}$$
$$p_t = \frac{\alpha_2\beta_1}{1-\alpha_1\beta_1}u_t + \frac{\beta_2}{1-\alpha_1\beta_1}m_t + \frac{\beta_1\varepsilon_{2t} + \varepsilon_{1t}}{1-\alpha_1\beta_1}.$$

Proceeding in a similar fashion, we can get the reduced form equation for w_t. The reduced forms are given by

$$p_t = \frac{\beta_2}{1-\alpha_1\beta_1}m_t + \frac{\alpha_2\beta_1}{1-\alpha_1\beta_1}u_t + \frac{\varepsilon_{1t} + \beta_1\varepsilon_{2t}}{1-\alpha_1\beta_1},$$

$$w_t = \frac{\alpha_1\beta_2}{1-\alpha_1\beta_1}m_t + \frac{\alpha_2}{1-\alpha_1\beta_1}u_t + \frac{\alpha_1\varepsilon_{1t} + \varepsilon_{2t}}{1-\alpha_1\beta_1}.$$

The above approach for creating the reduced form equations from the structural equations should suggest an approach to estimating the structural equation parameters. We could, in principle, conduct an OLS on the reduced form equations and then attempt to extract the structural equation parameters from the reduced form parameters. Another option is to first run OLS on the reduced form equations to get the predicted values of the endogenous variables. The predicted values of the endogenous variables can then be used in an OLS in the structural equations to estimate the parameters of interest. We will discuss estimation techniques in the next section. For now, we will move on to the system of equations in the more general case.

As shown in Greene (2003, p. 382), the general structure of the structural equations at time t with n endogenous and k exogenous variables can be written as

$$\alpha_{11}y_{t1} + \cdots + \alpha_{n1}y_{tn} + \beta_{11}x_{t1} + \cdots + \beta_{k1}x_{tk} = \varepsilon_{t1}$$
$$\alpha_{12}y_{t1} + \cdots + \alpha_{n2}y_{tn} + \beta_{12}x_{t1} + \cdots + \beta_{k2}x_{tk} = \varepsilon_{t2}$$
$$\vdots$$
$$\alpha_{1n}y_{t1} + \cdots + \alpha_{nn}y_{tn} + \beta_{1n}x_{t1} + \cdots + \beta_{kn}x_{tk} = \varepsilon_{tn}.$$

In matrix notation, the system of structural equations at time t can be written as $\mathbf{y}_t^T \mathbf{\Gamma} + \mathbf{x}_t^T \mathbf{B} = \mathbf{\varepsilon}_t^T$, where

$$\mathbf{y}_t^T = [y_{t1}, \ldots, y_{tm}],$$

$$\mathbf{\Gamma} = \begin{bmatrix} \alpha_{11} & \alpha_{12} & \cdots & \alpha_{1n} \\ \alpha_{21} & \alpha_{22} & \cdots & \alpha_{2n} \\ \vdots & \vdots & \ddots & \vdots \\ \alpha_{n1} & \alpha_{n2} & \cdots & \alpha_{nn} \end{bmatrix},$$

$$\mathbf{x}_t^T = [x_{t1}, \ldots, x_{tk}],$$

$$\mathbf{B} = \begin{bmatrix} \beta_{11} & \beta_{12} & \cdots & \beta_{1n} \\ \beta_{21} & \beta_{22} & \cdots & \beta_{2n} \\ \vdots & \vdots & \ddots & \vdots \\ \beta_{k1} & \beta_{k2} & \cdots & \beta_{kn} \end{bmatrix}, \text{and}$$

$$\mathbf{\varepsilon}_t^T = [\varepsilon_1, \ldots, \varepsilon_n].$$

Here, $\mathbf{\Gamma}$ is a $n \times n$ matrix of coefficients for the endogenous variables, $\mathbf{y}_t$ is a $n \times 1$ vector of endogenous variables, $\mathbf{B}$ is a $k \times n$ matrix of coefficients of the exogenous variables, $\mathbf{x}_t$ is a $k \times 1$ vector consisting of exogenous variables, and $\mathbf{\varepsilon}_t$ is a $n \times 1$ vector of disturbances.

Assuming that $\mathbf{\Gamma}$ is nonsingular, we can express the endogenous variables in reduced form as a function of the exogenous variables and the random disturbances. The reduced form system of equations at time period t is given by (Greene, 2003, p. 384)

$$\mathbf{y}_t^T = -\mathbf{x}_t^T \mathbf{B} \mathbf{\Gamma}^{-1} + \mathbf{\varepsilon}_t^T \mathbf{\Gamma}^{-1}$$
$$= \mathbf{x}_t^T \mathbf{\Pi} + \mathbf{v}_t^T,$$

where $\mathbf{\Pi}$ is a $k \times n$ matrix containing the parameters of the reduced form equations and $\mathbf{v}_t$ contains the disturbances of the reduced form equations

If we apply this to the wage–inflation model, we get

$$\mathbf{\Gamma} = \begin{bmatrix} 1 & -\alpha_1 \\ -\beta_1 & 1 \end{bmatrix}, \quad \mathbf{y}_t = \begin{bmatrix} p_t \\ w_t \end{bmatrix},$$

$$\mathbf{B} = \begin{bmatrix} -\beta_2 & 0 \\ 0 & -\alpha_2 \end{bmatrix}, \quad \mathbf{x}_t = \begin{bmatrix} m_t \\ u_t \end{bmatrix}, \quad \text{and } \mathbf{\varepsilon}_t = \begin{bmatrix} \varepsilon_{1t} \\ \varepsilon_{2t} \end{bmatrix}.$$

It is easily verified that the reduced form equation of the wage–inflation model at the tth observation is given by

$$[p_t \quad w_t] = -[m_t \quad u_t] \begin{bmatrix} -\beta_2 & 0 \\ 0 & -\alpha_2 \end{bmatrix} \begin{bmatrix} \frac{1}{1-\alpha_1\beta_1} & \frac{\alpha_1}{1-\alpha_1\beta_1} \\ \frac{\beta_1}{1-\alpha_1\beta_1} & \frac{1}{1-\alpha_1\beta_1} \end{bmatrix} + [\varepsilon_{1t} \quad \varepsilon_{2t}] \begin{bmatrix} \frac{1}{1-\alpha_1\beta_1} & \frac{\alpha_1}{1-\alpha_1\beta_1} \\ \frac{\beta_1}{1-\alpha_1\beta_1} & \frac{1}{1-\alpha_1\beta_1} \end{bmatrix},$$

which simplifies to the reduced form equations given earlier.

9.4 THE PROBLEM OF IDENTIFICATION

In the ideal case, we would estimate the parameters of the reduced-form equation and then use these to estimate the parameters of the structural-form equations. In most cases, however, the reduced form estimates do not provide direct estimates of the parameters for the structural equation. It turns out that they only provide estimates of functions of the structural equation parameters. The problem of *identification* deals with whether we can solve the reduced form equations for unique values of the parameters of the structural equations.

To lay the groundwork for whether the structural equation parameters can be estimated, consider the wage–price inflation system of equations. The reduced form parameters are

$$\Pi_{11} = \frac{\beta_2}{1-\alpha_1\beta_1},$$

$$\Pi_{12} = \frac{\beta_1\alpha_2}{1-\alpha_1\beta_1},$$

$$\Pi_{21} = \frac{\alpha_1\beta_2}{1-\alpha_1\beta_1},$$

$$\Pi_{22} = \frac{\alpha_2}{1-\alpha_1\beta_1}.$$

It should be obvious that the knowledge of the four reduced form parameters will allow us to estimate the four structural form parameters. That is, the ratio Π_{12}/Π_{22} gives us α_2 while the ratio Π_{21}/Π_{11} gives us α_1. The values of β_1 and β_2 can be extracted in a similar manner. Here, we say that the structural equations are identified. In the wage–price inflation system, we have four reduced form parameters and four structural parameters, and we could solve the reduced form parameter equations easily to extract the structural equation parameters.

9.4.1 Determining if the Structural Equations are Identified

Consider the following simple two-equation model with two endogenous and two exogenous variables

$$y_1 = \beta_0 y_2 + \beta_1 x_1 + \beta_2 x_2 + \varepsilon_1,$$
$$y_2 = \alpha_0 y_1 + \alpha_1 x_1 + \alpha_2 x_2 + \varepsilon_2.$$

It can be shown that the reduced form parameters are given by

$$\Pi_{11} = \frac{\beta_0\alpha_1 + \beta_1}{1-\beta_0\alpha_0},$$

$$\Pi_{12} = \frac{\beta_0\alpha_2 + \beta_2}{1-\beta_0\alpha_0},$$

$$\Pi_{21} = \frac{\beta_1\alpha_0 + \alpha_1}{1-\beta_0\alpha_0},$$

$$\Pi_{22} = \frac{\beta_2\alpha_0 + \alpha_2}{1-\beta_0\alpha_0}.$$

It should be obvious that the knowledge of the four reduced form parameters will not allow us to estimate the six structural equation parameters. However, assume that β_2 is 0. In this case, we can easily show that

$$\beta_0 = \Pi_{12}/\Pi_{22},$$

$$\beta_1 = \Pi_{11} - \frac{\Pi_{12}}{\Pi_{22}}\Pi_{21}.$$

Similarly, if we assume that $\alpha_2 = 0$, then it can be shown that

$$\alpha_0 = \Pi_{22}/\Pi_{12},$$

$$\alpha_1 = \Pi_{21} - \frac{\Pi_{22}}{\Pi_{12}}\Pi_{11}.$$

In the first case, we say that the structural equation for y_1 is identified and in the second that the structural equation for y_2 is identified.

It should therefore be clear that one approach to estimating the structural equation parameters is to put restrictions on the structural equations. In the example used, both structural equations were identified by eliminating the exogenous variable x_2. This methodology can be formalized.

When no restrictions were placed on the structural parameters of the above equations, the equations were not identified. In this case, no exogenous variables were removed and both structural equations had one endogenous variable on the right-hand side. When one exogenous variable was removed from the structural equations, we were able to estimate the structural equations.

In general, if we let k be the number of excluded exogenous variables in the structural equation and let m be the number of included endogenous variables in the structural equation, then the structural equation is identified if $k = m$. On the other hand, the structural equation is not identified if $k < m$. See Ashenfelter et al. (2003, pp. 223–226) for a good discussion of this. Greene (2003, pp. 390–392) extends this to the general case.

9.5   ESTIMATION OF SIMULTANEOUS EQUATION MODELS

The two-stage least squares method is the most commonly used method for estimating parameters in a simultaneous equation system. The approach involves first using OLS to estimate the reduced form equations. The predicted values of the endogenous variables are then used in an OLS regression of the identified structural form equation of interest to estimate the parameters.

A brief description of the method is summarized here. First, recall that each structural equation is written so that an endogenous variable on the left-hand side of the equation is expressed as a function of endogenous variables and the exogenous variables on the right-hand side of the equation. Thus, for a system with n equations, there are n endogenous variables. If we let $\mathbf{y}_j$ represent the endogenous variable for the j th equation, then we can write the structural equation as

$$\mathbf{y}_j = \mathbf{Y}_j^* \boldsymbol{\alpha}_j + \mathbf{X}_j \boldsymbol{\beta}_j + \boldsymbol{\varepsilon}_j \quad j = 1, \ldots, n.$$

Here, $\mathbf{Y}_j^*$ is a $T \times n_j$ matrix of n_j included endogenous variables on the right hand, $\boldsymbol{\alpha}_j$ is a $n_j \times 1$ vector of coefficients for $\mathbf{Y}_j^*$, $\mathbf{X}_j$ is a $T \times k_j^*$ matrix of included exogenous variables, $\boldsymbol{\beta}_j$ is the $k_j \times 1$ vector of coefficients for $\mathbf{X}_j$, and $\boldsymbol{\varepsilon}_j$ is the $T \times 1$ disturbance vector.

Let, $\mathbf{W}_j = [\, \mathbf{Y}_j^* \quad \mathbf{X}_j \,]$ and $\boldsymbol{\theta}_j = [\, \boldsymbol{\alpha}_j \quad \boldsymbol{\beta}_j \,]$, then the jth equation can be written as

$$\mathbf{y}_j = \mathbf{W}_j \boldsymbol{\theta}_j + \boldsymbol{\varepsilon}_j \quad j = 1, \ldots, n.$$

A naïve approach to analysis is to conduct an OLS of $\mathbf{y}_j$ on $\mathbf{W}_j$ to get

$$\hat{\boldsymbol{\theta}}_j = (\mathbf{W}_j^T \mathbf{W}_j)^{-1} \mathbf{W}_j^T \mathbf{y}_j.$$

As shown earlier, OLS estimates are biased and inconsistent. Greene (2003, p. 396) gives an expression of the OLS estimates and discusses the bias in the general setting.

The 2SLS method involves first conducting an OLS of the included endogenous variables $\mathbf{Y}_j^*$ on all the exogenous variables in the system of equations. If we denote the predicted values of $\mathbf{Y}_j^*$ as $\hat{\mathbf{Y}}_j^*$, then the 2SLS estimator is given by

$$\hat{\boldsymbol{\theta}}_{IV,j} = (\hat{\mathbf{W}}_j^T \hat{\mathbf{W}}_j)^{-1} \hat{\mathbf{W}}_j^T \mathbf{y}_j,$$

where $\hat{\mathbf{W}}_j = \lfloor \hat{\mathbf{Y}}_j^* \mathbf{X}_j \rfloor$. The asymptotic variance–covariance matrix for the 2SLS estimator is given by

$$Var(\hat{\boldsymbol{\theta}}_{IV,j}) = \hat{\sigma}_j \lfloor \hat{\mathbf{W}}_j^T \hat{\mathbf{W}}_j \rfloor,$$

where

$$\hat{\sigma}_j = \frac{(\mathbf{y}_j - \mathbf{W}_j \hat{\boldsymbol{\theta}}_{IV,j})^T (\mathbf{y}_j - \mathbf{W}_j \hat{\boldsymbol{\theta}}_{IV,j})}{T}.$$

See Greene (2003, pp. 398–399) for details.

We will illustrate computing the 2SLS estimators for a labor–wage equation. Ashenfelter et al. (2003, p. 233) gives a simultaneous system consisting of two equations, one for labor and one for wages, for the US agricultural labor market. The authors provide the equations and a description of the variables list that is summarized below.

$$ln(L_i) = \beta_0 + \beta_1 ln(W_i) + \beta_2 ln(Land_i) + \beta_3 ln(RE_i) + \beta_4 ln(Othexp_i) + \beta_5 ln(Sch_i) + \varepsilon_1$$
$$ln(W_i) = \alpha_0 + \alpha_1 ln(L_i) + \alpha_2 ln(Sch_i) + \alpha_3 ln(Othwag_i) + \varepsilon_2.$$

Here, $L =$ farm labor by state as measured by total number of worked days, $W =$ wages per hour of farm workers, $RE =$ research and development expenses, $Land =$ value of farmland capital, $Othexp =$ operating expenses besides labor, $Sch =$ median number of years of formal schooling of males, and $Othwag =$ weighted average of wages for nonfarm workers. Note that the variables L and W are endogenous while the other variables are exogenous. It is easy to show that the labor equation is exactly identified while the wage equation is overidentified.

The following statements can be used to conduct an OLS regression on the labor equation.

```
data SE;
     set SE;
     L=log(labor);
     W=log(wage);
     L_re=Log(RE);
     L_Land=log(land);
     L_Othexp=log(othexp);
     L_Sch=log(sch);
     L_Othwag=Log(othwag);
run;
proc reg data=SE;
     model L=W L_Land L_re L_Othexp L_Sch;
run;
```

Output 9.1 contains the analysis of the data using standard OLS. Recall that OLS gives biased and inconsistent estimators of the structural model parameters. Notice that the parameters for wages, value of land, R&D expenditure, and other nonlabor expense are highly significant. The coefficient for schooling is not significant. The sign for the wages coefficient makes intuitive sense. As wages increase, employers cut back on labor and therefore the number of labor days should decrease.

The following statements can be used to estimate the labor equation by the 2SLS method. The analysis results are given in Output 9.2. There are four main statements in the Proc Syslin procedure as it relates to the labor equation. First, the option 2SLS is used to request the two-stage instrumental variable estimator. Second, the endogenous statement lists out the two endogenous variables in the system. Next, we list out the instruments that will be used in the estimation. This is followed by the model statement.

```
proc syslin data=SE 2SLS;
     endogenous L W;
     instruments L_land L_RE L_Othexp L_Sch L_Othwag;
     Labor: model L = W L_Land L_RE L_Othexp L_Sch;
run;
```

The analysis indicates that the global F test for the model is significant (p value < 0.001). The value of the statistic is very close to the one obtained by OLS. As we saw in the OLS output, schooling is not significant whereas all the other model variables are significant. However, the sign associated with schooling is positive compared to negative in the OLS model. The sign for the intercept is also different while the other variables have the same sign as in the OLS model. The values of the coefficient of determination and the root MSE are very close to what was obtained from OLS.

Since the model used is a log–log model, we can interpret the coefficients as follows (taking research and development as an example): If the research and development expense increases by 10%, then farm labor increases by $0.10 \times 0.47 = 0.047 = 4.7\%$ days.

The 2SLS estimator is based on the assumption of homoscedastic disturbances. White's estimator for the variance covariance can be used if this assumption is violated. The resulting estimator is called the generalized method of moments (GMM) estimator. To see

The REG Procedure
Model: MODEL1
Dependent Variable: L

Number of Observations Read	39
Number of Observations Used	39

Analysis of Variance					
Source	DF	Sum of Squares	Mean Square	F Value	Pr > F
Model	5	11.01673	2.20335	45.12	<0.0001
Error	33	1.61160	0.04884		
Corrected Total	38	12.62833			

Root MSE	0.22099	R-Square	0.8724
Dependent Mean	10.29862	Adj R-Sq	0.8530
Coeff Var	2.14582		

Parameter Estimates					
Variable	DF	Parameter Estimate	Standard Error	t Value	Pr > |t|
Intercept	1	0.87573	1.94350	0.45	0.6552
W	1	-1.10577	0.28931	-3.82	0.0006
L_Land	1	0.31180	0.08373	3.72	0.0007
L_re	1	0.44460	0.11402	3.90	0.0004
L_Othexp	1	0.26454	0.11278	2.35	0.0252
L_Sch	1	-0.26232	0.60820	-0.43	0.6691

OUTPUT 9.1. OLS analysis of the labor equation.

The SYSLIN Procedure
Two-Stage Least Squares Estimation

Model	LABOR
Dependent Variable	L

Analysis of Variance					
Source	DF	Sum of Squares	Mean Square	F Value	Pr > F
Model	5	10.59918	2.119837	41.63	<0.0001
Error	33	1.680444	0.050923		
Corrected Total	38	12.62833			

Root MSE	0.22566	R-Square	0.86315
Dependent Mean	10.29862	Adj R-Sq	0.84242
Coeff Var	2.19117		

Parameter Estimates					
Variable	DF	Parameter Estimate	Standard Error	t Value	Pr > |t|
Intercept	1	-0.85913	3.306459	-0.26	0.7966
W	1	-1.44928	0.601228	-2.41	0.0217
L_Land	1	0.328677	0.089282	3.68	0.0008
L_re	1	0.468702	0.122089	3.84	0.0005
L_Othexp	1	0.249787	0.117341	2.13	0.0408
L_Sch	1	0.342098	1.111155	0.31	0.7601

OUTPUT 9.2. 2SLS analysis of the labor equation.

the construction of this estimator, first note that the 2SLS estimator can be written as (Greene, 2003, p. 399)

$$\hat{\boldsymbol{\theta}}_{IV,j} = [(\mathbf{W}_j^T\mathbf{X})(\mathbf{X}^T\mathbf{X})^{-1}(\mathbf{X}^T\mathbf{W}_j)]^{-1}(\mathbf{W}_j^T\mathbf{X})(\mathbf{X}^T\mathbf{X})^{-1}\mathbf{X}^T\mathbf{y}_j,$$

where $(\mathbf{X}^T\mathbf{X})$ is called the weight matrix. Replacing this with White's estimator

$$S_{0,j} = \sum_{t=1}^{T} \mathbf{x}_t\mathbf{x}_t^T\left(y_{jt}-\mathbf{w}_{jt}^T\hat{\boldsymbol{\theta}}_{IV,j}\right)^2$$

```
                  The MODEL Procedure

              ┌─────────────────────────┐
              │      Model Summary       │
              ├──────────────────────┬──┤
              │ Model Variables      │ 6│
              ├──────────────────────┼──┤
              │ Endogenous           │ 1│
              ├──────────────────────┼──┤
              │ Exogenous            │ 4│
              ├──────────────────────┼──┤
              │ Parameters           │ 6│
              ├──────────────────────┼──┤
              │ Equations            │ 1│
              ├──────────────────────┼──┤
              │ Number of Statements │ 1│
              └──────────────────────┴──┘
```

Model Variables	L W L_Land L_re L_Othexp L_Sch
Parameters	b0 b1 b2 b3 b4 b5
Equations	L

The Equation to Estimate is	
L =	F(b0(1), b1(W), b2(L_Land), b3(L_re), b4(L_Othexp), b5(L_Sch))
Instruments	1 L_Land L_re L_Othexp L_Sch L_Othwag

```
NOTE: At GMM Iteration 0 convergence assumed because OBJECTIVE=1.095301E-24 is almost zero (<1E-12).
```

```
                  The MODEL Procedure
                  GMM Estimation Summary
```

Data Set Options	
DATA=	SE

Minimization Summary	
Parameters Estimated	6
Kernel Used	PARZEN
l(n)	2.080717
Method	Gauss
Iterations	0

Final Convergence Criteria	
R	1
PPC	5.23E-12
RPC	.
Object	.
Trace(S)	0.050923
Objective Value	1.1E-24

Observations Processed	
Read	39
Solved	39

OUTPUT 9.3. GMM estimators for the labor equation.

The MODEL Procedure

Nonlinear GMM Summary of Residual Errors							
Equation	DF Model	DF Error	SSE	MSE	Root MSE	R-Square	Adj R-Sq
L	6	33	1.6804	0.0509	0.2257	0.8669	0.8468

Nonlinear GMM Parameter Estimates				
Parameter	Estimate	Approx Std Err	t Value	Approx Pr > |t|
b0	-0.85913	3.1700	-0.27	0.7881
b1	-1.44928	0.5905	-2.45	0.0196
b2	0.328677	0.0904	3.64	0.0009
b3	0.468702	0.1092	4.29	0.0001
b4	0.249787	0.1332	1.88	0.0696
b5	0.342098	1.2268	0.28	0.7821

Number of Observations		Statistics for System	
Used	39	Objective	1.095E-24
Missing	0	Objective*N	4.272E-23

OUTPUT 9.3. (*Continued*)

gives the GMM estimator (Greene, 2003, p. 401). Notice the similarity in the construction of White's estimator to the one used in Chapter 5. Also, notice the GMM estimator is constructed in three steps. First, a 2SLS is used to get $\hat{\theta}_{IV,j}$. In the second step, White's estimator is calculated. In the third step, this is used as a weight matrix to calculate the robust 2SLS estimator.

We will illustrate this method on the labor equation. We will use Proc Model to estimate the GMM estimator since Proc Syslin does not have the option to perform a GMM analysis. The following SAS code can be used. The analysis results are given in Output 9.3. The procedure statements start off with specifying the model. Note the coefficient names are not unique and can be changed. However, the variable names have to be identical to the ones used in the data set. Next, we specify the endogenous and exogenous variables. These are then followed by specifying the instrument variables and requesting a model using GMM estimation. Proc Model can fit the OLS and the 2SLS models. We used Proc Reg and Proc Syslin to minimize the output that is produced by using Proc Model. Notice that the parameter estimates and the model diagnostic statistics are very similar to the ones from 2SLS estimation.

```
Proc Model
     L=b0+b1*W+b2*L_Land+b3*L_RE+b4*L_Othexp+b5*L_Sch;
     Endogenous W;
     Exogenous L_Land L_RE L_Othexp L_Sch;
     Instruments L_Land L_RE L_Othexp L_Sch L_OTHWAG;
     Fit L/GMM;
Run;
```

9.6 HAUSMAN'S SPECIFICATION TEST

Hausman's specification test can be used to test whether an included exogenous variable in a simultaneous equation is endogenous. As with all specification tests by Hausman we used in Chapters 4 and 7, this test compares two estimators, both of which are consistent under the null hypothesis of exogeneity but only one is consistent under alternative hypothesis of endogeneity.

Details of the Hausman's specification test as it applies to simultaneous equation models can be found in Greene (2003, pp. 413–415). The analysis can easily be done in SAS by using the Proc Model procedure. The following statements (labor–wage data being used) can be used to conduct the Hausman's specification test comparing the OLS estimates with the 2SLS estimates. We did not include the output for the test as the OLS and 2SLS outputs were already provided. The p value for the test is around 0.90 and indicates that there is no gain in using the 2SLS model over the OLS model.

```
proc model data=SE;
      L = b0+b1*W+b2*L_Land+b3*L_RE+b4*L_Othexp+b5*L_Sch;
      W = a0+a1*L+a2*L_Sch+a3*L_Othwag;
      ENDOGENOUS L W;
      EXOGENOUS L_land L_RE L_Othexp L_Sch L_Othwag;
      fit L/ols 2sls hausman;;
      instruments L_land L_RE L_Othexp L_Sch L_Othwag;
run;
```

10

DISCRETE CHOICE MODELS

10.1 INTRODUCTION

The preceding chapters were focused on discussion of modeling techniques where the responses were continuous. In practice, we often encounter responses that are discrete. For example, direct marketing companies often model the response behavior of consumers receiving an offer to buy their products. Here, y, the response variable, equals 1 if the consumer responds to the mail offer, and it equals 0 otherwise. Direct marketing companies also build conversion models where again the response variable equals 1 if the consumer's inquiry about a mail offer results in a sale; the response variable equals 0 otherwise. Another example involves attrition models built by insurance companies that predict the likelihood that an existing consumer will cancel his or her policy to take up a new policy with a competitor. Here, the response variable equals 1 if the consumer cancels a policy, and it equals 0 otherwise. Attrition models can also be built by using duration models but the common approach in industry is to treat the attrition response as 0 or 1. A common theme in each example is the binary nature of the response variable. Of course, the response variable can assume more than two discrete values. This chapter deals with estimating parameters of models where the distribution of the response variable is not continuous but discrete. We will focus our attention on logistic regression with dichotomous responses, and Poisson regression.

By definition, the set of plausible values of a discrete random variable can be placed in a 1:1 correspondence with a finite or a countable infinite set. Some examples of discrete random variables are as follows:

1. The number of customers walking into a bank between 12 noon and 1 p. m. Here, the response variable can assume values 0, 1, 2, ...
2. The response of a consumer to a mailed offer for a new credit card or auto insurance. Here, the response variable can assume one of two possible values: 1, response; 0, no response.
3. The views on abortion of an individual can be measured as 1, strongly agree; 2, agree; 3, neutral; 4, disagree; and 5, strongly disagree. Here, the response variable is ordinal as the values have a natural rank.
4. The mode of transportation chosen by a commuter. The choices can be classified as 1, drive; 2, ride-share; 3, bus; and so on. Here, the values cannot be ordered, and therefore the random variable is called a discrete nominal variable.
5. A consumer's choice of a fast-food restaurant among several available brand names. This is similar to the example presented in (4).

10.2   BINARY RESPONSE MODELS

In this section, we will discuss techniques for modeling a response variable where there are two possible outcomes (0 and 1) and a set of explanatory variables $\mathbf{x}$, which is assumed to influence the response. To illustrate this, consider a response model where a person either responds to an offer in the mail ($y = 1$) or does not ($y = 0$). Furthermore, assume that a set of factors such as age, occupation, marital status, number of kids in the household, and so on, explains his or her decision and that are captured in a vector of explanatory variables $\mathbf{x}$. First note that the response variable is a Bernoulli random variable with mean

$$E(y|\mathbf{x}) = 1 \times P(y = 1|\mathbf{x}) + 0 \times P(y = 0|\mathbf{x}) = P(y = 1|\mathbf{x}) = P(\mathbf{x}),$$

where the probability function $P(\mathbf{x})$ denotes the dependence of the response variable on $\mathbf{x}$. Also note that

$$Var(y|\mathbf{x}) = E(y^2|\mathbf{x}) - [E(y|\mathbf{x})]^2 = P(\mathbf{x})(1 - P(\mathbf{x})).$$

We will denote the probability function as $P(\mathbf{x}, \boldsymbol{\beta})$, where $\boldsymbol{\beta}$ measure the impact of $\mathbf{x}$ on the probability of response $P(y = .|\mathbf{x})$. Using the change in notation, we can write

$$P(y = 1|\mathbf{x}) = P(\mathbf{x}, \boldsymbol{\beta}),$$
$$P(y = 0|\mathbf{x}) = 1 - P(\mathbf{x}, \boldsymbol{\beta}).$$

Our objective is to estimate $\boldsymbol{\beta}$ and given that $E(y|\mathbf{x}) = P(\mathbf{x}, \boldsymbol{\beta})$, a naïve approach to estimation of the parameters may start off by utilizing the traditional OLS method on the linear probability model

$$y_i = \mathbf{x}^T \boldsymbol{\beta} + \varepsilon_i, \quad i = 1, \dots, n.$$

However, as the next section illustrates, there are fundamental problems with OLS estimation when the response variable is dichotomous.

10.2.1   Shortcomings of the OLS Model

There are three main reasons why the OLS model should not be used to model discrete choice data (Agresti, 1990; Greene 2003). They are

1. *Nonnormal Disturbances:* Notice that the response variable y_i is binary and is either 0 or 1. Therefore, the disturbance ε_i is also binary and has only two possible outcomes: $\varepsilon_i = 1 - \mathbf{x}_i^T \boldsymbol{\beta}$ with probability $P(\mathbf{x}_i, \boldsymbol{\beta})$ and $\varepsilon_i = -\mathbf{x}_i^T \boldsymbol{\beta}$ with probability $1 - P(\mathbf{x}_i, \boldsymbol{\beta})$. Therefore, the error terms are not normally distributed. This poses problems in any inference on the model parameters (hypothesis tests, confidence intervals, etc.).

2. *Heteroscedastic error:* The linear probability model violates the assumption of homoscedastic disturbances. It is easy to see this by realizing that

$$Var(\varepsilon_i|\mathbf{x}_i) = Var(y_i|\mathbf{x}_i) = P(\mathbf{x}_i, \boldsymbol{\beta})(1 - P(\mathbf{x}_i, \boldsymbol{\beta})).$$

   Therefore, the variance of the disturbance depends on $E(y_i|\mathbf{x}_i) = P(\mathbf{x}_i, \boldsymbol{\beta})$ and as the mean changes so does the variance. Therefore, the homoscedasticity assumption is violated.

3. The conditional expectation, $E(y_i|\mathbf{x}_i) = P(\mathbf{x}_i, \boldsymbol{\beta})$, is a probability and it must be bounded by 0 and 1. In the linear probability model $E(y_i|\mathbf{x}_i) = \mathbf{x}_i^T \boldsymbol{\beta}$, which is defined over the range $(-\infty, \infty)$ and therefore does not guarantee that the conditional expectation will be bounded.

An alternative to the OLS model is to use weighted least squares (WLS) to account for the heteroscedasticity, with weights defined as $w_i = \sqrt{(\mathbf{x}_i^T \hat{\boldsymbol{\beta}}) \times (1 - \mathbf{x}_i^T \hat{\boldsymbol{\beta}})}$ with $\hat{\boldsymbol{\beta}}$ calculated from OLS. This is a two-step process where in step 1, OLS would be used to get an estimate of $\boldsymbol{\beta}$. The predicted value for y_i given by the first term under the square root in w_i would then be used to calculate the weights for each observation. The weights would then be used in step 2 to estimate $\boldsymbol{\beta}$. The WLS model thus becomes (Chapter 5, Myers, 1990, p. 316)

$$\frac{y_i}{w_i} = \left(\frac{\mathbf{x}_i^T}{w_i}\right)\boldsymbol{\beta} + \frac{\varepsilon_i}{w_i} \text{ or } y_i^* = \mathbf{x}_i^{*T}\boldsymbol{\beta} + \varepsilon_i^*.$$

The WLS would ensure homoscedasticity but would not ensure that the conditional mean $\mathbf{x}_i^T\boldsymbol{\beta}$ is bounded by [0,1].

Estimation of $\boldsymbol{\beta}$ involving more efficient techniques exist and will now be discussed. Specifically, we will transform the conditional mean by assuming either the probit or the logit distribution of the probability function $P(\mathbf{x}, \boldsymbol{\beta})$.

10.2.2 The Probit and Logit Models

As shown in Agresti (1990), by assuming a continuous distribution for $P(\mathbf{x}, \boldsymbol{\beta})$, the conditional expectation $\mathbf{x}_i^T\boldsymbol{\beta}$ can be bounded in the interval $[0, 1]$. There are two distributions that are commonly used: the normal and the logistic distribution.

If $P(\mathbf{x}, \boldsymbol{\beta})$ is taken to be the cumulative standard normal distribution, then the resulting model is a probit model. That is,

$$P(\mathbf{x}, \boldsymbol{\beta}) = F(\mathbf{x}^T\boldsymbol{\beta}),$$

where $F(t) = \int_{-\infty}^{t} f(s)ds$ with $f(s)$ being the probability density function of the standard normal distribution.

If

$$P(\mathbf{x}, \boldsymbol{\beta}) = \frac{e^{\mathbf{x}^T\boldsymbol{\beta}}}{1 + e^{\mathbf{x}^T\boldsymbol{\beta}}} = G(\mathbf{x}^T\boldsymbol{\beta})$$

then we have the logit model where $G(x)$ is the logistic cumulative function given by

$$G(x) = \frac{\exp(x)}{1 + \exp(x)} = \frac{1}{1 + \exp(-x)}.$$

It is trivial to see that in both cases, $P(\mathbf{x}, \boldsymbol{\beta})$ is bounded in the interval [0,1]. Agresti (1990, p. 105) discusses the complementary log–log model as another alternative to modeling binary response variables. However, the probit and logit distributions are the two most commonly used to model binary response variables and we therefore focus on estimation techniques involving these two distributions.

Notice that estimation using maximum likelihood methods will have to be used since the parameter $\boldsymbol{\beta}$ is now no longer a linear function of $P(\mathbf{x}, \boldsymbol{\beta})$. The standard approach is to use the OLS estimates as the initial values for the MLE and then to iterate until convergence.

10.2.3 Interpretation of Parameters

As discussed in Wooldridge (2002, p. 458), proper interpretation of the model parameters for the Probit and Logit models is crucial for successful implementation of the model. To see this, consider the formulation of the probit model where $P(\mathbf{x}, \boldsymbol{\beta}) = F(\mathbf{x}^T\boldsymbol{\beta})$ where $F(\mathbf{x}^T\boldsymbol{\beta})$ is the cumulative normal distribution. Taking derivatives of this with respect to $\mathbf{x}$ we get (Greene, 2003, p. 667)

$$\frac{\partial P(\mathbf{x}, \boldsymbol{\beta})}{\partial \mathbf{x}} = \frac{\partial F(\mathbf{x}^T\boldsymbol{\beta})}{\partial(\mathbf{x}^T\boldsymbol{\beta})}\boldsymbol{\beta} = f(\mathbf{x}^T\boldsymbol{\beta})\boldsymbol{\beta}$$

where $f(\bullet)$ is the normal probability density function. Therefore, the marginal effects depend on $\mathbf{x}$ via the density function. It can be shown that the derivative of the conditional expectation for the logit model is given by

$$\frac{\partial P(\mathbf{x}, \boldsymbol{\beta})}{\partial \mathbf{x}} = G(\mathbf{x}^T\boldsymbol{\beta})[1 - G(\mathbf{x}^T\boldsymbol{\beta})]\boldsymbol{\beta}.$$

Thus, the interpretation of $\boldsymbol{\beta}$ must be based on calculations done at a prespecified value of the explanatory variables $\mathbf{x}$. A common approach is to interpret the model parameters at the mean, $\bar{\mathbf{x}}$. If a variable in $\mathbf{x}$ contains a dummy variable, then the marginal effect of the variable can be computed as (Greene, 2003, p. 668; Wooldridge, 2002, p. 459)

$$P(y = 1|\bar{\mathbf{x}}_{(-d)}, x_d = 1) - P(y = 0|\bar{\mathbf{x}}_{(-d)}, x_d = 0)$$

where x_d denotes that the explanatory variables contains a dummy variable and $\bar{\mathbf{x}}_{(d)}$ is the mean vector of the other variables in the model.

10.2.4   Estimation and Inference

The method of maximum likelihood is used to estimate the parameters because both the Probit and the Logit models are nonlinear in $\boldsymbol{\beta}$. To start our discussion of model estimation, assume that we have an independent random sample $(y_i, \mathbf{x}_i)$, where $i = 1, \ldots, n$ from the Bernoulli distribution with probability $P(\mathbf{x}_i, \boldsymbol{\beta})$. Recall that the probability density function of a Bernoulli random variable, x, is given by $f(x) = p^x(1-p)^{1-x}$, where $x = 0, 1$, and p is the probability of success. Collecting all n observations and assuming independence gives the likelihood function

$$
\begin{aligned}
L(\boldsymbol{\beta}|\mathbf{y}, \mathbf{X}) &= \prod_{i=1}^{n} \Pr(y_i|\mathbf{x}_i, \boldsymbol{\beta}), \\
&= \prod_{i=1}^{n} \Pr(y_i = 0|\mathbf{x}_i, \boldsymbol{\beta})^{1-y_i} \Pr(y_i = 1|\mathbf{x}_i, \boldsymbol{\beta})^{y_i}, \\
&= \prod_{i=1}^{n} [1 - P(\mathbf{x}_i, \boldsymbol{\beta})]^{1-y_i} [P(\mathbf{x}_i, \boldsymbol{\beta})]^{y_i}.
\end{aligned}
$$

Taking the log of $L = L(\boldsymbol{\beta}|\mathbf{y}, \mathbf{X})$ yields

$$\ln(L) = \sum_{i=1}^{n} \{y_i \ln P(\mathbf{x}_i, \boldsymbol{\beta}) + (1-y_i) \ln [1 - P(\mathbf{x}_i, \boldsymbol{\beta})]\}.$$

It can easily be shown that the log-likelihoods for the probit and logit models are as follows:

$$\ln(L) = \sum_{i=1}^{n} \{y_i \ln F(\mathbf{x}_i^T \boldsymbol{\beta}) + (1-y_i) \ln [1 - F(\mathbf{x}_i^T \boldsymbol{\beta})]\} \text{ for the Probit model, and}$$

$$\ln(L) = \sum_{i=1}^{n} \{y_i \ln G(\mathbf{x}_i^T \boldsymbol{\beta}) + (1-y_i) \ln [1 - G(\mathbf{x}_i^T \boldsymbol{\beta})]\} \text{ for the Logit model.}$$

The maximum likelihood parameters are found by taking the derivative of the log-likelihood equations, setting them to zero and then solving for the unknown parameters. The first-order conditions are given by

$$L(\boldsymbol{\beta}) = \sum_{i=1}^{n} \left[\frac{y_i p_i}{P_i} + (1-y_i) \frac{-p_i}{1-P_i} \right] \mathbf{x}_i.$$

where p_i is the derivative of P_i with respect to $\mathbf{x}_i$.

Extending this to the Probit model, we get

$$L(\boldsymbol{\beta}) = \sum_{i=1}^{n} \frac{y_i - \Phi(\mathbf{x}_i^T \boldsymbol{\beta})}{\Phi(\mathbf{x}_i^T \boldsymbol{\beta}) \times [1 - \Phi(\mathbf{x}_i^T \boldsymbol{\beta})]} \phi(\mathbf{x}_i^T \boldsymbol{\beta}) \mathbf{x}_i,$$

whereas for the Logit model we have

$$L(\boldsymbol{\beta}) = \sum_{i=1}^{n} \left[y_i - \frac{\exp(\mathbf{x}_i^T \boldsymbol{\beta})}{1 + \exp(\mathbf{x}_i^T \boldsymbol{\beta})} \right] \mathbf{x}_i$$

As shown in Greene (2003, p. 671), it can be shown that the first-order condition for the Logit model can be written as

$$L(\boldsymbol{\beta}) = \sum_{i=1}^{n} (y_i - G_i) \mathbf{x}_i = 0.$$

The Hessian (second-order condition) for the Logit model is given by

$$\mathbf{H} = \frac{\partial \ln L}{\partial \boldsymbol{\beta} \partial \boldsymbol{\beta}^T} = -\sum_{i=1}^{n} G_i (1 - G_i) \mathbf{x}_i \mathbf{x}_i^T.$$

The Newton–Raphson method is used to obtain a solution for these likelihood equations. The Hessian calculated in the final iteration is the estimated covariance matrix of the coefficient vector.

10.2.5   The Newton–Raphson Method for the Logit Model

The Newton–Raphson algorithm to obtain a solution for the likelihood equation of the Logit model can be summarized as follows (Agresti, 1990):

1. Start with an initial set of estimates, $\hat{\boldsymbol{\beta}}_{t=0}$. Most often, the starting values are simply the OLS estimates. Here, $t = 0$ simply denotes the starting point.
2. Calculate the estimated value of the coefficient vector at time $t + 1$ as given by $\hat{\boldsymbol{\beta}}_{t+1} = \hat{\boldsymbol{\beta}}_t - \hat{\mathbf{H}}_t^{-1} \hat{\mathbf{g}}_t$, where $\hat{\mathbf{g}}_t$ is the gradient vector defined as $\hat{\mathbf{g}}_t = y_i - \hat{G}_t$ and $\mathbf{H}_t$ is the Hessian matrix. Here, $\hat{G}_t$ is the predicted probability of time t.
3. Iterate until the convergence criteria is reached. That is, until the difference of consecutive $\hat{\boldsymbol{\beta}}$ values are insignificant.

We will now illustrate the Newton–Raphson algorithm to calculate the parameters (assuming a Logit model) for an unemployment data set arising from a sample of 4877 blue collar workers who lost their jobs in the United States between 1982 and 1991 (McCall, 1995). All individuals in this study is assumed to have applied for unemployment benefits. Note that this data was also analyzed by Verbeek (2004, pp. 197–199). A description of the variables was given by the author and is summarized here. The variables in the data are as follows:

Y is the response variable and takes a value of 1 if the unemployed worker received unemployment benefits.

Age is the age of the subject.

Age2 is the square of the Age variable.

Tenure is the years of tenure at the last job.

Slack is an indicator variable that equals 1 if the subject was fired because of poor performance.

Abol is an indicator variable that equals 1 if the subject's position was eliminated.

Seasonal is an indicator variable that equals 1 if the subject was a temporary worker.

NWHITE is an indicator variable that equals 1 if the subject's race is nonwhite.

School12 is an indicator variable that equals 1 if the subject has more than 12 years of education.

Male is an indicator variable that equals 1 if the subject is a male.

SMSA is an indicator variable that equals 1 if the subject lives in a SMSA.

Married is an indicator variable that equals 1 if the subject is married.

DKIDS is an indicator variable that equals 1 if the subject has kids.

DYKIDS is an indicator variable that equals 1 if the subject has young kids.

YRDISP records the year when the job was lost. Here, 1982 = 1 and 1991 = 10.

RR is the replacement rate that is the ratio of the benefits received versus the last recorded weekly earnings.

RR2 is the square of RR.

Head is an indicator variable that equals 1 if the subject is the head of a household.

StateUR is the state unemployment rate.

StateMB is the maximum benefits available for a given state.

The following Proc IML code can be used.

```
* Read the data file and scale age squared. ;
      libname in "C:\Temp";
      data test;
            set in.unemp;
            age2=age2/10;
run;
proc iml;
* Read the data into matrices and calculate some constants. ;
            use test;
            read all var {'rr' 'rr2' 'age' 'age2' 'tenure'
            'slack' 'abol' 'seasonal' 'head' 'married' 'dkids'
            'dykids' 'smsa' 'nwhite' 'yrdispl' 'school12' 'male'
            'stateur' 'statemb'} into X;
            read all var {'y'} into Y;
            n=nrow(X);
            X=J(n,1,1)||X;
            c=ncol(X);
* Calculate an initial estimate of the parameter vector. ;
            Beta=inv(X'*X)*X'*y;
* Start the Newton-Raphson procedure. ;
            Col_One=J(4877,1,1);
            do index=1 to 5;
                  PI=exp(X*Beta)/(1+exp(X*Beta));
                  Temp_PI=Col_One-PI;
                  Diag_PI=Diag(PI#Temp_PI);
                  COV=inv(X'*Diag_PI*X);
                  Beta_New=Beta+COV*X'*(Y-PI);
                  DIFF=sum(abs(BETA_NEW-BETA));
                  print DIFF;
                  if DIFF<0.00001 then
                        do;
                              print 'The estimates of the coefficients are:';
                              print Beta_New;
                              SE=sqrt(vecdiag(COV));
                              print 'The estimated standard errors are:';
                              print SE;
                              stop;
                        end;
                        beta=beta_new;
                  end;
run;
```

The analysis results are given in Output 10.1. Notice that the test statistic value and the p values were not computed here. They can be easily computed by incorporating additional statements in the code provided. We will discuss the methodology when we interpret the output from Proc Logistic.

The parameter estimates and their standard errors are:

OUTPUT_TABLE		
	BETA	SE
INTERCEPT	-2.8005	0.6042
RR	3.0681	1.8682
RR2	-4.8906	2.3335
AGE	0.0677	0.0239
AGE2	-0.0597	0.0304
TENURE	0.0312	0.0066
SLACK	0.6248	0.0706
ABOL	-0.0362	0.1178
SEASONAL	0.2709	0.1712
HEAD	-0.2107	0.0812
MARRIED	0.2423	0.0794
DKIDS	-0.1579	0.0862
DYKIDS	0.2059	0.0975
SMSA	-0.1704	0.0698
NWHITE	0.0741	0.0930
YRDISPL	-0.0637	0.0150
SCHOOL12	-0.0653	0.0824
MALE	-0.1798	0.0875
STATEUR	0.0956	0.0159
STATEMB	0.0060	0.0010

OUTPUT 10.1. Proc IML output for the logistic model of unemployment data.

The following statements can be used to fit an OLS model. The analysis results are given in Output 10.2. Note that this is the incorrect method of analyzing data and we include it here simply for comparing the results to the results obtained by using logistic regression.

```
proc reg data=test;
     model y=rr rr2 age age2 tenure slack abol
     seasonal head married dkids dykids smsa nwhite
     yrdispl school12 male stateur statemb;
run;
```

The following statements can be used to fit a Logit model to the unemployment data. Note that the model statement contains the option "event=1" that forces the procedure to model the probability of a response. The option "descending" in the first line can be used to achieve the same results. The analysis results are given in Output 10.3.

```
proc logistic data=test;
     model y(event='1')=rr rr2 age age2 tenure slack abol
     seasonal head married dkids dykids smsa nwhite
     yrdispl school12 male stateur statemb;
run;
```

We will now provide details about the output produced by Proc Logistic.

The first table gives basic model information that includes the names of the data set and the response variable, the number of levels of the response variable (here two since the response is binary), and type of model (logit), and the optimization technique used by the program (Fisher's scoring).

The next three tables give information on the number of observations in the data set and the number of observations that were used to estimate the model. These numbers will be different if there are missing observations in the data set. The next table gives the response profile. There are 1542 nonresponders and 3335 responders. Therefore, out of the total number of individuals in the study, 3335 unemployed workers applied for and received unemployment insurance while the remaining 1542 applied for unemployment insurance but did not receive insurance.

```
The REG Procedure
Model: MODEL1
Dependent Variable: y
```

Number of Observations Read	4877
Number of Observations Used	4877

Analysis of Variance					
Source	DF	Sum of Squares	Mean Square	F Value	Pr > F
Model	19	70.55319	3.71333	18.33	<0.0001
Error	4857	983.90037	0.20257		
Corrected Total	4876	1054.45356			

Root MSE	0.45008	R-Square	0.0669
Dependent Mean	0.68382	Adj R-Sq	0.0633
Coeff Var	65.81857		

Parameter Estimates					
Variable	DF	Parameter Estimate	Standard Error	t Value	Pr > |t|
Intercept	1	−0.07687	0.12206	−0.63	0.5289
rr	1	0.62886	0.38421	1.64	0.1017
rr2	1	−1.01906	0.48095	−2.12	0.0342
age	1	0.01575	0.00478	3.29	0.0010
age2	1	−0.01459	0.00602	−2.43	0.0153
tenure	1	0.00565	0.00122	4.65	<0.0001
slack	1	0.12813	0.01422	9.01	<0.0001
abol	1	−0.00652	0.02483	−0.26	0.7928
seasonal	1	0.05787	0.03580	1.62	0.1060
head	1	−0.04375	0.01664	−2.63	0.0086
married	1	0.04860	0.01613	3.01	0.0026

Parameter Estimates					
Variable	DF	Parameter Estimate	Standard Error	t Value	Pr > |t|
dkids	1	−0.03051	0.01743	−1.75	0.0802
dykids	1	0.04291	0.01976	2.17	0.0299
smsa	1	−0.03520	0.01401	−2.51	0.0121
nwhite	1	0.01659	0.01871	0.89	0.3753
yrdispl	1	−0.01331	0.00307	−4.34	<0.0001
school12	1	−0.01404	0.01684	−0.83	0.4047
male	1	−0.03632	0.01781	−2.04	0.0415
stateur	1	0.01815	0.00308	5.88	<0.0001
statemb	1	0.00124	0.00020393	6.08	<0.0001

OUTPUT 10.2. Proc Reg output for the linear model of the unemployment data.

The Model Fit Statistics table gives three statistics that can be used to assess the fit of the model. The first column gives the model fit statistics for the model with only the intercept term while the second column adjusts the values to account for the addition of explanatory variables to the model. Smaller values of all three statistics are desirable.

The next table contains the chi-square test statistic value and the corresponding p value for the Global Null Hypothesis that none of the coefficients in the model are significant versus the alternative that at least one of the coefficients is significant. The chi-square value for the likelihood ratio is simply the difference between the log-likelihood values from the previous table. That is, it is computed by taking the difference $6086.056 - 5746.393 = 339.663$. The p value indicates that the overall model is highly significant.

The next table gives the parameter estimates, their standard errors along with the individual chi-square values and the associated p value. The Wald's chi-squared test statistic can be computed by taking the square of the ratio of the parameter estimate to its standard error. For example, the Wald's chi-squared value for the age variable is given by

$$\left(\frac{0.0677}{0.0239}\right)^2 = 8.02.$$

This can easily be programmed into Proc IML and then the PROBCHI function can be used to calculate the p values.

There are several variables that are significant in the model. As an example of how to interpret the variables consider the two state variables and the slack variable. All three are highly significant. The significance of the state variables indicate that the higher the state's unemployment rate and the higher the benefits allowed the more likely it is that an employed person will receive unemployment benefits. The variable slack indicates that a person whose job was terminated because of poor performance will have a higher likelihood of receiving employment benefits.

The next table gives the odds ratio estimates along with the corresponding 95% confidence intervals. The odds ratios are calculated by simply exponentiating the parameter estimates. For instance, the odds ratio for the variable slack is $\exp{(0.6248)} = 1.868$. This implies that a person who was fired for poor performance is 1.868 times more likely to receive unemployment benefits than a person who lost his or her job for some other reason (all other variables being held constant). The odds ratio for males indicates that they are 0.893 times as likely to get unemployment insurance benefits than females. That is, their likelihood of getting unemployment insurance benefits is lower than that of females.

The last table gives association of predicted probabilities and the observed responses. See the SAS 9.2 User's guide for a description of these statistics.

The LOGISTIC Procedure

Model Information	
Data Set	WORK.TEST
Response Variable	y
Number of Response Levels	2
Model	binary logit
Optimization Technique	Fisher's scoring

Number of Observations Read	4877
Number of Observations Used	4877

Response Profile		
Ordered Value	y	Total Frequency
1	0	1542
2	1	3335

Probability modeled is y=1.

Model Convergence Status
Convergence criterion (GCONV=1E-8) satisfied.

Model Fit Statistics		
Criterion	Intercept Only	Intercept and Covariates
AIC	6088.056	5786.393
SC	6094.548	5916.239
-2 Log L	6086.056	5746.393

OUTPUT 10.3. Logit model output for the unemployment data.

Testing Global Null Hypothesis: BETA=0			
Test	Chi-Square	DF	Pr > ChiSq
Likelihood Ratio	339.6629	19	<0.0001
Score	326.3187	19	<0.0001
Wald	305.0800	19	<0.0001

Analysis of Maximum Likelihood Estimates					
Parameter	DF	Estimate	Standard Error	Wald Chi-Square	Pr > ChiSq
Intercept	1	-2.8005	0.6042	21.4860	<0.0001
rr	1	3.0681	1.8682	2.6969	0.1005
rr2	1	-4.8906	2.3335	4.3924	0.0361
age	1	0.0677	0.0239	8.0169	0.0046
age2	1	-0.0597	0.0304	3.8585	0.0495
tenure	1	0.0312	0.00664	22.1189	<0.0001
slack	1	0.6248	0.0706	78.2397	<0.0001
abol	1	-0.0362	0.1178	0.0943	0.7588
seasonal	1	0.2709	0.1712	2.5042	0.1135
head	1	-0.2107	0.0812	6.7276	0.0095
married	1	0.2423	0.0794	9.3075	0.0023
dkids	1	-0.1579	0.0862	3.3552	0.0670
dykids	1	0.2059	0.0975	4.4601	0.0347
smsa	1	-0.1704	0.0698	5.9598	0.0146
nwhite	1	0.0741	0.0930	0.6349	0.4255
yrdispl	1	-0.0637	0.0150	18.0409	<0.0001
school12	1	-0.0653	0.0824	0.6270	0.4285
male	1	-0.1798	0.0875	4.2204	0.0399
stateur	1	0.0956	0.0159	36.1127	<0.0001
statemb	1	0.00603	0.00101	35.6782	<0.0001

OUTPUT 10.3. (*Continued*)

The following statements can be used to fit a binary Probit model. The analysis results are given in Output 10.4.

```
proc logistic data=test;
     model y(event='1')=rr rr2 age age2 tenure slack abol
     seasonal head married dkids dykids smsa nwhite
     yrdispl school12 male stateur statemb/l=Probit;
run;
```

Notice that the model statistics from the first few tables are very similar between the logit and probit models. The model parameters also have consistent signs. However, the magnitude of the parameter estimates are different.

The parameter estimates from the Probit model can be used to calculate the predicted probability that a person will receive unemployment insurance. For a given individual, the predicted probability of receiving unemployment insurance can be calculated by using

$$F(\mathbf{x}_i^T \hat{\boldsymbol{\beta}}) = F(-1.6999 + 1.8633rr - 2.9801rr2 + \cdots + 0.00364\ statemb),$$

where $F()$ is the cumulative normal distribution. The coefficients cannot be interpreted as the impact on predicted probability—all we can say is whether the predicted probabilities will increase or decrease based on the signs of the coefficients.

SAS can be used to generate a table of predicted probabilities by using the statement

$$\text{output out} = \textit{file-name} \text{ predicted} = \textit{y\_hat};$$

Here *file-name* is the temporary SAS dataset while *y_hat* is the variable name which will hold the predicted probabilities.

Odds Ratio Estimates			
Effect	Point Estimate	95% Wald Confidence Limits	
rr	21.500	0.552	836.911
rr2	0.008	<0.001	0.728
age	1.070	1.021	1.121
age2	0.942	0.888	1.000
tenure	1.032	1.018	1.045
slack	1.868	1.626	2.145
abol	0.964	0.766	1.215
seasonal	1.311	0.937	1.834
head	0.810	0.691	0.950
married	1.274	1.090	1.489
dkids	0.854	0.721	1.011
dykids	1.229	1.015	1.487
smsa	0.843	0.736	0.967
nwhite	1.077	0.898	1.292
yrdispl	0.938	0.911	0.966
school12	0.937	0.797	1.101
male	0.835	0.704	0.992
stateur	1.100	1.067	1.135
statemb	1.006	1.004	1.008

Association of Predicted Probabilities and Observed Responses			
Percent Concordant	65.7	Somers' D	0.318
Percent Discordant	33.9	Gamma	0.320
Percent Tied	0.4	Tau-a	0.138
Pairs	5142570	c	0.659

OUTPUT 10.3. (*Continued*)

10.2.6 Comparison of Binary Response Models for the Unemployment Data

The marginal effects for the OLS, Probit, and Logit models were given in Section 10.2.3. Consider comparing the marginal effects from the three models on the unemployment data. We will calculate the marginal effect of getting unemployment benefits for a male, with more than 12 years of education. Both the variables Male and the one recording more than 12 years of education are binary variables with 1 for both implying that the person is a male with over 12 years of education. Table 10.1 summarizes the calculations. Notice that the conditional probability for the OLS is negative, which points to one of the issues with the linear probability model that was discussed earlier.

10.3 POISSON REGRESSION

Often, we are interested in modeling responses that count some phenomenon of interest. For example, we may be interested in modeling the number of patents filed by a firm with respect to its R&D investment (Verbeek, 2004, p. 215). As discussed by the

TABLE 10.1. Comparing Marginal Effect of the Three Models

Model	Constant	> 12 years of Education	Male	$x_i^T\beta$	Marginal Effect
LMP	−0.077	−0.014	−0.036	−0.127	−0.127
Probit	−1.7	−0.042	−0.107	−1.849	0.032
Logit	−2.8	−0.065	−0.18	−3.045	0.045

author, the number of patents can assume values starting at 0 (for no patents filed) to some large number. Typically, there is no upper bound for the count variables. These random variables are modeled by using the Poisson distribution.

A Poisson random variable, y, with mean λ has a probability density function given by

$$f(y) = \frac{\exp(-\lambda)\lambda^y}{y!}, \qquad y = 0, 1, 2, \ldots.$$

In Poisson regression, the dependent variable for observation i ($i = 1, \ldots, n$) is assumed to follow a Poisson distribution with mean λ_i, which is a function of explanatory variables and unknown coefficients. That is,

$$f(y_i) = \frac{\exp(-\lambda_i)\lambda_i^{y_i}}{y_i!}, \qquad y_i = 0, 1, 2, \ldots,$$

where $\lambda_i = \exp(\mathbf{x}_i^T \boldsymbol{\beta})$. Therefore, the density function of y_i can be written as

$$f(y_i) = \frac{\exp(-\mathbf{x}_i^T \boldsymbol{\beta})(\mathbf{x}_i^T \boldsymbol{\beta})^{y_i}}{y_i!}, \quad y_i = 0, 1, 2, \ldots.$$

It is trivial to show that the mean and variance of a Poisson random variable are equal. That is, $E(y_i) = Var(y_i) = \lambda_i$. For the Poisson regression model, we can write the conditional expectation and conditional variance of y_i given $\mathbf{x}_i$ as

$$E(y_i|\mathbf{x}_i) = Var(y_i|\mathbf{x}_i) = \exp(\mathbf{x}_i^T \boldsymbol{\beta}).$$

The Poisson regression model is therefore a nonlinear model, and estimation of the unknown parameters is done using maximum likelihood methods.

The LOGISTIC Procedure

Model Information	
Data Set	WORK.TEST
Response Variable	y
Number of Response Levels	2
Model	binary probit
Optimization Technique	Fisher's scoring

Number of Observations Read	4877
Number of Observations Used	4877

Response Profile		
Ordered Value	y	Total Frequency
1	0	1542
2	1	3335

Probability modeled is y=1.

Model Convergence Status
Convergence criterion (GCONV=1E-8) satisfied.

Model Fit Statistics		
Criterion	Intercept Only	Intercept and Covariates
AIC	6088.056	5788.142
SC	6094.548	5917.987
-2 Log L	6086.056	5748.142

OUTPUT 10.4. Probit model output for the unemployment data.

Testing Global Null Hypothesis: BETA=0			
Test	Chi-Square	DF	Pr > ChiSq
Likelihood Ratio	337.9143	19	<0.0001
Score	326.3187	19	<0.0001
Wald	316.7543	19	<0.0001

Analysis of Maximum Likelihood Estimates					
Parameter	DF	Estimate	Standard Error	Wald Chi-Square	Pr > ChiSq
Intercept	1	-1.6999	0.3630	21.9367	<0.0001
rr	1	1.8633	1.1293	2.7221	0.0990
rr2	1	-2.9801	1.4119	4.4549	0.0348
age	1	0.0422	0.0143	8.6975	0.0032
age2	1	-0.0377	0.0181	4.3397	0.0372
tenure	1	0.0177	0.00385	21.1488	<0.0001
slack	1	0.3755	0.0424	78.4751	<0.0001
abol	1	-0.0223	0.0719	0.0964	0.7562
seasonal	1	0.1612	0.1041	2.3987	0.1214
head	1	-0.1248	0.0491	6.4656	0.0110
married	1	0.1455	0.0478	9.2568	0.0023
dkids	1	-0.0966	0.0518	3.4700	0.0625
dykids	1	0.1236	0.0586	4.4437	0.0350
smsa	1	-0.1002	0.0418	5.7290	0.0167
nwhite	1	0.0518	0.0558	0.8599	0.3538
yrdispl	1	-0.0385	0.00905	18.0755	<0.0001
school12	1	-0.0416	0.0497	0.6985	0.4033
male	1	-0.1067	0.0527	4.0947	0.0430
stateur	1	0.0568	0.00943	36.2915	<0.0001
statemb	1	0.00364	0.000606	36.0192	<0.0001

Association of Predicted Probabilities and Observed Responses			
Percent Concordant	65.7	Somers' D	0.318
Percent Discordant	33.9	Gamma	0.319
Percent Tied	0.4	Tau-a	0.137
Pairs	5142570	c	0.659

OUTPUT 10.4. (*Continued*)

10.3.1 Interpretation of the Parameters

As was the case with the parameters in the Logit and Probit models, the parameters in the Poisson model are interpreted by calculating the marginal effects with respect to the explanatory variables. That is,

$$\frac{\partial E(y_i \mid \mathbf{x}_i)}{\partial \mathbf{x}} = \lambda_i \boldsymbol{\beta}$$

$$= \exp(\mathbf{x}_i^T \boldsymbol{\beta}) \boldsymbol{\beta},$$

which implies that the interpretation of the model depends on both $\boldsymbol{\beta}$ and the explanatory variables.

10.3.2 Maximum Likelihood Estimation

Maximum likelihood estimation can be used to estimate the parameters in a Poisson regression model. A brief description of the steps is provided below.

Assume that we have a random sample $(y_i, \mathbf{x}_i)$, $i = 1, \ldots, n$, from a Poisson distribution with mean λ_i. The joint likelihood function for the n observations is given by

$$L = \prod_{i=1}^{n} \frac{\exp(-\lambda_i)\lambda_i^{y_i}}{y_i!}.$$

Taking the log of the likelihood function and simplifying gives

$$\ln(L) = \sum_{i=1}^{n} [-\lambda_i + y_i \mathbf{x}_i^T \boldsymbol{\beta} - \log(y_i!)].$$

The first-order condition involves taking the derivative of the log-likelihood with respect to $\boldsymbol{\beta}$ and setting it equal to $\mathbf{0}$. That is,

$$S(\boldsymbol{\beta}) = \frac{\partial \log(L)}{\partial \boldsymbol{\beta}} = \sum_{i=1}^{n} [y_i - \lambda_i] \mathbf{x}_i = \mathbf{0}$$

$$= \sum_{i=1}^{n} [y_i - \exp(-\mathbf{x}_i^T \boldsymbol{\beta})] \mathbf{x}_i = \mathbf{0}.$$

The Hessian is given by

$$\frac{\partial^2 \log(L)}{\partial \boldsymbol{\beta} \partial \boldsymbol{\beta}^T} = -\sum_{i=1}^{n} \exp(-\mathbf{x}_i^T \boldsymbol{\beta}) \mathbf{x}_i \mathbf{x}_i^T.$$

The Newton–Raphson method can be used to estimate $\boldsymbol{\beta}$. The asymptotic variance–covariance matrix of the parameters is given by

$$\left[\sum_{i=1}^{n} \hat{\lambda}_i \mathbf{x}_i \mathbf{x}_i^T \right]^{-1}.$$

We will now show how to conduct Poisson regression in SAS. We will make use of Cincera's (1997) patent data, which was also analyzed by Verbeek (2004, pp. 215–217). The data consist of 181 international manufacturing firms. As described by the author, for each firm, their annual expenditures on research and development (R&D), the industrial sector it operates in, the country of its registered office, and the total number of patent applications for a number of consecutive years is recorded. The variable names in the model and their descriptions are summarized below:

P91=The number of patents filed in the year 1991,
LR91=The research and development expenses in 1991,
AEROSP=An indicator variable that is 1 if the company is in the aerospace industry and 0 otherwise,
CHEMIST=An indicator variable that is 1 if the company is in the chemical industry and 0 otherwise,
COMPUTER=An indicator variable that is 1 if the company is in the computer industry and 0 otherwise,
MACHINES=An indicator variable that is 1 if the company is in the heavy machine manufacturing industry and 0 otherwise,
VEHICLES=An indicator variable that is 1 if the company is in the auto industry and 0 otherwise,
US=An indicator that is 1 if the company is in US and 0 otherwise,
JAPAN=An indicator variable if the company is in Japan and 0 otherwise.
We will use Proc Genmod to conduct the analysis. The following statements can be used to fit the Poisson regression model.

Proc Genmod is very useful for fitting models that belong to the class of Generalized Linear Models where the distributions can be any member of the exponential family of distributions. The procedure can therefore be used for OLS, Logit, and Probit models. The advantage of this procedure over Proc Logistic is that interaction terms can be incorporated directly into the model statement whereas in Proc Logistic the interaction terms have to be added in the data step module of the code.

The GENMOD Procedure

Model Information		
Data Set	WORK.PATENT	
Distribution	Poisson	
Link Function	Log	
Dependent Variable	P91	P91

Number of Observations Read	181
Number of Observations Used	181

Criteria For Assessing Goodness Of Fit			
Criterion	DF	Value	Value/DF
Deviance	172	9081.9013	52.8018
Scaled Deviance	172	9081.9013	52.8018
Pearson Chi-Square	172	10391.9101	60.4181
Scaled Pearson X2	172	10391.9101	60.4181
Log Likelihood		54225.8240	

Algorithm converged.

Analysis Of Parameter Estimates				Wald 95% Confidence Limits			
Parameter	DF	Estimate	Standard Error			Chi-Square	Pr > ChiSq
Intercept	1	−0.8737	0.0659	−1.0028	−0.7446	175.94	<0.0001
LR91	1	0.8545	0.0084	0.8381	0.8710	10381.6	<0.0001
AEROSP	1	−1.4219	0.0956	−1.6093	−1.2344	221.00	<0.0001
CHEMIST	1	0.6363	0.0255	0.5862	0.6863	621.25	<0.0001
COMPUTER	1	0.5953	0.0233	0.5496	0.6411	650.70	<0.0001
MACHINES	1	0.6890	0.0383	0.6138	0.7641	322.76	<0.0001

Analysis Of Parameter Estimates				Wald 95% Confidence Limits			
Parameter	DF	Estimate	Standard Error			Chi-Square	Pr > ChiSq
VEHICLES	1	−1.5297	0.0419	−1.6117	−1.4476	1335.01	<0.0001
JAPAN	1	0.2222	0.0275	0.1683	0.2761	65.29	<0.0001
US	1	−0.2995	0.0253	−0.3491	−0.2499	140.14	<0.0001
Scale	0	1.0000	0.0000	1.0000	1.0000		

The scale parameter was held fixed.

OUTPUT 10.5. Poisson regression of the patent data.

The main statement here is the one that specifies the distribution and the log link function. The analysis output is given in Output 10.5.

```
proc genmod data=patent;
    model p91=lr91 aerosp chemist computer machines vehicles
    japan us/dist=poisson link=log;
run;
```

We will now provide details about the output produced by Proc Genmod.

The first table of the output gives the model information including the distribution specified, the link function, and the dependent variable. The log link function is used by default when the Poisson distribution is specified. The "link" function references the function that is used to relate the conditional expectation and the explanatory variables. For the Poisson model, recall that the conditional mean was given by

$$E(y_i|\mathbf{x}_i) = \exp(\mathbf{x}_i^T\boldsymbol{\beta}),$$
$$\ln(E(y_i|\mathbf{x}_i)) = \mathbf{x}_i^T\boldsymbol{\beta}$$

so that the link function that is used is the LOG function.

The next table gives various statistics for assessing the model fit. Small values for the goodness of fit statistics are desired. Large ratios may indicate that the model is misspecified or that the model suffers from overdispersion.

The final table gives the parameter estimates, along with their standard errors, confidence intervals, and chi-squared test statistic value. Recall that in Poisson regression we model $E(y_i|\mathbf{x}_i)$ versus $\exp(\mathbf{x}_i^T\boldsymbol{\beta})$. Therefore, the regression coefficient can be interpreted by first exponentiating the parameter estimates. For example, the parameter estimate for the variable *aerospace* is -1.4219 giving $\exp(-1.4219) = 0.241$. The coefficient of this variable compares the number of patent applications received for the aerospace and nonaerospace industries. The value 0.241 is the predicted ratio of the number of patents filed for the aerospace companies to the numbers filed by the nonaerospace companies. Therefore, based on the estimated model, the number of patents filed by the aerospace companies is predicted to be lower than the number filed by the nonaerospace companies assuming all other explanatory variables are held constant.

11

DURATION ANALYSIS

11.1 INTRODUCTION

Often, we are interested in modeling the duration or time between two events. For example, we may be interested in the time between the start and the end of a strike, the time it takes an unemployed person to find a job, the time to failure of a machine component, or the recidivism duration of an ex-convict. In each of these examples, the data set consists of a response variable that records the time or duration between the events of interest. Due to the nature of the study, the data set usually consists of a mixture of complete and censored observations. For example, in the recidivism study conducted by Chung, Schmidt, and Witte (1991), a random sample of 1,445 former inmates released from prison between July 1, 1977 and June 30, 1978 was collected using April 1, 1984 as the end point of the study. The study found that 552 former inmates were arrested again for different violations. Their duration or time response was therefore recorded. Duration measurements for the remaining 893 individuals were not recorded and were censored. That is, at the time the study concluded, these individuals had not been arrested for any violation since their release. Note that the censored times for these individuals will vary due to the staggered entry of the subjects into the study.

As another example, consider an auto and home insurance company that may be interested in the time between a policy holder's start and termination date. The company's objective may be to understand the attrition behavior of policy holders who cancel their policies and move to a competitor company. As with the recidivism study, the time window for the study is usually predefined. Therefore, there may be several policy holders for whom the event of interest (attrition) is not observed. That is, a few policy holders may attrite after the conclusion of the study; their time to attrite response is therefore censored.

The use of duration models, although fairly recent in economics, is well established in engineering and biomedical fields. For example, reliability engineers are often interested in the time it takes for a machine component to fail. They may use this information to optimize preventive maintenance strategies on the machine. Pharmaceutical companies conduct clinical trials where patients are administered a new drug and then are followed for a predefined length of time in order to evaluate the drug's effectiveness. In both these fields, the event of interest is the "time to failure," and the data are comprised of censored observations as well as observations for which the actual duration of the event of interest is observed.

11.2 FAILURE TIMES AND CENSORING

There are three conditions required to determine failure times accurately in any duration analysis study. First, the time window of the study must be clearly defined. That is, the origin and the termination points must be defined. In most duration analysis studies, the subjects will have a staggered entry. Therefore, the starting point may not be the same for all subjects. However, there are

Applied Econometrics Using the SAS® System, by Vivek B. Ajmani
Copyright © 2009 John Wiley & Sons, Inc.

instances where the starting point is the same for all subjects. Kennan (1984) gives an example of a major strike affecting a certain type of industry where most of the subjects have a common starting point. In the recidivism study, the time window for the study was the release period from July 1, 1977 to June 30, 1978 with a study termination date of April 1, 1984. Notice that the observation period ranged from 70 to 81 months. Therefore, the censored times are also in the range of 70–81 months. In an attrition modeling study I was involved with, the time window of the study was set from January 1, 2006 to July 31, 2006. The study was terminated on June 1, 2007. The time lag between the time window and the observation point or termination point is to allow various factors (known or unknown) to influence the subject into either taking an action (recidivism or attrition) or not taking an action.

Second, the measurement scale must be clearly understood. In the recidivism and attrition studies, the response variable is the number of days to the event of interest. In an example to study automobile reliability, the response variable may be the number of miles recorded before the car breaks down. In a reliability study on light bulbs, we are interested in the number of hours until the light bulb fails.

Third, the meaning of "failure" or "success" must be clearly understood. In the recidivism example, "failure" or "success" occurs when a former inmate is arrested again. In the attrition study, a "failure" or "success" occurs when the policy holder attrites. In an employment study a "failure" or "success" is observed when the unemployed person is employed again.

To define censoring more formally, assume that we have n subjects in the duration analysis study. In the absence of censoring, the ith subject has a duration time denoted by t_i. That is, the event of interest was observed for this subject in the study's time window. Assume that the duration analysis study window for the ith subject has length c_i. Here, without loss of generality, we are assuming that the subjects enter the study in a staggered fashion. If the event of interest was not observed for the ith subject when the study was terminated, then the duration time recorded for this subject will be c_i. The duration time can therefore be denoted by $y_i = \min(t_i, c_i)$. In the recidivism and attrition examples, the data are referred to as right censored data since for the censored observations the duration times are assumed to occur after the study termination date.

11.3 THE SURVIVAL AND HAZARD FUNCTIONS

We now turn our attention to understanding the survival and hazard rate functions for the data collected from duration analysis studies. First, assume that the random variable T (the duration time) has a continuous probability distribution $f(t)$. The cumulative distribution function (CDF) $F(t)$ is therefore given by

$$F(t) = P(T \leq t) = \int_0^t f(\tau)d\tau.$$

The CDF measures the probability that a subject will have a duration of time less than or equal to t. In the recidivism study, we could interpret this as the probability that a former inmate will be arrested again by time t. In the attrition study, this could be interpreted as the probability that a policy holder will attrite by time t.

We are usually interested in the probability that the duration for a subject will exceed t. That is, we are typically interested in the survival function. The survival function is defined as the probability that a subject will survive past time t. That is,

$$S(t) = P(T \geq t) = 1 - P(T > t) = 1 - F(t) = = \int_t^\infty f(\tau)d\tau.$$

We are also interested in the instantaneous failure rate or the hazard rate. To see the form of the hazard function, first define the probability that a subject will survive a small time interval $[t, t + \Delta t]$ given that he or she has survived past t as $P(t \leq T < t + \Delta t | T \geq t)$.

The hazard function is given by

$$\lambda(t) = \lim_{\Delta t \to 0} \frac{P(t \leq T < t + \Delta t | T \geq t)}{\Delta t}.$$

It can be shown that the hazard function is a ratio of the probability density function to the survival function. That is,

$$\lambda(t) = \frac{f(t)}{S(t)}.$$

To see this, first notice that by using the fact $P(A|B) = P(A \cap B)/P(B)$, we can write the numerator of the hazard function as

$$P(t \leq T < t + \Delta t | T \geq t) = \frac{P(t \leq T < t + \Delta t)}{P(T \geq t)} = \frac{F(t + \Delta t) - F(t)}{1 - F(t)} = \frac{F(t + \Delta t) - F(t)}{S(t)}.$$

Next, realize that by definition

$$\lambda(t) = \lim_{\Delta t \to 0} \frac{F(t + \Delta t) - F(t)}{\Delta t} = \frac{\partial}{\partial t} F(t).$$

Again, by definition

$$\frac{\partial}{\partial t} F(t) = f(t)$$

and we have our result.

A few more results relating the hazard, survival, and probability density functions are of interest follow. First note that by definition, the survival function can be written as

$$f(t) = -\frac{\partial}{\partial t} S(t)$$

so that

$$\lambda(t) = -\frac{1}{S(t)} \frac{\partial}{\partial t} S(t) = -\frac{\partial}{\partial t} \log[S(t)].$$

The cumulative hazard rate function $\Lambda(t)$ can be written as

$$\Lambda(t) = \int_0^t \lambda(\tau) d\tau = \int_0^t -\frac{\partial}{\partial \tau} \log[S(\tau)] dt = \int_0^t -\frac{\partial}{\partial \tau} \log[1 - F(\tau)] dt.$$

It is trivial to show that the above expression simplifies to

$$\Lambda(t) = -\log[1 - F(t)].$$

Upon further simplification, we get

$$F(t) = 1 - e^{-\Lambda(t)}.$$

Differentiating both sides of the above expression gives

$$f(t) = \frac{\partial}{\partial t} \Lambda(t) = e^{-\Lambda(t)} = \lambda(t) e^{-\Lambda(t)}.$$

Therefore, the probability distribution function can be completely characterized by the knowledge of the hazard function. It is for this reason that the analysis of duration data is based on understanding the hazard function.

Defining the hazard rate as $\partial \lambda(t) / \partial t$ we see that

1. If $\partial \lambda / \partial t = 0$, then the hazard rate is constant and independent of time. Using the attrition example, this implies that the probability of a policy holder's attrition is constant over the life of the policy holder's tenure with the company.

2. If $\partial\lambda/\partial t > 0$, then the hazard rate is increasing with time. As an example, Greene (2003) discusses an unemployment study where the time between a subject's unemployment and employment is modeled. The longer the subject is unemployed, the more likely the subject is to be employed. As discussed by the author, the longer the unemployment lasts, the more is the likelihood that the subject will take up any job.

3. If $\partial\lambda/\partial t < 0$, then the hazard rate is decreasing with time. Using the attrition example, the longer a policy holder is with the company, the less likely he or she is to attrite. Similarly, it can be hypothesized that the longer a former convict is free, the less likely he or she is to be arrested.

   Readers familiar with the "bathtub" curve in reliability analysis will recall that the first part of the curve represents ("infant mortality") a decreasing hazard rate, the second part of the curve represents constant hazard rate, and the last part of the curve represents ("wear-out" failures) increasing hazard rate.

We will use the RECID data from Chung, Schmidt, and Witte (1991) to show how SAS can be used to plot the survival and hazard functions. The data set consists of the duration time for each ex-convict along with information on race, alcohol problems, number of prior convictions, age, and soon. The complete list of variable description can be downloaded from Jeffrey Wooldridge's web site at Michigan State University.

The Proc Lifetest can be used to conduct preliminary analysis of the duration data. It is useful for plotting the survival and hazard functions and also for comparing survival and hazard functions across groups. Two methods of estimating the survival function are included: the Kaplan Meier method by default and the Life Table method (use option `method=life`). See Allison (1995) for more details on these procedures. At the minimum, the following statements should be used:

```
proc lifetest data=duration;
    time l_durat2*censored(1);
run;
```

Here, `duration` is the temporary SAS data set, and the command `time` is used to generate the survival function table. The response variable is `l_durat` and the censoring variable is `censored`. A value of 1 for `censored` denotes a censored observation. The analysis creates a survival table using the Kaplan Meier method. A partial output is given in Output 11.1.

The first part of the output contains the survival table while the second part of the output contains basic summary statistics. The first table lists the duration times at which failures occurred, the survival estimates, the failure estimates (1-survival estimates), and the survival standard error that is given by

$$\sqrt{\frac{S(t_i)(1-S(t_i))}{n}},$$

where $S(t_i)$ and $1-S(t_i)$ are the survival and failure estimates, respectively. The total number of subjects is given denoted by n. The final two columns give the number of subjects who failed and the number left.

Notice that there were eight failures in the first month. The survival estimate is therefore given by

$$n_{\text{left}}/n_{\text{total}} \Rightarrow 1437/1445 = 0.9945.$$

The failure estimate is therefore 1–0.9945 or 0.0055. The standard error is calculated as follows

$$\sqrt{\frac{0.9945 \times 0.0055}{1445}} = 0.00195.$$

The last three tables give the summary statistics. Note that the 25th percentile is estimated to be at 27 months. This implies that the survival rate estimate is 75% at 27 months, and 25% of the ex-convicts are expected to be arrested again by the 27th month.

The Kaplan Meier method is recommended for small data sets and for highly accurate measurements. When the data set is large and the accuracy of the duration times are in question then the life table method is recommended. Adding the

The LIFETEST Procedure

l_durat2	Survival	Failure	Survival Standard Error	Number Failed	Number Left
0.0000	1.0000	0	0	0	1445
1.0000	.	.	.	1	1444
1.0000	.	.	.	2	1443
1.0000	.	.	.	3	1442
1.0000	.	.	.	4	1441
1.0000	.	.	.	5	1440
1.0000	.	.	.	6	1439
1.0000	.	.	.	7	1438
1.0000	0.9945	0.00554	0.00195	8	1437
2.0000	.	.	.	9	1436
2.0000	.	.	.	10	1435
2.0000	.	.	.	11	1434
2.0000	.	.	.	12	1433
2.0000	.	.	.	13	1432
2.0000	.	.	.	14	1431
2.0000	.	.	.	15	1430
2.0000	.	.	.	16	1429
2.0000	.	.	.	17	1428
2.0000	.	.	.	18	1427
2.0000	.	.	.	19	1426
2.0000	.	.	.	20	1425
2.0000	.	.	.	21	1424
2.0000	.	.	.	22	1423
2.0000	0.9841	0.0159	0.00329	23	1422
3.0000	.	.	.	24	1421
3.0000	.	.	.	25	1420
3.0000	.	.	.	26	1419
3.0000	.	.	.	27	1418
3.0000	.	.	.	28	1417
3.0000	.	.	.	29	1416
3.0000	.	.	.	30	1415
3.0000	.	.	.	31	1414
3.0000	.	.	.	32	1413
3.0000	.	.	.	33	1412
3.0000	.	.	.	34	1411
3.0000	.	.	.	35	1410
3.0000	.	.	.	36	1409
3.0000	0.9744	0.0256	0.00416	37	1408

Table title: Product-Limit Survival Estimates

OUTPUT 11.1. Proc Lifetest analysis of the RECID data using the Kaplan Meier method.

`method=life` option to the first line of the Proc Life procedure statements invokes the analysis using the life table technique. The output is given in Output 11.2.

The analysis takes the range of duration data and divides them into intervals. The number of intervals can be easily adjusted by using the "width" option. The first table gives information on the number of ex-convicts arrested in the given time intervals, the effective (or entering) sample sizes for each interval along with the estimate of the conditional probability of failure, its standard error, and the survival estimate. The conditional probabilities are calculated by simply dividing the number of arrests by the effective sample size for each interval. The survival estimate is simply 1—the conditional probability of failure. The standard errors are calculated using the formula provided earlier; however, the sample size used is the effective sample size of the interval.

The second table gives the failure rate and the survival estimate standard error. See Lee (1992) for the formulas used in the calculations.

The LIFETEST Procedure

l_durat2		Survival	Failure	Survival Standard Error	Number Failed	Number Left
Product-Limit Survival Estimates						
81.0000	*	.	.	.	552	1
81.0000	*	.	.	.	552	0

The marked survival times are censored observations.

Summary Statistics for Time Variable l_durat2

		95% Confidence Interval	
Percent	Point Estimate	[Lower	Upper)
Quartile Estimates			
75	.	.	.
50	.	.	.
25	27.0000	24.0000	31.0000

Mean	Standard Error
56.7032	0.7402

The mean survival time and its standard error were underestimated because the largest observation was censored and the estimation was restricted to the largest event time.

Total	Failed	Censored	Percent Censored
Summary of the Number of Censored and Uncensored Values			
1445	552	893	61.80

OUTPUT 11.1. (*Continued*)

The last table gives the probability of failure (PDF), the hazard estimate, and their standard errors for each interval. The PDF is calculated by dividing the number of subjects arrested in each interval by the interval width and the total number of subjects in the study. For example, the PDF for the interval [40, 50] is calculated as follows:

$$\frac{43}{10 \times 1445} = 0.00298.$$

The hazard estimate is calculated as follows (Lee, 1992):

$$\frac{n_a}{w \times (n_{in} - n_a/2)}.$$

Here, n_a represents the number arrested in the time interval, w is the width of the interval, and n_m is the number of subjects coming into the interval. Using the same interval as above, we get

$$\frac{43}{10 \times (1004 - 43/2)} = 0.004377.$$

The number entering into the interval is calculated by subtracting the cumulative arrests up to the interval of interest from the total number of subjects in the study. See Lee (1992) for the formulas of the standard errors of both the PDF and the hazard estimate.

The LIFETEST Procedure

Life Table Survival Estimates							
Interval							
[Lower,	Upper)	Number Failed	Number Censored	Effective Sample Size	Conditional Probability of Failure	Conditional Probability Standard Error	Survival
0	10	136	0	1445.0	0.0941	0.00768	1.0000
10	20	146	0	1309.0	0.1115	0.00870	0.9059
20	30	105	0	1163.0	0.0903	0.00840	0.8048
30	40	54	0	1058.0	0.0510	0.00677	0.7322
40	50	43	0	1004.0	0.0428	0.00639	0.6948
50	60	37	0	961.0	0.0385	0.00621	0.6651
60	70	23	0	924.0	0.0249	0.00513	0.6394
70	80	8	776	513.0	0.0156	0.00547	0.6235
80	90	0	117	58.5	0	0	0.6138

The LIFETEST Procedure

Interval					
[Lower,	Upper)	Failure	Survival Standard Error	Median Residual Lifetime	Median Standard Error
0	10	0	0	.	.
10	20	0.0941	0.00768	.	.
20	30	0.1952	0.0104	.	.
30	40	0.2678	0.0116	.	.
40	50	0.3052	0.0121	.	.
50	60	0.3349	0.0124	.	.
60	70	0.3606	0.0126	.	.
70	80	0.3765	0.0127	.	.
80	90	0.3862	0.0130	.	.

The LIFETEST Procedure

Interval		Evaluated at the Midpoint of the Interval			
[Lower,	Upper)	PDF	PDF Standard Error	Hazard	Hazard Standard Error
0	10	0.00941	0.000768	0.009877	0.000846
10	20	0.0101	0.000793	0.011812	0.000976
20	30	0.00727	0.000683	0.009455	0.000922
30	40	0.00374	0.000499	0.005238	0.000713
40	50	0.00298	0.000447	0.004377	0.000667
50	60	0.00256	0.000416	0.003926	0.000645
60	70	0.00159	0.000329	0.002521	0.000526
70	80	0.000972	0.000342	0.001572	0.000556
80	90	0	.	0	.

Summary of the Number of Censored and Uncensored Values			
Total	Failed	Censored	Percent Censored
1445	552	893	61.80

OUTPUT 11.2. Proc Lifetest analysis of the RECID data using the Life Table method.

Proc Lifetest can be used to plot the survival and hazard functions by using the "`plot=`" function as shown in the following statements.

```
proc lifetest data=duration method=life plot=(s h);
     time l_durat2*censored(1);
run;
```

In the above *s* denotes the survival function plot and *h* denotes the hazard function plot. Note that only the survival function is available when using the Kaplan Meier method. Figure 11.1 contains the Kaplan Meier survival function plot while Figures 11.2 and 11.3 contain the lifetime survival and hazard plots.

As an example of how to interpret the output, notice that from the lifetime survival function plot, we see that the survival probability at 20 months is a little over 75%. In terms of the example used, this indicates that the probability an ex-convict will be arrested within 20 months of release is under 25%. The hazard function plot indicates that the hazard rate is the highest in the 10–20 month window after release and that the risk rapidly decreases with time.

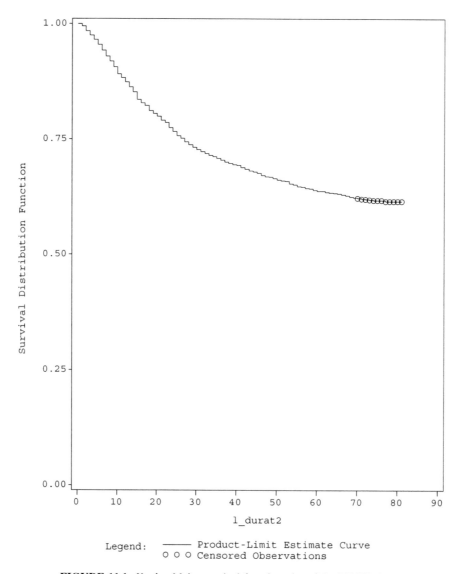

FIGURE 11.1. Kaplan Meier survival function plot of the RECID data.

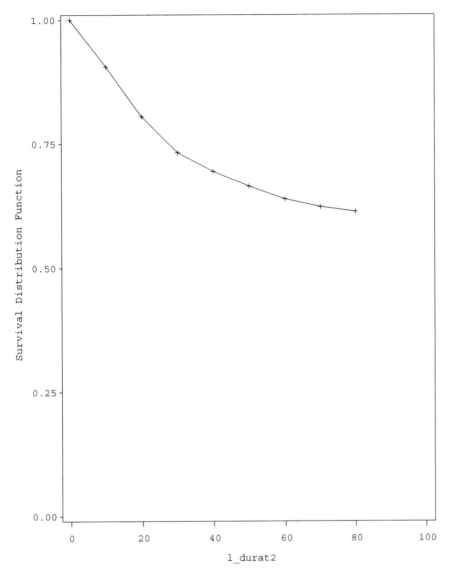

FIGURE 11.2. Lifetime survival function plot of the RECID data.

Using the option $plot=(ls)$ produces a plot of the cumulative hazard function versus time. This plot can be used to check the rate at which the hazard is changing. The reader is asked to verify that the cumulative hazard plot for the RECID data is increasing at a gradual rate.

Proc Lifetime can also be used to compare two or more groups with respect to their survival and hazard functions. Consider testing the survival functions of married ex-convicts to unmarried ex-convicts. The hypothesis tested is as follows:

$$H_0 : S_m(t) = S_{um}(t),$$
$$H_1 : S_m(t) > S_{um}(t).$$

Here, $S_m(t)$ and $S_{um}(t)$ are the survival functions of the married and unmarried ex-convicts. The following statements can be used to conduct the test in SAS. The analysis results are given in Output 11.3.

```
proc lifetest data=duration method=life plots=(s);
     time l_durat2*censored(1);
     strata l_marry;
run;
```

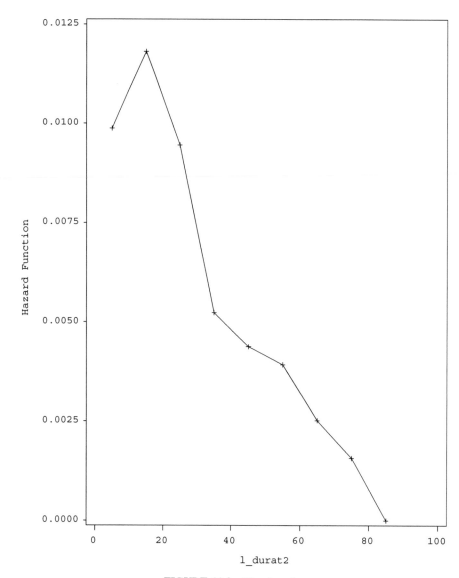

FIGURE 11.2. (*Continued*).

The first part of the output consists of the life tables for the unmarried ex-convicts. This is followed by the life tables for the married ex-convicts. The survival function plot indicates that the survivorship probability is higher for married subjects compared to unmarried subjects. The last part of the output contains test statistics from the Log-Rank, Wilcoxon, and the Likelihood-Ratio tests. See Lee (1992, pp. 104–122) for details on these tests. All three test statistics are highly significant indicating that the two survival functions are different. This analysis can easily be extended to more than two groups.

Proc Lifetest is usually used in the preliminary stages to understand the data and to isolate factors that appear to have an impact on the response of interest. In the later sections, we will introduce Proc Lifereg and Proc Phreg that are used to conduct regression analysis on data with censored observations. We now move on to discuss the different distributions that may be used to model duration data.

11.4 COMMONLY USED DISTRIBUTION FUNCTIONS IN DURATION ANALYSIS

There are three distribution functions commonly used in duration analysis studies.

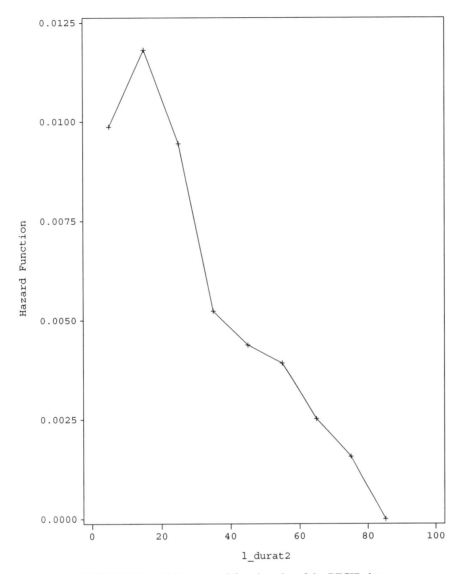

FIGURE 11.3. Lifetime hazard function plot of the RECID data.

11.4.1 The Exponential Distribution

The general formula for the probability density function of the exponential distribution is $f(t) = \theta^{-1} \exp(-t/\theta)\, \theta > 0$, where θ is the scale parameter. The cumulative density, survival, and hazard functions are given by $f(t) = 1 - \exp(-t/\theta), S(t) = \exp(-t/\theta)$, and $\lambda(t) = 1/\theta$, respectively.

The hazard and survival function plots for the exponential distribution are given in Figures 11.4 and 11.5 (*Source*: NIST). Notice that the hazard function is constant and independent of time. The survival function graph indicates the classic exponential decay behavior.

11.4.2 The Weibull Distribution

The general formula for the probability density function of the two-parameter Weibull distribution is $f(t) = \gamma \alpha^\gamma t^{\gamma-1} \exp(-(\alpha t))^\gamma \alpha$, $\gamma > 0$, where γ is the shape parameter and α is the scale parameter (Casella and Berger, 1990). The cumulative density, survival, and hazard functions are given by $F(t) = 1 - \exp(-(\alpha t)^\gamma)$, $S(t) = \exp(-(\alpha t)^\gamma)$, and $\lambda(t) = \gamma \alpha^\gamma t^{\gamma-1}\ \lambda(t) = \gamma \alpha^\gamma t^{\gamma-1}\alpha, \gamma > 0$. The hazard and survival plots for different values of the shape parameters are given in Figures 11.6 and 11.7 (Source: NIST). Notice that the Weibull distribution offers more flexibility in modeling the hazard function. The dependency of the hazard function on time

The LIFETEST Procedure

Stratum 1: l_marry = 0						
Life Table Survival Estimates						
Interval						
[Lower,	Upper)	Number Failed	Number Censored	Effective Sample Size	Conditional Probability of Failure	Conditional Probability Standard Error	Survival
0	10	118	0	1076.0	0.1097	0.00953	1.0000
10	20	115	0	958.0	0.1200	0.0105	0.8903
20	30	76	0	843.0	0.0902	0.00986	0.7835
30	40	47	0	767.0	0.0613	0.00866	0.7128
40	50	30	0	720.0	0.0417	0.00745	0.6691
50	60	30	0	690.0	0.0435	0.00776	0.6413
60	70	13	0	660.0	0.0197	0.00541	0.6134
70	80	7	552	371.0	0.0189	0.00706	0.6013
80	90	0	88	44.0	0	0	0.5900

Interval					
[Lower,	Upper)	Failure	Survival Standard Error	Median Residual Lifetime	Median Standard Error
0	10	0	0	.	.
10	20	0.1097	0.00953	.	.
20	30	0.2165	0.0126	.	.
30	40	0.2872	0.0138	.	.
40	50	0.3309	0.0143	.	.
50	60	0.3587	0.0146	.	.
60	70	0.3866	0.0148	.	.
70	80	0.3987	0.0149	.	.
80	90	0.4100	0.0152	.	.

Interval		Evaluated at the Midpoint of the Interval			
[Lower,	Upper)	PDF	PDF Standard Error	Hazard	Hazard Standard Error
0	10	0.0110	0.000953	0.011603	0.001066
10	20	0.0107	0.000942	0.012771	0.001188
20	30	0.00706	0.000781	0.009441	0.001082
30	40	0.00437	0.000623	0.006321	0.000922
40	50	0.00279	0.000502	0.004255	0.000777
50	60	0.00279	0.000502	0.004444	0.000811
60	70	0.00121	0.000333	0.001989	0.000552
70	80	0.00113	0.000426	0.001905	0.00072
80	90	0	.	0	.

OUTPUT 11.3. Testing survival functions of married ex-convicts to unmarried ex-convicts in the RECID data set.

The LIFETEST Procedure

	Stratum 2: l_marry = 1
	Life Table Survival Estimates
	Interval						
[Lower,	Upper)	Number Failed	Number Censored	Effective Sample Size	Conditional Probability of Failure	Conditional Probability Standard Error	Survival
0	10	18	0	369.0	0.0488	0.0112	1.0000
10	20	31	0	351.0	0.0883	0.0151	0.9512
20	30	29	0	320.0	0.0906	0.0160	0.8672
30	40	7	0	291.0	0.0241	0.00898	0.7886
40	50	13	0	284.0	0.0458	0.0124	0.7696
50	60	7	0	271.0	0.0258	0.00964	0.7344
60	70	10	0	264.0	0.0379	0.0117	0.7154
70	80	1	224	142.0	0.00704	0.00702	0.6883
80	90	0	29	14.5	0	0	0.6835

	Interval				
[Lower,	Upper)	Failure	Survival Standard Error	Median Residual Lifetime	Median Standard Error
0	10	0	0	.	.
10	20	0.0488	0.0112	.	.
20	30	0.1328	0.0177	.	.
30	40	0.2114	0.0213	.	.
40	50	0.2304	0.0219	.	.
50	60	0.2656	0.0230	.	.
60	70	0.2846	0.0235	.	.
70	80	0.3117	0.0241	.	.
80	90	0.3165	0.0244	.	.

	Interval	colspan="4"	Evaluated at the Midpoint of the Interval
[Lower,	Upper)	PDF	PDF Standard Error	Hazard	Hazard Standard Error
0	10	0.00488	0.00112	0.005	0.001178
10	20	0.00840	0.00144	0.00924	0.001658
20	30	0.00786	0.00140	0.009493	0.001761
30	40	0.00190	0.000710	0.002435	0.00092
40	50	0.00352	0.000960	0.004685	0.001299
50	60	0.00190	0.000710	0.002617	0.000989
60	70	0.00271	0.000845	0.003861	0.001221
70	80	0.000485	0.000483	0.000707	0.000707
80	90	0	.	0	.

	Summary of the Number of Censored and Uncensored Values
Stratum	l_marry	Total	Failed	Censored	Percent Censored
1	0	1076	436	640	59.48
2	1	369	116	253	68.56
Total		1445	552	893	61.80

OUTPUT 11.3. (*Continued*)

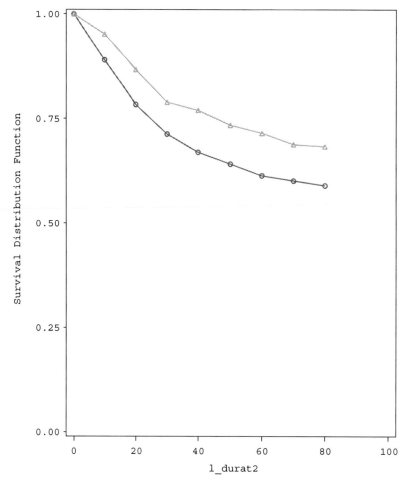

STRATA: ⊖–⊖–⊖ l_marry=0 △–△–△ l_marry=1

The LIFETEST Procedure

Testing Homogeneity of Survival Curves for l_durat2 over Strata

	Rank Statistics	
l_marry	Log-Rank	Wilcoxon
0	34.111	42523
1	−34.111	−42523

Covariance Matrix for the Log-Rank Statistics		
l_marry	0	1
0	108.307	−108.307
1	−108.307	108.307

Covariance Matrix for the Wilcoxon Statistics		
l_marry	0	1
0	1.5103E8	−1.51E8
1	−1.51E8	1.5103E8

Test of Equality over Strata			
Test	Chi-Square	DF	Pr > Chi-Square
Log-Rank	10.7432	1	0.0010
Wilcoxon	11.9728	1	0.0005
−2Log(LR)	12.2349	1	0.0005

OUTPUT 11.3. (*Continued*).

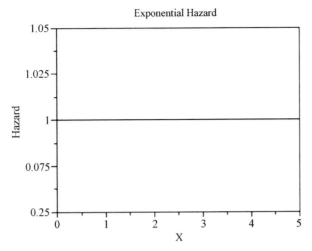

FIGURE 11.4. Hazard function of the exponential distribution.

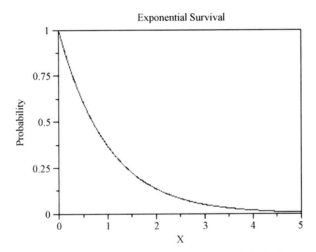

FIGURE 11.5. Survival function of the exponential distribution.

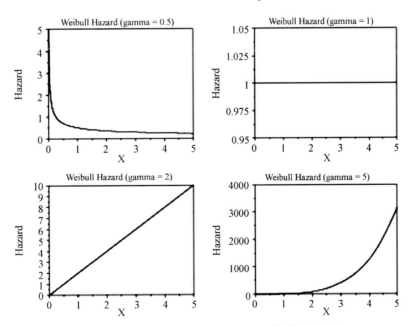

FIGURE 11.6. Hazard functions for the Weibull distribution.

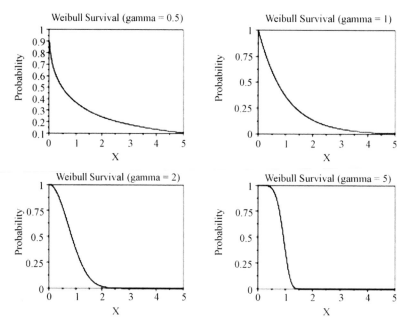

FIGURE 11.7. Survival functions for the Weibull distribution.

is negative for $\gamma < 1$ and positive for $\gamma > 1$. The exponential distribution is a special case of the Weibull distribution when $\gamma = 1$. The Weibull hazard rate is therefore constant when $\gamma = 1$.

11.4.3   The Lognormal Distribution

A variable T is lognormally distributed if $\ln(T)$ is normally distributed where ln refers to the natural logarithm function. The general formula for the probability density function of the lognormal distribution is (Casella and Berger, 1990)

$$f(t) = \frac{\exp^{[-\ln(t/m)^2/2\sigma^2]}}{\sigma\sqrt{2\pi t}} m, \qquad \sigma > 0,$$

where σ is the shape parameter and m is the scale parameter. The cumulative density function is given by

$$F(t) = \Phi\left[\frac{\ln(t)}{\sigma}\right], \qquad \sigma > 0,$$

and the survival and hazard functions are given by

$$S(t) = 1 - \Phi\left[\frac{\ln(t)}{\sigma}\right], \qquad \sigma > 0,$$

and

$$\lambda(t) = \frac{(1/t\sigma)\phi(\ln(t)/\sigma)}{\Phi\left(\frac{-\ln(t)}{\sigma}\right)}, \qquad \sigma > 0,$$

respectively.

The hazard and survival functions for different values of the shape parameters are given in Figures 11.8 and 11.9 (Source: NIST).

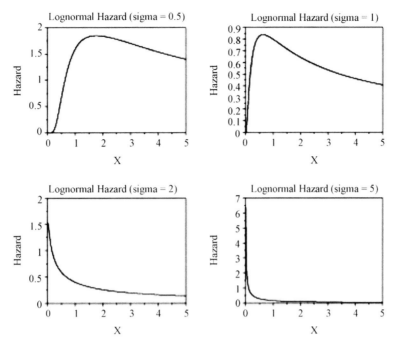

FIGURE 11.8. Hazard functions for the lognormal distribution.

Maximum likelihood estimation, which takes censoring into account, is used to estimate the parameters of these distributions. Details on the technique can be found in Meeker and Escobar (1998, pp. 153–159).

Proc Lifereg can be used to assess the goodness of fit of the distributions on the collected data. The resulting output also includes the maximum likelihood estimates of the parameters of the distributions. We will revisit this procedure in our discussion of regression analysis on duration data. For now, we will assess the goodness of fit of the distributions on the RECID data set. The

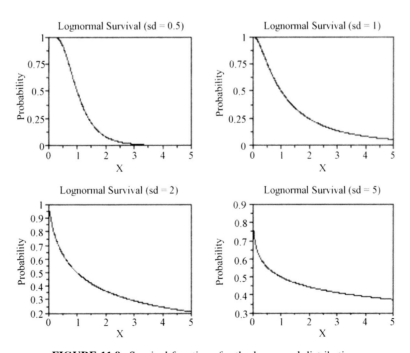

FIGURE 11.9. Survival functions for the lognormal distribution.

following statements can be used to fit a normal distribution to the duration response variable. Proc Lifereg invokes the procedure on the temporary SAS data set "duration." The model statement lists out the response and the censoring variable. A value of 1 indicates that the record is censored. The option "d" is used to specify the distribution. Changing the value of the option to exponential, Weibull, lnormal will give analysis results for the other distributions that were discussed in this section. The option "probplot" creates a probability plot while the option "inset" writes the parameter estimates on the probability plot. Note that there are several other options that can be used to enhance the graphs. See the *SAS Users Guide 9.2* from SAS Institute, Inc. for more information. Outputs 11.4–11.7 contain goodness of fit information as well as the probability plots for the various distributions. Output 11.4 indicates that the normal distribution does not fit the data well while Output 11.6 indicates that the log-normal distribution fit is the best.

```
Proc Lifereg data=duration;
    Model l_durat*censored(1)/d=normal;
    Probplot;
    Inset;
Run;
```

11.5   REGRESSION ANALYSIS WITH DURATION DATA

The discussion so far was limited to the estimation of parameters in the absence of exogenous variables. Going back to the recidivism study, there may be various factors a effecting a person's relapse into criminal behavior. For instance, a person's age, education, marital status, race, and number of prior convictions may all play a role in influencing the behavior of the ex-convict. This section deals with the introduction of regressors' analysis on duration data. There are two types of regression models that are

The LIFEREG Procedure

Model Information	
Data Set	WORK.DURATION
Dependent Variable	l_durat2
Censoring Variable	censored
Censoring Value(s)	1
Number of Observations	1445
Noncensored Values	552
Right Censored Values	893
Left Censored Values	0
Interval Censored Values	0
Name of Distribution	Normal
Log Likelihood	-3584.355095

Number of Observations Read	1445
Number of Observations Used	1445

Algorithm converged.

Analysis of Parameter Estimates							
Parameter	DF	Estimate	Standard Error	95% Confidence Limits		Chi-Square	Pr > ChiSq
Intercept	1	88.2733	2.3096	83.7466	92.7999	1460.81	<0.0001
Scale	1	60.1504	2.1312	56.1151	64.4758		

OUTPUT 11.4. Normal distribution fit for the RECID data.

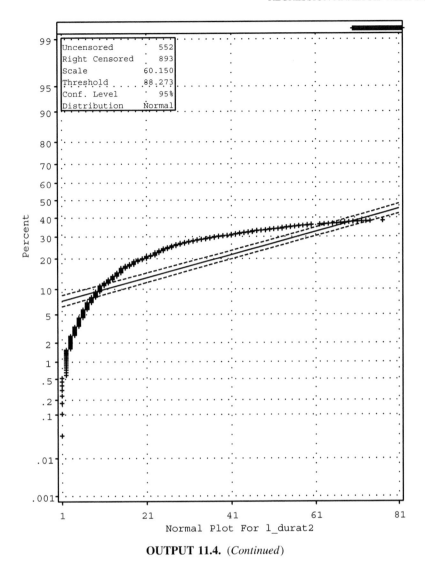

OUTPUT 11.4. (*Continued*)

used to model duration data—the parametric regression model and Cox proportional hazard regression model. Proc Lifereg can be used to fit the first set of models while Proc Phreg can be used to fit the second type of models.

To see the general form of these models, let t_i denote the duration time for the ith subject and let $\mathbf{x}_i$ denote the set of explanatory variables assumed to influence the duration time, t_i. The models fit by using Proc Lifereg are of the form

$$\ln(t_i) = \mathbf{x}_i^T\boldsymbol{\beta} + \sigma\varepsilon_i$$
$$t_i = \exp(\mathbf{x}_i^T\boldsymbol{\beta} + \sigma\varepsilon_i),$$

where ε_i is the disturbance with mean 0 and unit variance. The variance of disturbance is estimated via the parameter σ. These models are referred to as Accelerated Failure Time models with the covariates assumed to influence the failure rate. The choice of the distribution for ε_i leads to the different types of models. For example, if we choose ε_i to be normally distributed (or $\ln(t_i)$ to be normally distributed), then we have a log-normal model.

The covariates vector, $\mathbf{x}_i$, includes a constant term and a set of time-invariant regressors. Note that the regressor set may contain time-dependent regressors. However, Proc Lifereg cannot be used to accommodate these. Proc Phreg, on the other hand, can handle both time-invariant and time-dependent explanatory variables. Maximum likelihood methods are used to estimate the model parameters with the Newton–Raphson method used for optimization. Details on this can be found in Allison (1995), Lee (1992), and Meeker and Escobar (1998).

The LIFEREG Procedure

Model Information	
Data Set	WORK.DURATION
Dependent Variable	Log(l_durat2)
Censoring Variable	censored
Censoring Value(s)	1
Number of Observations	1445
Noncensored Values	552
Right Censored Values	893
Left Censored Values	0
Interval Censored Values	0
Name of Distribution	Exponential
Log Likelihood	-1739.894437

Number of Observations Read	1445
Number of Observations Used	1445

Algorithm converged.

Analysis of Parameter Estimates							
Parameter	DF	Estimate	Standard Error	95% Confidence Limits		Chi-Square	Pr > ChiSq
Intercept	1	4.9764	0.0426	4.8930	5.0598	13670.0	<0.0001
Scale	0	1.0000	0.0000	1.0000	1.0000		
Weibull Scale	1	144.9511	6.1695	133.3497	157.5618		
Weibull Shape	0	1.0000	0.0000	1.0000	1.0000		

Lagrange Multiplier Statistics		
Parameter	Chi-Square	Pr > ChiSq
Scale	35.9078	<0.0001

OUTPUT 11.5. Exponential distribution fit for the RECID data.

We will illustrate how Proc Lifereg can be used to estimate Accelerated Failure Time models. The following minimal set of statements should be used. The results of the analysis are given in Output 11.8.

```
Proc Lifereg data=duration;
     Model l_durat*censored(1)=l_black l_drugs l_workprg l_priors l_tserved
     l_felon l_alcohol l_marry l_educ l_age/d=lnormal;
Run;
```

Proc Lifereg invokes the procedure on the temporary SAS data set *duration*. The second statement lists out the model relating the explanatory variables on the right-hand side to the duration time *l_durat*. Note that the response variable has not been transformed and is in the original units of measurements (months). The log transformation is done by Proc Lifereg. The *d* option can be used to select the distribution. We decided on the log-normal distribution from the distributional analysis done in Section 11.4. More details on the different options that can be used with this procedure can be found in Allison (1995) and *SAS Users Guide 9.2* from SAS Institute, Inc.

The first table gives basic model information: the model name, the response variable, the censoring ID, the total number of observations, the number of censored and uncensored observations, the assumed distribution, and the log-likelihood value.

The next table gives the Wald's chi-square values and associated *p* values for the model parameters. Notice that the variables recording work programs and education are not significant. The Wald's chi-square values can be calculated by taking the

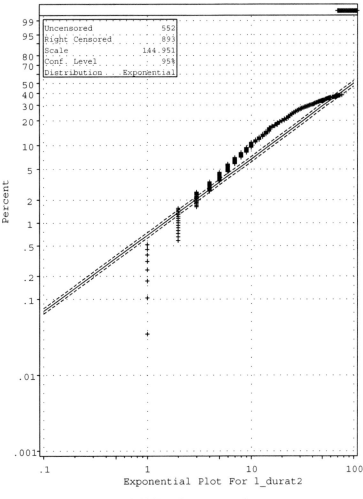

OUTPUT 11.5. (*Continued*).

square of the ratio of the parameter estimates to their standard errors. The next table gives the actual parameter estimates, their standard errors, 95% confidence intervals, the chi-square values and the associated p values. The 95% confidence intervals are calculated as

$$\hat{\beta} z_{\alpha/2} \times s.e(\hat{\beta})$$

where $z_{\alpha/2}$ is the $100 \times (1-\alpha)$ percentile from the standard normal distribution.

Notice that the estimate scale parameter is 1.81 indicating a decreasing hazard rate (see Figure 11.8).

The signs of the model parameters indicate that in general race, alcohol, and drug problems all appear to shorten the duration (or arrest) times. Marriage and whether a person was convicted of a felony both appear to lengthen the duration (or arrest) time. Using the fact that we have a semi-log model, we may interpret the coefficients as follows: Holding all other variables constant, for the `l_tserved` variable that records the time served in months, each additional month of time served, reduces the duration until the next arrest by about 2%. As another example, we may interpret the coefficient for the `l_drugs` variable as follows: Holding all other variables constant, the duration time until the next arrest for ex-convicts with drug problems is about 30% shorter than for those without drug problems.

As discussed in Allison (1995), we can also use the exponential transformation to interpret the coefficients. For dummy variables, the transformation is simply $\exp(\hat{\beta})$. For example, for the dummy variable `l_drugs` $\exp(-0.2982) = 0.74$. This implies that holding all other variables constant, the duration time until the next arrest for ex-convicts with drug problems is about 26% shorter than the duration time of ex-convicts without drug problems. For continuous variables, the transformation used is $100 \times \exp(\hat{\beta}-1)$. For example, for the variable `l_tserved`, we have $100 \times \exp(-0.0193-1) = 2\%$.

The LIFEREG Procedure

Model Information	
Data Set	WORK.DURATION
Dependent Variable	Log(l_durat2)
Censoring Variable	censored
Censoring Value(s)	1
Number of Observations	1445
Noncensored Values	552
Right Censored Values	893
Left Censored Values	0
Interval Censored Values	0
Name of Distribution	Lognormal
Log Likelihood	-1680.426985

Number of Observations Read	1445
Number of Observations Used	1445

Algorithm converged.

Analysis of Parameter Estimates							
Parameter	DF	Estimate	Standard Error	95% Confidence Limits		Chi-Square	Pr > ChiSq
Intercept	1	4.8454	0.0755	4.6975	4.9933	4121.78	<0.0001
Scale	1	1.9688	0.0684	1.8392	2.1077		

OUTPUT 11.6. Weibull distribution fit for the RECID data.

Not surprisingly, both sets of interpretations lead to exactly the same conclusion.

Proc Phreg can be used to fit the Cox's (1972) Proportional Hazard Models to duration data. These models make no distributional assumptions on the data and can be used to incorporate time variant regressors. The general form of the model is given by

$$h_i(t) = \lambda_0(t) \, \exp(\mathbf{x}_i^T \boldsymbol{\beta}),$$

where $h_i(t)$ denotes the hazard for the ith subject at time t, $\lambda_0(t)$ is the baseline hazard, and $\mathbf{x}_i$ is the vector of explanatory variables that may include both time-dependent and time-invariant variables. The name proportional hazard is derived from the fact that for two subjects in the study, the ratio $h_i(t)/h_j(t)$ is time invariant. That is,

$$\frac{h_i(t)}{h_j(t)} = \exp\left(\mathbf{x}_i^T \boldsymbol{\beta} - \mathbf{x}_j^T \boldsymbol{\beta}\right).$$

The method of partial likelihood is used to estimate the parameters of these models. More details can be found in Allison (1995), Lee (1992), and Meeker and Escobar (1998). We will illustrate how Proc Phreg can be used to estimate Cox's Proportional Hazard models. The following minimal set of statements should be used. The results of the analysis are given in Output 11.9.

```
Proc Phreg data=duration;
     Model l_durat*censored(1)=l_black l_drugs l_workprg l_priors l_tserved l_felon
     l_alcohol l_marry l_educ l_age;
Run;
```

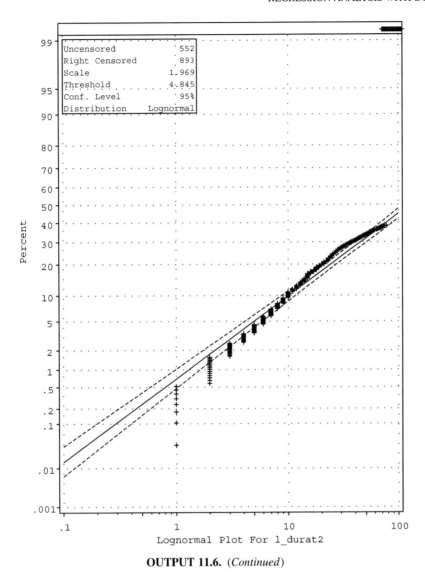

OUTPUT 11.6. (*Continued*)

Proc Phreg invokes the procedure on the temporary SAS data set "`duration.`" The "`model`" statement is identical to the statement used in Proc Lifereg with the exception of the missing "`d`" option. More details on the different options that can be used with this procedure can be found in Allison (1995) and *SAS Users Guide 9.2* from SAS Institute, Inc.

The first few tables list out basic information about the model. As discussed in Allison (1995), the partial likelihood method uses the ranks of the duration times. There may be instance where there are ties in the ranks. For instance, in the example used here, there were eight ex-convicts who were arrested again in the first month. Their ranks in the calculations are therefore equal. SAS offers three options for handling ties in the data. The default is the Breslow method. See both Allison (1995) and the *SAS Users Guide 9.2* from SAS Institute for more details.

The model fit statistics indicates a highly significant model. The last table gives information on the parameter estimates. The variables recording education, marriage, and work program participation are not significant. Note that the signs of the coefficients here are different from those obtained using Proc Lifereg. This is because, Proc Phreg models the log of the hazard while Proc Lifereg models the log of the survival. The chi-square test statistic values are calculated in the usual way and the hazard ratio is simply $\exp(\hat{\beta})$. The interpretations can be done by using the fact that a semi-log model was used for the data. For instance, holding all other variables constant, the hazard of arrest for ex-convicts with drug problems is about 28% higher than the hazard of arrest of ex-convicts without drug problems. As another example, holding all other variables constant, for each additional month of time served increases the hazard of arrest by about 1.3%.

The LIFEREG Procedure

Model Information	
Data Set	WORK.DURATION
Dependent Variable	Log(l_durat2)
Censoring Variable	censored
Censoring Value(s)	1
Number of Observations	1445
Noncensored Values	552
Right Censored Values	893
Left Censored Values	0
Interval Censored Values	0
Name of Distribution	Weibull
Log Likelihood	-1715.771096

Number of Observations Read	1445
Number of Observations Used	1445

Algorithm converged.

Analysis of Parameter Estimates							
Parameter	DF	Estimate	Standard Error	95% Confidence Limits		Chi-Square	Pr > ChiSq
Intercept	1	5.2245	0.0705	5.0863	5.3627	5490.88	<0.0001
Scale	1	1.3004	0.0516	1.2030	1.4057		
Weibull Scale	1	185.7613	13.0971	161.7861	213.2895		
Weibull Shape	1	0.7690	0.0305	0.7114	0.8312		

OUTPUT 11.7. Long Normal distribution fit for the RECID data.

It may be of interest to see how optimization techniques such as the Newton–Raphson and the BHHH methods work. Program 18 in Appendix E contains Proc IML code to conduct analysis of the strike duration data from Kennan (1985, pp. 14–16) using both the optimization techniques. As stated in Greene (2003, p. 800), the strike data contains the number of days for 62 strikes that started in June of years 1968–1976. The data set also contains an explanatory variable measures deviation of production due to various seasonal trends (Greene, 2003).

The Proc IML code analyzes the data using the Newton–Raphson and the BHHH methods and was written by Thomas B. Fomby from the Department of Economics at the Southern Methodist University in 2005. The code has been reprinted with permission from Thomas Fomby. As is always the case, the optimization techniques require a starting value. Using Proc Reg on the log of the response variable against the explanatory variable gives values of 3.2 and −9.2 for the intercept and slope coefficient. The starting values have been set at 4 for the intercept and −9 for the slope coefficient. The scale value is set at 1, the scale value for the exponential distribution.

The analysis results are given in Output 11.10. Notice that the convergence of the Newton–Raphson algorithm occurs at the sixth iteration with values of 0.9922 for the scale parameter, 3.78 for the intercept, and −9.33 for the slope coefficient. Also, note that the Wald's test for constant hazard cannot be rejected, indicating that the exponential model may be used to model the strike duration data. The BHHH algorithm converges much later but with parameter values similar to the ones from the Newton–Raphson method. Output 11.11 contains the analysis results of the strike data using Proc Lifereg using the exponential distribution. Notice that the p value from LM test for the scale parameter indicates that we cannot reject the hypothesis that the scale parameter is 1, indicating that the exponential distribution was a good choice for the data.

Since the model is a semi-log model, we can interpret the coefficient for "eco" as follows: A one-unit change in production is expected to decrease the average length of the strikes by 9.33%.

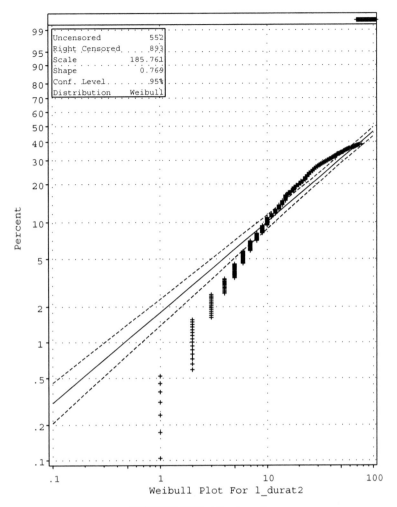

OUTPUT 11.7. (*Continued*).

The LIFEREG Procedure

Model Information	
Data Set	WORK.DURATION
Dependent Variable	Log(l_durat2)
Censoring Variable	censored
Censoring Value(s)	1
Number of Observations	1445
Noncensored Values	552
Right Censored Values	893
Left Censored Values	0
Interval Censored Values	0
Name of Distribution	Lognormal
Log Likelihood	-1597.058956

Number of Observations Read	1445
Number of Observations Used	1445

Algorithm converged.

Type III Analysis of Effects			
Effect	DF	Wald Chi-Square	Pr > ChiSq
l_black	1	21.3548	<0.0001
l_drugs	1	5.0457	0.0247
l_workprg	1	0.2717	0.6022
l_priors	1	40.9108	<0.0001
l_tserved	1	42.1368	<0.0001
l_felon	1	9.3649	0.0022
l_alcohol	1	19.3817	<0.0001
l_marry	1	5.9350	0.0148
l_educ	1	0.8144	0.3668
l_age	1	41.6081	<0.0001

Analysis of Parameter Estimates							
Parameter	DF	Estimate	Standard Error	95% Confidence Limits		Chi-Square	Pr > ChiSq
Intercept	1	4.0994	0.3475	3.4182	4.7805	139.14	<0.0001
l_black	1	-0.5427	0.1174	-0.7729	-0.3125	21.35	<0.0001
l_drugs	1	-0.2982	0.1327	-0.5583	-0.0380	5.05	0.0247
l_workprg	1	-0.0626	0.1200	-0.2978	0.1727	0.27	0.6022
l_priors	1	-0.1373	0.0215	-0.1793	-0.0952	40.91	<0.0001
l_tserved	1	-0.0193	0.0030	-0.0252	-0.0135	42.14	<0.0001
l_felon	1	0.4440	0.1451	0.1596	0.7284	9.36	0.0022
l_alcohol	1	-0.6349	0.1442	-0.9176	-0.3522	19.38	<0.0001
l_marry	1	0.3407	0.1398	0.0666	0.6148	5.94	0.0148
l_educ	1	0.0229	0.0254	-0.0269	0.0727	0.81	0.3668
l_age	1	0.0039	0.0006	0.0027	0.0051	41.61	<0.0001
Scale	1	1.8105	0.0623	1.6924	1.9368		

OUTPUT 11.8.

The PHREG Procedure

Model Information	
Data Set	WORK.DURATION
Dependent Variable	l_durat2
Censoring Variable	censored
Censoring Value(s)	1
Ties Handling	BRESLOW

Number of Observations Read	1445
Number of Observations Used	1445

Summary of the Number of Event and Censored Values			
Total	Event	Censored	Percent Censored
1445	552	893	61.80

Convergence Status
Convergence criterion (GCONV=1E-8) satisfied.

Model Fit Statistics		
Criterion	Without Covariates	With Covariates
-2 LOG L	7788.360	7632.760
AIC	7788.360	7652.760
SBC	7788.360	7695.895

Testing Global Null Hypothesis: BETA=0			
Test	Chi-Square	DF	Pr > ChiSq
Likelihood Ratio	155.6005	10	<0.0001
Score	168.7868	10	<0.0001
Wald	166.1585	10	<0.0001

The PHREG Procedure

Analysis of Maximum Likelihood Estimates						
Variable	DF	Parameter Estimate	Standard Error	Chi-Square	Pr > ChiSq	Hazard Ratio
l_black	1	0.43257	0.08838	23.9547	<0.0001	1.541
l_drugs	1	0.27558	0.09786	7.9297	0.0049	1.317
l_workprg	1	0.08403	0.09081	0.8562	0.3548	1.088
l_priors	1	0.08759	0.01348	42.2354	<0.0001	1.092
l_tserved	1	0.01296	0.00168	59.1317	<0.0001	1.013
l_felon	1	-0.28284	0.10616	7.0989	0.0077	0.754
l_alcohol	1	0.43063	0.10572	16.5922	<0.0001	1.538
l_marry	1	-0.15490	0.10921	2.0117	0.1561	0.857
l_educ	1	-0.02133	0.01945	1.2028	0.2728	0.979
l_age	1	-0.00358	0.0005223	47.0211	<0.0001	0.996

OUTPUT 11.9.

Calculation of Unrestricted MLE estimates using Hessian-Based Newton-Raphson
Method

Iteration steps

			RESULT		
ITER	SIGMA	BETA1	BETA2	G1	G2
1	0.9537711	3.7467595	-9.443068	-3.175459	-12.59413
2	0.9869113	3.7769746	-9.345545	6.0076113	3.5681181
3	0.9921146	3.7797375	-9.332412	0.6819702	0.3315151
4	0.9922036	3.7797742	-9.332198	0.0110044	0.0048339
5	0.9922037	3.7797742	-9.332198	2.9191E-6	1.207E-6
6	0.9922037	3.7797742	-9.332198	1.984E-13	8.085E-14
0	0	0	0	0	0
0	0	0	0	0	0
0	0	0	0	0	0
0	0	0	0	0	0

	RESULT	
G3	CRIT	LNLU
-0.169208	0.5124223	-98.77871
0.0551014	0.1073396	-97.45512
0.0053577	0.0143944	-97.28771
0.0000803	0.0002348	-97.28542
2.0407E-8	6.1706E-8	-97.28542
1.465E-15	5.601E-15	-97.28542
0	0	0
0	0	0
0	0	0
0	0	0

Unrestricted Log-likelihood =

LNLU
-97.28542

The Maximum Likelihood Estimates: Hessian-Based Newton-Raphson Iteration

THETA	
SIGMA	0.9922037
BETA1	3.7797742
BETA2	-9.332198

Asymptotic Covariance Matrix-From Hessian

OUTPUT 11.10. Duration analysis of the strike data using Thomas Fomby's Proc IML code.

COV			
	SIGMA	BETA1	BETA2
SIGMA	0.0099437	-0.004184	-0.00209
BETA1	-0.004184	0.0186872	-0.094236
BETA2	-0.00209	-0.094236	8.6292303

Standard errors:

SE_SIGMA_H
0.0997181

SE_BETA1_H
0.1367012

SE_BETA2_H
2.9375552

Asymptotic Covariance Matrix-From bhhh

COVBH3			
	SIGMA	BETA1	BETA2
SIGMA	0.0145537	-0.002197	-0.038556
BETA1	-0.002197	0.0191359	-0.109777
BETA2	-0.038556	-0.109777	8.7277958

Standard errors:

SE_SIGMA_B
0.1206388

SE_BETA1_B
0.1383325

SE_BETA2_B
2.9542843

Wald test of hypothesis of constant hazard (sigma=1)

	WALD	CRITVAL	PVAL
Results of Wald test Using Hessian	0.0032527	3.8414588	0.9545197

	WALD	CRITVAL	PVAL
Results of Wald test Using BHHH	0.0031764	3.8414588	0.9550554

OUTPUT 11.10. (*Continued*)

```
Maximum Likelihood Estimation of Restricted Model
```

```
****************************************************
```

Iteration steps

			RESULT			
ITER	BETA1	BETA2	G1	G2	CRIT	LNLR
1	4	-9	-12.59413	-0.169208	0.4878992	-98.77871
2	3.7494458	-9.418651	1.7610376	0.0293685	0.0867187	-97.3134
3	3.7761427	-9.336144	0.0244856	0.0005357	0.0023563	-97.28845
4	3.7765119	-9.333816	5.1415E-6	1.6029E-7	9.1067E-7	-97.28844
5	3.7765119	-9.333815	2.572E-13	1.125E-14	7.292E-14	-97.28844
0	0	0	0	0	0	0
0	0	0	0	0	0	0
0	0	0	0	0	0	0
0	0	0	0	0	0	0
0	0	0	0	0	0	0

The Maximum Likelihood Estimates-Restricted Model

BETA1	BETA2
3.7765119	-9.333815

Asymptotic Covariance Matrix-From Hessian of Restricted Model

COV		
	BETA1	BETA2
BETA1	0.0171936	-0.096573
BETA2	-0.096573	8.761006

	LM	CRITVAL	PVAL
Results of LM test Using Hessian	0.0061132	3.8414588	0.9376792

	LM	CRITVAL	PVAL
Results of LM test Using BHHH	0.0089884	3.8414588	0.9244679

	LR	CRITVAL	PVAL
Results of LR test	0.0060429	3.8414588	0.9380379

```
Calculation of Unrestricted MLE estimates using BHHH-Based Newton-Raphson
Method
```

Iteration steps

OUTPUT 11.10. (*Continued*)

		RESULT				RESULT		
ITER	SIGMA	BETA1	BETA2	G1	G2	G3	CRIT	LNL
1	0.9898141	3.778817	-9.315595	-0.593324	0.4162062	0.0064475	0.0369484	-97.28985
2	0.9937401	3.7803749	-9.344186	0.2894756	0.1131911	-0.00063	0.0289013	-97.28582
3	0.99118	3.7793555	-9.324198	-0.183216	-0.070137	0.0005733	0.0201776	-97.28559
4	0.9928738	3.7800404	-9.337409	0.123469	0.0482311	-0.000373	0.013337	-97.28549
5	0.9917598	3.7795943	-9.328734	-0.080232	-0.030943	0.0002455	0.0087576	-97.28545
6	0.9924954	3.7798909	-9.334469	0.0534042	0.0207745	-0.000162	0.00579	-97.28543
7	0.9920109	3.7796964	-9.330694	-0.034992	-0.013536	0.0001068	0.0038108	-97.28543
8	0.9923306	3.7798251	-9.333186	0.023168	0.0089956	-0.00007	0.0025157	-97.28542
9	0.9921199	3.7797404	-9.331545	-0.015235	-0.005901	0.0000464	0.0016575	-97.28542
10	0.9922589	3.7797964	-9.332628	0.0100633	0.0039042	-0.000031	0.0010935	-97.28542
11	0.9921672	3.7797595	-9.331914	-0.006628	-0.002569	0.0000202	0.0007208	-97.28542
12	0.9922277	3.7797838	-9.332385	0.0043735	0.0016961	-0.000013	0.0004753	-97.28542
13	0.9921878	3.7797678	-9.332074	-0.002882	-0.001117	$8.7761E{-}6$	0.0003134	-97.28542
14	0.9922141	3.7797784	-9.332279	0.0019011	0.0007372	$-5.787E{-}6$	0.0002067	-97.28542
15	0.9921968	3.7797714	-9.332144	-0.001253	-0.000486	$3.8156E{-}6$	0.0001363	-97.28542
16	0.9922082	3.779776	-9.332233	0.0008265	0.0003205	$-2.516E{-}6$	0.0000898	-97.28542
17	0.9922007	3.779773	-9.332174	-0.000545	-0.000211	$1.6589E{-}6$	0.0000592	-97.28542
18	0.9922056	3.779775	-9.332213	0.0003593	0.0001393	$-1.094E{-}6$	0.0000391	-97.28542
19	0.9922023	3.7797737	-9.332187	-0.000237	-0.000092	$7.2126E{-}7$	0.0000258	-97.28542
20	0.9922045	3.7797745	-9.332204	0.0001562	0.0000606	$-4.756E{-}7$	0.000017	-97.28542
21	0.9922031	3.779774	-9.332193	-0.000103	-0.00004	$3.1358E{-}7$	0.0000112	-97.28542
22	0.992204	3.7797743	-9.3322	0.0000679	0.0000263	$-2.068E{-}7$	$7.3837E{-}6$	-97.28542
23	0.9922034	3.7797741	-9.332196	-0.000045	-0.000017	$1.3634E{-}7$	$4.8686E{-}6$	-97.28542
24	0.9922038	3.7797742	-9.332199	0.0000295	0.0000114	$-8.99E{-}8$	$3.2102E{-}6$	-97.28542
25	0.9922035	3.7797741	-9.332197	-0.000019	$-7.549E{-}6$	$5.9275E{-}8$	$2.1167E{-}6$	-97.28542
26	0.9922037	3.7797742	-9.332198	0.0000128	$4.9776E{-}6$	$-3.908E{-}8$	$1.3957E{-}6$	-97.28542
27	0.9922036	3.7797742	-9.332197	$-8.465E{-}6$	$-3.282E{-}6$	$2.5771E{-}8$	$9.2029E{-}7$	-97.28542
28	0.9922037	3.7797742	-9.332198	$5.5817E{-}6$	$2.1641E{-}6$	$-1.699E{-}8$	$6.0681E{-}7$	-97.28542
29	0.9922036	3.7797742	-9.332197	$-3.68E{-}6$	$-1.427E{-}6$	$1.1205E{-}8$	$4.0011E{-}7$	-97.28542
30	0.9922037	3.7797742	-9.332198	$2.4268E{-}6$	$9.4089E{-}7$	$-7.388E{-}9$	$2.6382E{-}7$	-97.28542
31	0.9922036	3.7797742	-9.332198	$-1.6E{-}6$	$-6.204E{-}7$	$4.8714E{-}9$	$1.7396E{-}7$	-97.28542
32	0.9922037	3.7797742	-9.332198	$1.0551E{-}6$	$4.0907E{-}7$	$-3.212E{-}9$	$1.147E{-}7$	-97.28542
33	0.9922036	3.7797742	-9.332198	$-6.957E{-}7$	$-2.697E{-}7$	$2.1179E{-}9$	$7.5632E{-}8$	-97.28542
34	0.9922037	3.7797742	-9.332198	$4.5872E{-}7$	$1.7785E{-}7$	$-1.397E{-}9$	$4.9869E{-}8$	-97.28542
35	0.9922036	3.7797742	-9.332198	$-3.025E{-}7$	$-1.173E{-}7$	$9.208E{-}10$	$3.2882E{-}8$	-97.28542
36	0.9922037	3.7797742	-9.332198	$1.9944E{-}7$	$7.7325E{-}8$	$-6.07E{-}10$	$2.1682E{-}8$	-97.28542
37	0.9922036	3.7797742	-9.332198	$-1.315E{-}7$	$-5.099E{-}8$	$4.003E{-}10$	$1.4296E{-}8$	-97.28542
38	0.9922037	3.7797742	-9.332198	$8.671E{-}8$	$3.3619E{-}8$	$-2.64E{-}10$	$9.4266E{-}9$	-97.28542
39	0.9922036	3.7797742	-9.332198	$-5.717E{-}8$	$-2.217E{-}8$	$1.741E{-}10$	$6.2156E{-}9$	-97.28542
40	0.9922037	3.7797742	-9.332198	$3.7699E{-}8$	$1.4616E{-}8$	$-1.15E{-}10$	$4.0984E{-}9$	-97.28542
41	0.9922037	3.7797742	-9.332198	$-2.486E{-}8$	$-9.638E{-}9$	$7.568E{-}11$	$2.7024E{-}9$	-97.28542

OUTPUT 11.10. (*Continued*)

			RESULT				RESULT	
ITER	SIGMA	BETA1	BETA2	G1			G3	CRIT	LNL
42	0.9922037	3.7797742	-9.332198	1.639E-8	6.3548E-9	-4.99E-11	1.7819E-9	-97.28542
43	0.9922037	3.7797742	-9.332198	-1.081E-8	-4.19E-9	3.29E-11	1.1749E-9	-97.28542
44	0.9922037	3.7797742	-9.332198	7.1261E-9	2.7629E-9	-2.17E-11	7.747E-10	-97.28542
45	0.9922037	3.7797742	-9.332198	-4.699E-9	-1.822E-9	1.43E-11	5.108E-10	-97.28542
46	0.9922037	3.7797742	-9.332198	3.0982E-9	1.2012E-9	-9.43E-12	3.368E-10	-97.28542
47	0.9922037	3.7797742	-9.332198	-2.043E-9	-7.92E-10	6.219E-12	2.221E-10	-97.28542
48	0.9922037	3.7797742	-9.332198	1.347E-9	5.222E-10	-4.1E-12	1.464E-10	-97.28542
49	0.9922037	3.7797742	-9.332198	-8.88E-10	-3.44E-10	2.704E-12	9.656E-11	-97.28542
0	0	0	0	0	0	0	0	0
0	0	0	0	0	0	0	0	0
0	0	0	0	0	0	0	0	0
0	0	0	0	0	0	0	0	0
0	0	0	0	0	0	0	0	0
0	0	0	0	0	0	0	0	0
0	0	0	0	0	0	0	0	0
0	0	0	0	0	0	0	0	0
0	0	0	0	0	0	0	0	0
0	0	0	0	0	0	0	0	0
0	0	0	0	0	0	0	0	0

OUTPUT 11.10. (*Continued*)

The LIFEREG Procedure

Model Information	
Data Set	WORK.STRIKE
Dependent Variable	Log(dur)
Number of Observations	62
Noncensored Values	62
Right Censored Values	0
Left Censored Values	0
Interval Censored Values	0
Name of Distribution	Exponential
Log Likelihood	-97.28844102

Number of Observations Read	62
Number of Observations Used	62

Algorithm converged.

Type III Analysis of Effects			
Effect	DF	Wald Chi-Square	Pr > ChiSq
eco	1	9.9441	0.0016

Analysis of Parameter Estimates							
Parameter	DF	Estimate	Standard Error	95% Confidence Limits		Chi-Square	Pr > ChiSq
Intercept	1	3.7765	0.1311	3.5195	4.0335	829.50	<0.0001
eco	1	-9.3338	2.9599	-15.1351	-3.5325	9.94	0.0016
Scale	0	1.0000	0.0000	1.0000	1.0000		
Weibull Shape	0	1.0000	0.0000	1.0000	1.0000		

Lagrange Multiplier Statistics		
Parameter	Chi-Square	Pr > ChiSq
Scale	0.0061	0.9377

OUTPUT 11.11. Analysis of the strike data using Proc Lifereg.

12

SPECIAL TOPICS

Chapters 1 through 11 discussed basic econometric analysis using SAS. This chapter introduces additional analytical methods within the context of what was covered in the previous chapters.

12.1 ITERATIVE FGLS ESTIMATION UNDER HETEROSCEDASTICITY

In Section 5.6, we introduced FGLS estimation where we assumed that the variance of the disturbances is a function of one or more explanatory variables. For example, we assumed that $\sigma_i^2 = \sigma^2 z_i^\alpha$, where $z_i = income$. The estimation was done over two steps, where in step 1, the OLS residuals were used in a regression with $\log(z_i)$ to get an estimate of α. The weights using α were calculated resulting in the two-step FGLS estimator.

We can very easily iterate the two-step estimation process to convergence. The method involves recomputing the residuals using the first set of FGLS estimators and then using these residuals to recompute the FGLS estimates. The iteration continues until the difference between the most recent FGLS estimates does not differ from the estimates computed in the previous stage. Program 9 in Appendix E gives IML code to carry out these computations on the credit card data set, which was used in Chapter 5 with $z_i = income$. The analysis results are given in Output 12.1. As discussed in Greene (2003), the asymptotic properties of the iterated FGLS are similar to those of the FGLS.

12.2 MAXIMUM LIKELIHOOD ESTIMATION UNDER HETEROSCEDASTICITY

To motivate our discussion of maximum likelihood estimation, first consider the joint distribution of $y_i = \mathbf{x}_i^T \boldsymbol{\beta} + \varepsilon_i$, where $i = 1, \ldots, n$, assuming that $\varepsilon_i \sim iid\, N(0, \sigma_i^2)$, where $\sigma_i^2 = \sigma^2 f_i(\boldsymbol{\alpha})$. Note that the observations y_i are independently distributed since the disturbances ε_i are assumed to be independently distributed. The joint distribution is therefore given by (Casella and Berger, 1990; Greene, 2003, p. 228–229)

$$\prod_{i=1}^{n} f(y_i | \boldsymbol{\beta}, \mathbf{x}_i, \sigma_i^2) = (2\pi)^{-n/2} \prod_{i=1}^{n} \sigma_i^2 \exp\left(-\frac{1}{2} \sum_{i=1}^{n} \frac{(y_i - \mathbf{x}_i^T \boldsymbol{\beta})^2}{\sigma_i^2} \right).$$

Applied Econometrics Using the SAS® System, by Vivek B. Ajmani
Copyright © 2009 John Wiley & Sons, Inc.

The value of alpha is	ALPHA_S 1.7622762

	ITER	
Convergence was obtained in	18	iterations.

The estimates of the coefficients are

STAT_TABLE		
	BHAT	SE
INT	−130.384	143.9658
AGE	−2.7754	3.9523
OWNRENT	59.1258	60.5929
INCOME	169.7363	75.6177
INCOME2	−8.5995	9.2446

OUTPUT 12.1. Iterative FGLS estimators for the credit card expenditure data.

Taking the log of the likelihood after substituting $\sigma^2 f_i(\boldsymbol{\alpha})$ for σ_i^2 and ε_i for $y_i - \mathbf{x}_i^T \boldsymbol{\beta}$ and rearranging the terms involving $\boldsymbol{\alpha}$, we get

$$\text{Long-likelihood} = -\frac{n}{2}\left[\log(2\pi) + \log\sigma^2\right] - \frac{1}{2}\sum_{i=1}^{n}\left[\log f_i(\boldsymbol{\alpha}) + \frac{1}{\sigma^2}\frac{1}{f_i(\boldsymbol{\alpha})}\varepsilon_i^2\right].$$

Our objective is to find values of $\boldsymbol{\alpha}, \boldsymbol{\beta}$, and σ^2 that maximize this log-likelihood function. Taking derivative of this function with respect to $\boldsymbol{\alpha}, \boldsymbol{\beta}$, and σ^2, we get (Greene, 2003, p. 229)

$$S(\boldsymbol{\beta}) = \sum_{i=1}^{n}\mathbf{x}_i\frac{\varepsilon_i}{\sigma^2 f_i(\boldsymbol{\alpha})},$$

$$S(\sigma^2) = \sum_{i=1}^{n}\left(\frac{1}{2\sigma^2}\right)\left(\frac{\varepsilon_i^2}{\sigma^2 f_i(\boldsymbol{\alpha})} - 1\right),$$

$$S(\boldsymbol{\alpha}) = \sum_{i=1}^{n}\left(\frac{1}{2}\right)\left(\frac{\varepsilon_i^2}{\sigma^2 f_i(\boldsymbol{\alpha})} - 1\right)\frac{1}{f_i(\boldsymbol{\alpha})}\frac{\partial f_i(\boldsymbol{\alpha})}{\partial \boldsymbol{\alpha}}.$$

The values of these derivatives equate to zero at the maximum likelihood estimators of $\boldsymbol{\alpha}, \boldsymbol{\beta}$, and σ^2. We will consider two cases for $\boldsymbol{\alpha}$. In the first case, we assume that $\boldsymbol{\alpha}$ has a single parameter. In the second case, we will discuss estimation when $\boldsymbol{\alpha}$ has more than one parameter. The estimation process in the first case is straightforward and is outlined in the following steps:

1. Take a range of values for α.
2. For each value of α from Step 1, compute the GLS estimator of $\boldsymbol{\beta}$ using weights defined by $f_i(\alpha)$.
3. Compute the Generalized Sums of Squares (GSS) $\hat{\sigma}^2$ for each $(\alpha, \hat{\boldsymbol{\beta}})$ pair. The expression for $\hat{\sigma}^2$ can be derived by equating $S(\sigma^2)$ to zero and solving for σ^2. The GSS is given by

$$\hat{\sigma}^2 = \frac{1}{n}\sum_{i=1}^{n}\frac{(y_i - x_i^T\hat{\boldsymbol{\beta}})^2}{f_i(\alpha)}.$$

4. Finally, calculate the value of the log-likelihood equation at $(\alpha, \hat{\boldsymbol{\beta}}, \sigma^2)$. A plot of the log-likelihood values versus α can be used to locate the optimal value of α. Weighted least squares can then be performed by using weights defined by $f_i(\alpha)$.

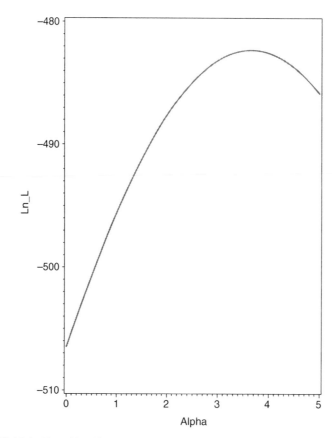

FIGURE 12.1. Plot of log-likelihood at various values of α for the credit card data set.

We will illustrate the computations involved using the credit card data set that was used in Chapter 5. We define $f_i(\alpha) = income^{\alpha}$ and then take a range of values of α and follow steps 1 through 4 to calculate the maximum likelihood estimator of $\boldsymbol{\beta}$ under heteroscedasticity.

The plot of the log-likelihood versus α appears in Figure 12.1. Program 10 in Appendix E contains the complete code to create the plot. Note that the log-likelihood is maximized around 3.6. The exact value of α that maximizes the log-likelihood function can easily be found by various techniques using SAS.

The exact value of α that maximizes the log-likelihood is 3.651 and the value of the log-likelihood at this value of α is -482.324. To get the MLE estimates of $\boldsymbol{\beta}$, use Proc Reg with weights $= 1/income^{3.651}$. The output from this analysis is given in Output 12.2. Note that the standard errors that appear in this output are not based on the GSS. To get the correct standard errors, use Proc IML to first compute the GSS using the optimal values of α and $\boldsymbol{\beta}$. Then use the fact that the variance–covariance matrix for $\hat{\boldsymbol{\beta}}$ is given by $Var(\hat{\boldsymbol{\beta}}|\mathbf{X}) = \hat{\sigma}^2(\mathbf{X}^T\hat{\boldsymbol{\Omega}}^{-1}(\alpha)\mathbf{X})^{-1}$. The correct standard errors are (113.06, 2.76, 43.51, 81.04, and 13.43). Program 11 in Appendix E contains the complete IML code to conduct this analysis.

12.3   HARVEY'S MULTIPLICATIVE HETEROSCEDASTICITY

The previous section dealt with MLE estimates assuming that $\boldsymbol{\alpha}$ has a single parameter. As shown in Greene (2003, pp. 232–235), the case where $\boldsymbol{\alpha}$ has more than one parameter is a straightforward extension of the maximum likelihood estimation used for the previous case. Harvey's model of multiplicative heteroscedasticity can be used to calculate the estimates of the model parameters. In the procedure, $\mathbf{z}_i^T = [1, \mathbf{q}_i^T]$, where $\mathbf{q}_i$ consists of the variables suspected of causing heteroscedasticity. In the credit card example, $\mathbf{q}_i^T = [income, incomesq]$. An iterated scoring method is used to estimate the parameters in the model $\log(\sigma_i^2) = \log(\sigma^2) + \boldsymbol{\alpha}^T\mathbf{z}_i$. The intercept at convergence can be used to estimate σ^2. Details of the scoring method and the formulas are provided in Greene (2003) and are summarized here.

```
                    The REG Procedure
                    Model: MODEL1
             Dependent Variable: AvgExp AvgExp
```

Number of Observations Read	72
Number of Observations Used	72

```
                    Weight: wt
```

Analysis of Variance					
Source	DF	Sum of Squares	Mean Square	F Value	Pr > F
Model	4	13826	3456.42517	5.35	0.0008
Error	67	43278	645.94103		
Corrected Total	71	57104			

Root MSE	25.41537	R-Square	0.2421
Dependent Mean	128.76438	Adj R-Sq	0.1969
Coeff Var	19.73789		

Parameter Estimates						
Variable	Label	DF	Parameter Estimate	Standard Error	t Value	Pr > |t|
Intercept	Intercept	1	−19.26287	117.20218	−0.16	0.8699
Age	Age	1	−1.70608	2.85938	−0.60	0.5527
OwnRent	OwnRent	1	58.10399	45.10486	1.29	0.2021
Income	Income	1	75.98559	84.00591	0.90	0.3690
Income_Sq		1	4.38904	13.92426	0.32	0.7536

OUTPUT 12.2. Regression analysis of the credit card expenditure data using optimal value of α.

1. Estimate $\boldsymbol{\beta}$ using OLS and calculate $\log(e_i^2)$.
2. Regress $\log(e_i^2)$ versus $\mathbf{z}_i$ to get estimates of the intercept and $\boldsymbol{\alpha}$.
3. Estimate σ_i^2 with $\exp(\log(\sigma^2) + \boldsymbol{\alpha}^T\mathbf{z}_i)$.
4. Use FGLS to estimate $\boldsymbol{\beta}$.
5. Update both $\log(\sigma^2)$ and $\boldsymbol{\beta}$. The formulas are given in Greene (2003, p. 234).
6. Stop the iteration process if the differences between the estimated values across the two periods are negligible.

Program 12 in Appendix E gives the complete code for conducting this analysis. The code was written to analyze the credit card data set from Greene (2003). The following weight function was used:

$$\sigma_i^2 = \exp(\log(\sigma^2) + \alpha_1 income + \alpha_2 incomesq).$$

The analysis results are given in Output 12.3.

An estimate of σ^2 is given by $\hat{\sigma}^2 = \exp(-0.042997) = 0.957914$. Again, we can use this along with the expression of the variance–covariance matrix of $\hat{\boldsymbol{\beta}}$ to generate the standard errors of the regression coefficients.

12.4 GROUPWISE HETEROSCEDASTICITY

In this section, we will discuss groupwise heteroscedasticity. That is, the case where the homoscedasticity assumption is violated because of the unequal variance of the disturbances between the groups. To motivate the discussion, consider analyzing the airlines data from Greene (2003) using the model

$$\ln(C) = \beta_1 + \beta_2 \ln(Q) + \beta_3 LF + \beta_4 \ln(PF) + \varepsilon.$$

	ALPHA
The value of alpha is	−0.042997
	5.3553578
	−0.563225

	I	
Convergence was obtained in	67	iterations.

The estimates of the coefficients are

STAT_TABLE		
	BHAT	SE
INT	−58.4173	60.7722
AGE	−0.3763	0.5383
OWNRENT	33.3545	36.3445
INCOME	96.8136	31.1198
INCOME2	−3.7999	2.5688

OUTPUT 12.3. Maximum likelihood estimates using a multivariate value of α.

This data set was used extensively to illustrate basic panel data models in Chapter 7. The least squares residuals was plotted for each airline and appear in Figures 12.2 and 12.3.

Airlines 3, 4, and 6 exhibit more variability in the disturbances than the other airlines. We therefore suspect that the model suffers from groupwise heteroscedasticity. A formal test to check this can be conducted by using the likelihood ratio test given by (Greene, 2003, p. 236).

$$\chi_0^2 = n\ln(s^2) - \sum_{i=1}^{K} n_i s_i^2.$$

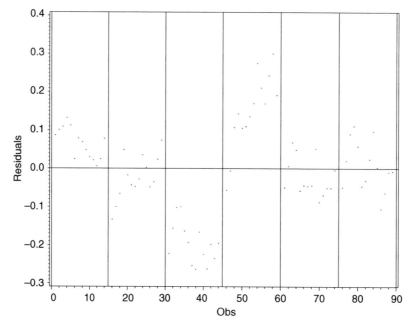

FIGURE 12.2. Time series plot of least squares residuals of individual airlines.

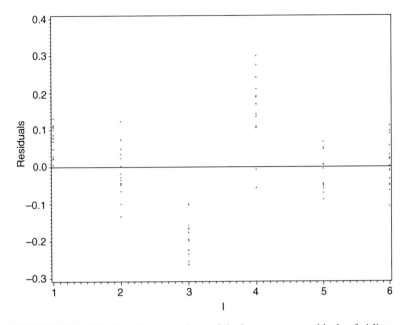

FIGURE 12.3. Side-by-side comparison of the least squares residuals of airlines.

This test statistic is a slight modification of the Bartlett test statistic to compare variances (Snedecor and Cochran, 1983). Here, s^2 is the mean square error (MSE) when the data set is pooled across the K groups and s_i^2 are the group-specific mean square errors. Under the null hypothesis of homoscedasticity, the test statistic χ_0^2 has a chi-squared distribution with $K-1$ degrees of freedom. For the model given above, the value of χ_0^2 equals 107.4 and the null hypothesis of homoscedasticity is therefore rejected.

Another test that can be used to check for groupwise heteroscedasticity is the test by Bartlett (Snedecor and Cochran, 1983), which is given by

$$\chi_0^2 = \frac{(n-K)\ln(s^2) - \sum_{i=1}^{K}(n_i-1)\ln(s_i^2)}{1 + \frac{1}{3(K-1)}\left(\left[\sum_{i=1}^{K}\frac{1}{n_i-1}\right] - \frac{1}{n-K}\right)}.$$

The term in the numerator is the Bartlett test statistic for comparing variances and is very similar to the test statistic given above.

In general, a group wise heteroscedasticity model is characterized by a common coefficients vector (or slope vector) across the K groups but different within-group disturbance variances. That is, a model with the form (Greene, 2003, p. 235)

$$y_i = \mathbf{x}_i^T\boldsymbol{\beta} + \varepsilon_i, \quad i = 1, \ldots, n,$$
$$Var(\varepsilon_{ik}|\mathbf{x}_{iK}) = \sigma_K^2, \quad i = 1, \ldots, n_K.$$

If the within-group variances within each group are known, then one can use GLS estimation to calculate an estimate of the least squares parameter. However, in most cases this will be unknown and estimation will need to be done using FGLS. That is, the least squares estimates can be computed using

$$\hat{\boldsymbol{\beta}} = \left[\sum_{i=1}^{K}\frac{1}{\hat{\sigma}_i^2}\mathbf{X}_i^T\mathbf{X}_i\right]^{-1}\left[\sum_{i=1}^{K}\frac{1}{\hat{\sigma}_i^2}\mathbf{X}_i^T\mathbf{y}_i\right].$$

The within-group residuals vector, $\mathbf{e}_i$ can be used to calculate $\hat{\sigma}_i^2$ ($i = 1, \ldots, K$). That is,

$$\hat{\sigma}_i^2 = \frac{\mathbf{e}_i^T\mathbf{e}_i}{n_i}, \quad i = 1, \ldots, K.$$

The following statements can be used to analyze the airlines data set under the groupwise heteroscedasticity assumption for the model

$$\ln(cost_{it}) = \beta_1 + \beta_2\ln(output_{it}) + \beta_3 LoadFactor_{it} + \beta_4\ln(FuelPrice_{it})$$
$$+ \alpha_2 Firm_2 + \alpha_3 Firm_3 + \alpha_4 Firm_4 + \alpha_5 Firm_5 + \alpha_6 Firm_6 + \varepsilon_{it}.$$

We are assuming that a temporary SAS data set "airline" with the appropriate transformations and dummy variables has already been created.

Step 1: Conduct an OLS to calculate the residuals. The following statements can be used.

```
proc reg data=airline;
       model LnC=LnQ LnPF LF delta2 delta3 delta4 delta5 delta6;
       output out=resid r=resid;
run;
```

Step 2: Calculate the within-group estimate of the disturbance variance. The following statements can be used.

```
data get_resid;
       set resid;
       temp=resid*resid;
run;
proc univariate data=get_resid noprint;
       var temp;
       by i;
       output out=out sum=sum n=n;
run;
data get_var;
   set out;
   var=sum/n;
run;
```

Step 3: Merge the original data set with the data set that contains the within-group variances, calculate the weights, and estimate the parameters.

```
data final_analysis;
       merge airline(in=a) get_var(in=b);
       by i;
       if a and b;
       weight=1/var;
run;
proc reg data=final_analysis;
       model LnC=LnQ LnPF LF delta2 delta3 delta4 delta5 delta6;
       weight weight;
run;
```

The results of the analysis are given in Output 12.4. The first part of the output includes the traditional OLS model without adjusting for the different within-group disturbance variances.

Another estimation method involves treating this model as a form of Harvey's multiplicative heteroscedasticity model with z_i equal to the set of $K - 1$ dummy variables. Program 13 in Appendix E contains the IML code for analyzing the airlines data using the Harvey's multiplicative heteroscedasticity approach (Output 12.5). Here, z_i contains $K - 1$ dummy variables.

```
                    The REG Procedure
                     Model: MODEL1
                 Dependent Variable: LnC
```

Number of Observations Read	90
Number of Observations Used	90

Analysis of Variance					
Source	DF	Sum of Squares	Mean Square	F Value	Pr > F
Model	8	113.74827	14.21853	3935.80	<0.0001
Error	81	0.29262	0.00361		
Corrected Total	89	114.04089			

Root MSE	0.06011	R-Square	0.9974
Dependent Mean	13.36561	Adj R-Sq	0.9972
Coeff Var	0.44970		

Parameter Estimates						
Variable	Label	DF	Parameter Estimate	Standard Error	t Value	Pr > |t|
Intercept	Intercept	1	9.70594	0.19312	50.26	<0.0001
LnQ		1	0.91928	0.02989	30.76	<0.0001
LnPF		1	0.41749	0.01520	27.47	<0.0001
LF	LF	1	−1.07040	0.20169	−5.31	<0.0001
delta2		1	−0.04124	0.02518	−1.64	0.1054
delta3		1	−0.20892	0.04280	−4.88	<0.0001
delta4		1	0.18456	0.06075	3.04	0.0032
delta5		1	0.02405	0.07990	0.30	0.7641
delta6		1	0.08706	0.08420	1.03	0.3042

Number of Observations Read	90
Number of Observations Used	90

```
                    Weight: weight
```

Analysis of Variance					
Source	DF	Sum of Squares	Mean Square	F Value	Pr > F
Model	8	46047	5755.82865	5526.84	<0.0001
Error	81	84.35595	1.04143		
Corrected Total	89	46131			

Root MSE	1.02051	R-Square	0.9982
Dependent Mean	13.47897	Adj R-Sq	0.9980
Coeff Var	7.57109		

Parameter Estimates						
Variable	Label	DF	Parameter Estimate	Standard Error	t Value	Pr > |t|
Intercept	Intercept	1	9.94232	0.16229	61.26	<0.0001
LnQ		1	0.92577	0.02678	34.57	<0.0001
LnPF		1	0.40561	0.01255	32.32	<0.0001
LF	LF	1	−1.21631	0.18559	−6.55	<0.0001
delta2		1	−0.04603	0.02376	−1.94	0.0562
delta3		1	−0.20210	0.03615	−5.59	<0.0001
delta4		1	0.19055	0.05516	3.45	0.0009
delta5		1	0.03717	0.07044	0.53	0.5992
delta6		1	0.09459	0.07436	1.27	0.2070

OUTPUT 12.4. Groupwise heteroscedasticity estimators for the airlines data.

	I	
Convergence was obtained in	12	iterations

The estimates of the coefficients are

ESTIMATES_GLS		
	BETA	SE
INTERCEPT	10.0570	0.1343
LNQ	0.9283	0.0227
LF	-1.2892	0.1638
LNPF	0.4000	0.0108
D2	-0.0487	0.0237
D3	-0.1996	0.0308
D4	0.1921	0.0499
D5	0.0419	0.0594
D6	0.0963	0.0631

The values of alpha are

ESTIMATES_ALPHA		
	ALPHA	SE_ALPHA
ALPHA1	-7.0882	0.3651
ALPHA2	2.0073	0.5164
ALPHA3	0.7581	0.5164
ALPHA4	2.3855	0.5164
ALPHA5	0.5300	0.5164
ALPHA6	1.0530	0.5164

OUTPUT 12.5. Groupwise heteroscedasticity estimators for the airlines data using Harvey's Multiplicative heteroscedasticity approach.

12.5 HAUSMAN–TAYLOR ESTIMATOR FOR THE RANDOM EFFECTS MODEL

Basic panel data models including both fixed and random effects models have been discussed in Chapter 7. A fundamental assumption in random effects models is that the unobserved subject-specific heterogeneity is independent of the observed explanatory variables. In reality, it is rare that this assumption holds. For example, both Baltagi (2005, p. 128) and Greene (2003, p. 305) give an example of a study where the interest is to gauge the impact of years of schooling on earnings. It is well known that a subject's motivation and desire (both assumed unobserved) are highly correlated to academic success and, therefore, to the subject's number of years of formal schooling. The random effects model cannot be used here since the independence assumption between the unobserved heterogeneity (motivation, desire) and observed explanatory variable (number of years of schooling) is correlated. As discussed in Greene (2003), most often these models have explanatory variables that are time invariant. That is, we may be interested in drivers such as gender, race, marital status, and so on with respect to their impact on earning's potential. However, fixed effects models cannot incorporate time-invariant explanatory variables as they are "swept" from the model. Suppose that the researcher wants to include the time-invariant explanatory variables in the model. In this case, the fixed effects model will not allow the estimation of the parameters of these time-invariant explanatory variables.

Hausman and Taylor (1981) introduced estimation techniques for the random effects model where the unobserved subject-specific heterogeneity is correlated with the observed explanatory variables and where there are time-invariant explanatory variables in the model.

Hausman and Taylor's general approach is to first partition the observed and unobserved explanatory variables into two sets. In each set, one set of variables are exogenous while the other set of variables are endogenous. Using the notation from Greene (2003, p. 303), we can write the general form of the model as

$$y_{it} = \mathbf{x}_{1it}^T\boldsymbol{\beta}_1 + \mathbf{x}_{2it}^T\boldsymbol{\beta}_2 + \mathbf{z}_{1i}^T\boldsymbol{\alpha}_1 + z_{2i}^T\boldsymbol{\alpha}_2 + \varepsilon_{it} + u_i, \quad i = 1, \ldots, n, t = 1, \ldots, T.$$

Here,

1. $\mathbf{x}_{1it}^{T}$ has k_1 observed explanatory variables that are time dependent and exogenous with respect to u_i.
2. $\mathbf{z}_1$ has l_1 observed individual-specific variables that are time independent and exogenous with respect to u_i.
3. $\mathbf{x}_{2it}^{T}$ has k_2 observed explanatory variables that are time dependent and endogenous with respect to u_i.
4. $\mathbf{z}_2$ has l_2 observed individual-specific variables that are time independent and endogenous with respect to u_i.

The assumptions about the random disturbances of this model are given in Greene (2003, p. 303). Hausman and Taylor proposed an instrumental variables approach to estimate the parameters of the general model in the presence of the endogenous variables. The Hausman and Taylor's approach can be outlined as follows (Baltagi, 2005, p. 126; Greene, 2003, p. 304).

We can estimate $\boldsymbol{\beta}_1$ and $\boldsymbol{\beta}_2$ by using the within-group estimator. However, the time-invariant explanatory variables are "swept" from the model and so we cannot estimate $\boldsymbol{\alpha}_1$ and $\boldsymbol{\alpha}_2$.

Note that in the within-group model, the time-invariant disturbance term, u_i, is swept from the model as well and, therefore, both sets of deviations $(\mathbf{x}_{1it}-\bar{\mathbf{x}}_{1i.})$ and $(\mathbf{x}_{2it}-\bar{\mathbf{x}}_{2i.})$ are independent of u_i. Hausman and Taylor recommended that these $k_1 + k_2$ deviations be used as instruments to estimate $\boldsymbol{\alpha}_1$ and $\boldsymbol{\alpha}_2$. Next, since $\mathbf{z}_1$ is also exogenous, additional l_1 instruments are available for estimation. At this stage, the number of instruments is less than the number of parameters that need to be estimated. That is, we have $k_1 + k_2 + l_1$ instruments and $k_1 + k_2 + l_1 + l_2$ parameters. As stated in Greene (2003), Hausman and Taylor show that as long as $k_1 \geq l_2$, the k_1 group means $\bar{\mathbf{x}}_1$ can also be used as instruments. The complete set of instruments is, therefore, given by

$$\left(\mathbf{x}_{1it}-\bar{\mathbf{x}}_{1i.}\right), \left(\mathbf{x}_{2it}-\bar{\mathbf{x}}_{2i.}\right), \mathbf{z}_1, \bar{\mathbf{x}}_1.$$

The following steps can then be taken to estimate $\boldsymbol{\alpha}$ and $\boldsymbol{\beta}$:

1. Estimate $\boldsymbol{\beta}_1$ and $\boldsymbol{\beta}_2$ using $\mathbf{x}_1$ and $\mathbf{x}_2$ via the within-group model. Estimate σ_ε^2, the variance of ε_{it}, using the residuals from this analysis.
2. Use the estimates of $\boldsymbol{\beta}_1$ and $\boldsymbol{\beta}_2$ from step 1 to get the within-group residuals. That is, calculate $\bar{e}_{i.} = \bar{y}_{i.} - \bar{\mathbf{x}}_{i.}\mathbf{b}$ where $\mathbf{b} = (\mathbf{b}_1^T, \mathbf{b}_2^T)$. Instrumental variable regression is then used to regress the residual group means against $\mathbf{z}_1$ and $\mathbf{z}_2$ using as instruments $\mathbf{z}_1$ and $\mathbf{x}_1$ to provide an estimate for $\boldsymbol{\alpha}_1$ and $\boldsymbol{\alpha}_2$.
3. Use the mean square errors from steps 1 and 2 to estimate σ_u^2, the variance of u_i, by using the formula $\sigma_u^2 = \sigma^{*2} - \sigma_\varepsilon^2/T$. Next, define the weights that will be used in FGLS estimation as

$$\theta = \sqrt{\frac{\sigma_\varepsilon^2}{\sigma_\varepsilon^2 + T\sigma_u^2}}.$$

4. Calculate the weighted instrumental variable estimator using the weights from step 3. To proceed, first consider the row vector $\mathbf{w}_{it}^{T} = (\mathbf{x}_{1it}^{T}, \mathbf{x}_{2it}^{T}, \mathbf{z}_{1i}^{T}, \mathbf{z}_{2i}^{T})$. The transformed variables using the weights are given by the row vectors (Greene, 2003, p. 305)

$$\mathbf{w}_{it}^{*T} = \mathbf{w}_{it}^{T} - (1-\hat{\theta})\bar{\mathbf{w}}_{i.}^{T},$$
$$y_{it}^{*} = y_{it} - (1-\hat{\theta})\bar{y}_{i.}.$$

The instrumental variables are given by the row vector

$$\mathbf{v}_{it}^{T} = \left\lfloor \left(\mathbf{x}_{1it}-\bar{\mathbf{x}}_{1i}\right)^{T} \quad \left(\mathbf{x}_{2it}-\bar{\mathbf{x}}_{2i}\right)^{T} \quad \mathbf{z}_{1i}^{T} \quad \bar{\mathbf{x}}_{1i}^{T}\right)\rfloor.$$

We can then stack the row vectors defined above to form $nt \times (2k_1 + k_2 + l_1)$ matrices $\mathbf{W}^*$ and $\mathbf{V}$. Let $\mathbf{Y}^*$ be the $nT \times 1$ vector of transformed responses. The instrumental variables estimator is, therefore, given by

$$(\hat{\boldsymbol{\beta}}^T, \hat{\boldsymbol{\alpha}}^T)_{IV}^T = \left[(\mathbf{W}^{*T}\mathbf{V})(\mathbf{V}^T\mathbf{V})^{-1}(\mathbf{V}^T\mathbf{W}^*) \right]^{-1} \left[(\mathbf{W}^{*T}\mathbf{V})(\mathbf{V}^T\mathbf{V})^{-1}(\mathbf{W}^{*T}\mathbf{Y}^*) \right] \text{ (Baltagi, 2005, p. 126; Greene, 2003, p. 305).}$$

The standard errors of the coefficients can be calculated using the approaches discussed in the previous chapters.

We will now illustrate the Hausman and Taylor method on the PSID return to schooling data based on a panel of 595 individuals observed over the period 1976–1982. The data were analyzed by Cornwell and Rupert (1988) and then again by Baltagi (2005). As described in Baltagi (2005, p. 128), the analysis involves regressing the log of wage on years of education (ED), weeks worked (WKS), years of full-time work experience (EXP), occupation (OCC = 1, an indicator that the person is in a full-collar occupation), location indicators SOUTH (1 if the person resides in the South), SMSA (1 if the person resides in a standard metropolitan resident area), industry (IND = 1 if the person works in a manufacturing industry), marital status (MS = 1 if the person is married), sex and race (FEM = 1 indicates that the person is a female, BLK = 1 indicates that the person is a black individual), and union coverage (UNION = 1 if the person belongs to a union). We will compare the random effects and LSDV model results with the Hausman and Taylor model. *Taking an identical approach to analyzing the data as the author, the following four groups of variables are first defined.*

$\mathbf{X}_1$ = OCC, SOUTH, SMSA, IND.
$\mathbf{X}_2$ = EXP, EXP^2, WKS, MS, UNION.
$\mathbf{Z}_1$ = FEM, BLK.
$\mathbf{Z}_2$ = ED.

The following statements can be used to fit a random effects model to the data set. The analysis results are given in Output 12.6.

```
proc panel data=wages;
     id people year;
     model lwage=EXP EXPSQ WKS OCC IND SOUTH SMSA MS FEM UNION
     ED BLK/ranone;
run;
```

The variables EXP, EXPSQ, OCC, MS, FEM, UNION, ED, and BLK are all significant while the variables WKS, IND, SOUTH, and SMSA are not significant. Since this is a semi-log model, we can interpret the coefficient for return to schooling as follows: an additional year of schooling results in an 10.7% wage gain. Note that the test statistic for Hausman's test cannot be computed here because the fixed effects model eliminates the model FEM, BLK, and ED. If we eliminate these variables from the model and rerun the random effects model using Proc Panel, we get a Hausman's test statistic value of 541.87 with 7 degrees of freedom that is highly significant. The correct degrees of freedom should be 9 since three explanatory variables were eliminated. Baltagi (2005, p. 128) gives the Hausman's test statistic value that is much larger than the one obtained by the Proc Panel procedure and has the correct degrees of freedom. We conducted the Hausman's test using Proc IML with the within-groups and random effects estimator and calculated the test statistics to be higher than the one reported by the author. Nevertheless, the Hausman test statistic is rejected that justifies the use of an instrumental variables approach.

As stated earlier, the random effects model does not take into account any possible correlation between the explanatory variables and the unobserved individual subject-specific effects. The within-group estimators can be calculated by making use of the code provided in Chapter 7—the code is also provided in the computations of the Hausman–Taylor estimator. The results from the analysis are given in Output 12.7.

As stated earlier, the within-group model sweeps the individual effects from the model resulting in a consistent estimator of the parameters associated with the time-dependent explanatory variables. However, the approach does not allow us to estimate the parameters of the time-invariant effects.

The following steps can be taken to obtain the coefficients estimates under the Hausman–Taylor approach. Note that Proc IML is used in conjunction with other SAS procedures. Also note that the results are slightly off from the results presented in Baltagi (2005, Table 7.4, p. 129). The differences are, however, very small.

```
                    The PANEL Procedure
        Fuller and Battese Variance Components (RanOne)
                Dependent Variable: LWAGE LWAGE
```

Model Description	
Estimation Method	RanOne
Number of Cross Sections	595
Time Series Length	7

Fit Statistics			
SSE	149.3005	DFE	4152
MSE	0.0360	Root MSE	0.1896
R-Square	0.4284		

Variance Component Estimates	
Variance Component for Cross Sections	0.100553
Variance Component for Error	0.023102

Hausman Test for Random Effects		
DF	m Value	Pr > m
0	.	.

Parameter Estimates						
Variable	DF	Estimate	Standard Error	t Value	Pr > |t|	Label
Intercept	1	4.030811	0.1044	38.59	<0.0001	Intercept
EXP	1	0.087726	0.00281	31.27	<0.0001	EXP
expsq	1	-0.00076	0.000062	-12.31	<0.0001	
WKS	1	0.000954	0.000740	1.29	0.1971	WKS
OCC	1	-0.04293	0.0162	-2.65	0.0081	OCC
IND	1	0.00381	0.0172	0.22	0.8242	IND
SOUTH	1	-0.00788	0.0281	-0.28	0.7795	SOUTH
SMSA	1	-0.02898	0.0202	-1.43	0.1517	SMSA
MS	1	-0.07067	0.0224	-3.16	0.0016	MS
FEM	1	-0.30791	0.0572	-5.38	<0.0001	FEM
UNION	1	0.058121	0.0169	3.45	0.0006	UNION
ED	1	0.10742	0.00642	16.73	<0.0001	ED
BLK	1	-0.21995	0.0660	-3.33	0.0009	BLK

OUTPUT 12.6. Random effects model for the wages data.

1. The group means for the response variable and all explanatory variables are calculated using Proc Univariate. The output is stored in a temporary SAS data set called summary.

```
proc univariate data=wages noprint;
        var occ south smsa ind exp expsq wks ms union fem blk
        ed lwage;
        by people;
        output out=summary mean=m_occ m_south m_smsa m_ind
        m_exp m_expsq m_wks m_ms m_union m_fem m_blk m_ed
        m_lwage;
run;
```

```
                    The REG Procedure
                    Model: MODEL1
                Dependent Variable: t_lwage
```

Number of Observations Read	4165
Number of Observations Used	4165

No intercept in model. R-Square is redefined.

Analysis of Variance

Source	DF	Sum of Squares	Mean Square	F Value	Pr > F
Model	9	158.38388	17.59821	889.03	<0.0001
Error	4156	82.26732	0.01979		
Uncorrected Total	4165	240.65119			

Root MSE	0.14069	R-Square	0.6581
Dependent Mean	8.52992E-18	Adj R-Sq	0.6574
Coeff Var	1.649418E18		

Model is not full rank. Least-squares solutions for the parameters are not unique. Some statistics will be misleading. A reported DF of 0 or B means that the estimate is biased.

The following parameters have been set to 0, since the variables are a linear combination of other variables as shown.

t_fem =	0
t_blk =	0
t_ed =	0

Parameter Estimates

Variable	DF	Parameter Estimate	Standard Error	t Value	Pr > |t|
t_occ	1	-0.02148	0.01276	-1.68	0.0924
t_south	1	-0.00186	0.03175	-0.06	0.9533
t_smsa	1	-0.04247	0.01798	-2.36	0.0182
t_ind	1	0.01921	0.01430	1.34	0.1792
t_exp	1	0.11321	0.00229	49.49	<0.0001
t_expsq	1	-0.00041835	0.00005054	-8.28	<0.0001
t_wks	1	0.00083595	0.00055509	1.51	0.1321
t_ms	1	-0.02973	0.01757	-1.69	0.0908
t_union	1	0.03278	0.01381	2.37	0.0177
t_fem	0	0	.	.	.
t_blk	0	0	.	.	.
t_ed	0	0	.	.	.

OUTPUT 12.7. Within-group effects model for the wages data.

The summary statistics are then merged with the original data set to create a data set where the observations on each variable are deviations from the group means. This is accomplished with the following statements.

```
data LSDV_Step;
     merge wages(in=a) summary(in=b);
     by people;
     if a and b;
     t_occ=occ-m_occ;
     t_south=south-m_south;
     t_smsa=smsa-m_smsa;
     t_ind=ind-m_ind;
```

```
        t_exp=exp-m_exp;
        t_expsq=expsq-m_expsq;
        t_wks=wks-m_wks;
        t_ms=ms-m_ms;
        t_union=union-m_union;
        t_fem=fem-m_fem;
        t_blk=blk-m_blk;
        t_ed=ed-m_ed;
        t_lwage=lwage-m_lwage;
    run;<?}j?>
```

2. Proc IML is now used to calculate the within-group estimates and the within-group mean residuals. The group mean residuals from this step are used as a dependent variable in an instrumental variable regression against z_1 and z_2 with instruments x_1 and z_1.

```
proc iml;
    use wages;
    use LSDV_Step;
    read all var{'t_occ', 't_south', 't_smsa', 't_ind',
    't_exp', 't_expsq', 't_wks', 't_ms', 't_union'} into
    X;
    read all var{'t_lwage'} into Y;
    beta=inv(X'*X)*X'*Y;
    summary var{occ south smsa ind exp expsq wks ms union
    lwage} class{people} stat{mean} opt{save};
    Y_M=lwage;
    X_M=occ||south||smsa||ind||
    exp||expsq||wks||ms||union;
    e=Y_M-X_M*beta;
    create e_data from e;
    append from e;
run;
```

3. A new data set is created with the within-group mean residuals and the explanatory variables for the purpose of doing the instrumental variables regression. The following statements can be used.

```
data e_data;
    set e_data;
    people=_n_;
run;
data step2;
    merge wages(in=a) e_data(in=b);
    by people;
    if a and b;
    rename col1=e_mean;
run;
```

The instrumental variable regression is done by using Proc Model. The following statements can be used. The analysis results are given in Output 12.8.

```
proc model data=step2;
    endo ed;
    instruments fem blk occ south smsa ind;
    e_mean=beta2*fem+beta3*blk+beta4*ed;
```

The MODEL Procedure

Model Summary	
Model Variables	4
Endogenous	3
Parameters	3
Equations	1
Number of Statements	1

Model Variables	FEM BLK ED e_mean
Parameters	beta2 beta3 beta4
Equations	e_mean

The Equation to Estimate is	
e_mean =	F(beta2(FEM), beta3(BLK), beta4(ED))
Instruments	1 FEM BLK OCC SOUTH SMSA IND

NOTE: At 2SLS Iteration 1 CONVERGE=0.001 Criteria Met.

The MODEL Procedure
2SLS Estimation Summary

Data Set Options	
DATA=	STEP2

Minimization Summary	
Parameters Estimated	3
Method	Gauss
Iterations	1

Final Convergence Criteria	
R	0
PPC	0
RPC(beta4)	3549.228
Object	0.992621
Trace(S)	1.24545
Objective Value	0.160006

Observations Processed	
Read	4165
Solved	4165

OUTPUT 12.8. Proc model output (preliminary step) to the Hausman and Taylor estimates for the wages data.

The MODEL Procedure

Nonlinear 2SLS Summary of Residual Errors							
Equation	DF Model	DF Error	SSE	MSE	Root MSE	R-Square	Adj R-Sq
e_mean	3	4162	5183.6	1.2454	1.1160	–0.1664	–0.1670

Nonlinear 2SLS Parameter Estimates				
Parameter	Estimate	Approx Std Err	t Value	Approx Pr > |t|
beta2	–0.12485	0.0560	–2.23	0.0258
beta3	0.056205	0.0679	0.83	0.4080
beta4	0.358572	0.00143	250.24	<0.0001

Number of Observations		Statistics for System	
Used	4165	Objective	0.1600
Missing	0	Objective*N	666.4242

OUTPUT 12.8. (*Continued*)

```
    fit e_mean/2sls;
run;
```

4. The final step is to calculate the weighted instrumental variables estimator. From the within-groups analysis, an estimate of σ_ε^2 is 0.0231 and from the instrumental regression analysis, an estimate of σ^{*2} is 1.2454. These values can be used to calculate an estimate of σ_u^2. That is,

$$
\begin{aligned}
\hat{\sigma}_u^2 &= \hat{\sigma}^{*2} - \sigma_\varepsilon^2/T \\
&= 1.2454 - 0.0231/7 \\
&= 1.2421.
\end{aligned}
$$

An estimate of the weight, $\hat{\theta}$, can now be derived as follows:

$$
\begin{aligned}
\hat{\theta} &= \sqrt{\frac{\hat{\sigma}_\varepsilon^2}{\hat{\sigma}_\varepsilon^2 + T\hat{\sigma}_u^2}} \\
&= \sqrt{\frac{0.0231}{0.0231 + 7 \times 1.2421}} \\
&= 0.051476.
\end{aligned}
$$

The calculation is carried out in the following data step statements in SAS. The variables are also transformed using this weight in the following statements.

```
data Final_Step;
    merge wages(in=a) summary(in=b);
    by people;
    if a and b;
    sigmae=0.0231;
```

```
        sigmau=1.2421;
        Theta=sqrt(sigmae/(sigmae+7*sigmau));
        t_occ=occ-(1-theta)*m_occ;
        t_south=south-(1-theta)*m_south;
        t_smsa=smsa-(1-theta)*m_smsa;
        t_ind=ind-(1-theta)*m_ind;
        t_exp=exp-(1-theta)*m_exp;
        t_expsq=expsq-(1-theta)*m_expsq;
        t_wks=wks-(1-theta)*m_wks;
        t_ms=ms-(1-theta)*m_ms;
        t_union=union-(1-theta)*m_union;
        t_fem=fem-(1-theta)*m_fem;
        t_blk=blk-(1-theta)*m_blk;
        t_ed=ed-(1-theta)*m_ed;
        t_lwage=lwage-(1-theta)*m_lwage;
        s_occ=occ-m_occ;
        s_south=south-m_south;
        s_smsa=smsa-m_smsa;
        s_ind=ind-m_ind;
        s_exp=exp-m_exp;
        s_expsq=expsq-m_expsq;
        s_wks=wks-m_wks;
        s_ms=ms-m_ms;
        s_union=union-m_union;
        s_fem=fem-m_fem;
        s_blk=blk-m_blk;
        s_ed=ed-m_ed;
    run;
```

Proc IML is then used to calculate the Hausman–Taylor's estimates for the earnings equation. The analysis results are given in Output 12.9. The model indicates that an additional year of schooling results in a 13.73% wage gain. This is significantly different from the estimate obtained from the random effects model.

```
proc iml;
* Read the data into matrices.;
    use final_step;read all
    var{'t_occ','t_south','t_smsa','t_ind','t_exp',
    't_exsq','t_wks','t_ms','t_union',
    't_fem','t_blk','t_ed'} into W;
    read all var{t_lwage} into Y;
    W=J(4165,1,0.051408)||W;
    read all
    var{'s_occ','s_south','s_smsa','s_ind','s_exp',
    's_expsq','s_wks','s_ms','s_union',
    'fem','blk','m_occ','m_south','m_smsa','m_ind'} into
    V;
* Calculate the Hausman and Taylor estimates and standard
errors.;
    HT=inv((W'*V)*inv(V'*V)*(V'*W))*((W'*V)*inv(V'*V)*(V'
    *y));
    MSE=(y-W*HT)'*(y-W*HT)/(4165);
    SE=SQRT(vecdiag(MSE*inv((W'*V)*inv(V'*V)*(V'*W))));
run;
```

The Hausman & Taylor Estimates are

TABLE1		
	BHAT	SE
INTERCEPT	2.2098	0.2712
OCC	-0.0209	0.0135
SOUTH	0.0052	0.0319
SMSA	-0.0419	0.0189
IND	0.0152	0.0150
EXP	0.1132	0.0024
EXP_SQ	-0.0004	0.0001
WKS	0.0008	0.0006
MS	-0.0298	0.0186
UNION	0.0328	0.0146
FEM	-0.1291	0.1481
BLK	-0.2852	0.1798
ED	0.1373	0.0282

OUTPUT 12.9. Hausman and Taylor estimates of the wages equation.

12.6 ROBUST ESTIMATION OF COVARIANCE MATRICES IN PANEL DATA

The panel data models discussed in Chapter 7 were based on the assumption of homoscedastic disturbances. This section extends the discussion to heteroscedasticity in panel data models. We will focus our attention on the robust estimation of the covariance matrix for fixed effects models and will use the Proc Panel procedure to calculate various robust estimates of the covariance matrix. We illustrate the various techniques by revisiting the cost of US airlines data set from Greene (2003).

The HCCME option in Proc Model can be adjusted to generate robust estimates of the variance–covariance matrix. The various options are given below (The Panel Procedure, p. 58, SAS Institute, Inc.). Also see the discussion on heteroscedasticity in Chapter 5.

If we do not specify the HCCME option, then the analysis will default the OLS estimate of the covariance matrix. The OLS output for the airlines data has been given in Chapter 7.

HCCME = 0: This yields the White's estimator

$$\frac{1}{nT} \sum_{i=1}^{nT} \hat{\varepsilon}_i^2 \mathbf{x}_i \mathbf{x}_i^T.$$

HCCME = 1: This yields the first version of the Davidson and MacKinnon (1993) estimator where the end result of the White's estimator is scaled up by a factor of $nT/(nT - K)$:

$$\frac{1}{nT} \sum_{i=0}^{nT} \left(\frac{nT}{nT-K} \hat{\varepsilon}_i^2 \mathbf{x}_i \mathbf{x}_i^T \right).$$

HCCME = 2: This yields the second version of the Davidson and MacKinnon (1993) estimator where the White's estimator is adjusted by the diagonals of the hat matrix:

$$\hat{h}_i = \mathbf{X}_i (\mathbf{X}^T \mathbf{X})^{-1} \mathbf{X}_i^T.$$

The estimator is given by

$$\frac{1}{nT} \sum_{i=0}^{nT} \frac{\hat{\varepsilon}_i^2}{1 - \hat{h}_i} \mathbf{x}_i \mathbf{x}_i^T.$$

Obs	_TYPE_	_NAME_	LnQ	LnPF	LF
1	OLS	LnQ	0.000893416	-0.000317817	-0.00188
2	OLS	LnPF	-0.000317817	0.000231013	-0.00077
3	OLS	LF	-0.001884262	-0.000768569	0.04068
4	HCCME0	LnQ	0.000365016	-0.000125245	-0.00031
5	HCCME0	LnPF	-0.000125245	0.000183132	-0.00169
6	HCCME0	LF	-0.000306158	-0.001690757	0.04692
7	HCCME1	LnQ	0.000405574	-0.000139161	-0.00034
8	HCCME1	LnPF	-0.000139161	0.000203480	-0.00188
9	HCCME1	LF	-0.000340176	-0.001878619	0.05214
10	HCCME2	LnQ	0.000397411	-0.000134190	-0.00034
11	HCCME2	LnPF	-0.000134190	0.000192432	-0.00178
12	HCCME2	LF	-0.000337505	-0.001783683	0.04989
13	HCCME3	LnQ	0.000435062	-0.000144422	-0.00038
14	HCCME3	LnPF	-0.000144422	0.000202471	-0.00188
15	HCCME3	LF	-0.000378817	-0.001881047	0.05310
16	HCCME4	LnQ	0.000870151	0.000062860	-0.00794
17	HCCME4	LnPF	0.000062860	0.000301454	-0.00164
18	HCCME4	LF	-0.007938291	-0.001642741	0.14797

OUTPUT 12.10. HCCME estimators for the airlines data set.

HCCME = 3: This yields an estimator that is similar to the second version of the Davidson and MacKinnon's estimator. The adjustment is now based on $(1-\hat{h}_i)^2$ instead of $(1-\hat{h}_i)$.

$$\frac{1}{nT}\sum_{i=0}^{nT}\frac{\hat{\varepsilon}_i^2}{(1-\hat{h}_i)^2}\mathbf{x}_i\mathbf{x}_i^T.$$

HCCME = 4: This yields the Arellano (1987) version of the White's estimator for panel data. The general idea involves calculating the White's estimator for each cross section in the panel ($i = 1, \ldots, n$) and then taking the average of the n estimates of the covariance matrices. See the Proc Panel Procedure (p. 58) for more details on this estimator.

Section 5.5 included a SAS program to print the various robust covariance matrices under heteroscedasticity. The code can easily be adjusted to generate the robust covariance matrices in the panel setting.

The results of the analysis are given in Output 12.10. The diagonal elements in each covariance matrix give the variance estimates for the parameters. Notice the similarity between White's estimator and the Davidson and McKinnon's estimators. The OLS estimators and the Arellano's version of the White's estimator are different from these.

12.7   DYNAMIC PANEL DATA MODELS

We now turn our attention to dynamic panel data models. That is, models that are characterized by lagged variables on the right-hand side of the model. The general form of these models is given by Baltagi (2005, pp. 134–142) and Verbeek (2004, pp. 360–366)

$$y_{i,t} = \rho y_{i,t-1} + \mathbf{x}_{i,t}^T\boldsymbol{\beta} + \alpha_i + \varepsilon_{i,t} \quad i = 1, \ldots, n; t = 1, \ldots, T$$

where $y_{i,t}$ is a 1×1 scalar dependent variable, $\mathbf{x}_{i,t}$ is a $k \times 1$ vector of explanatory variables, and ρ and $\boldsymbol{\beta}$ are 1×1 and $k \times 1$ parameters that need to be estimated. The term α_i is the unobserved subject-specific heterogeneity and $\varepsilon_{i,t}$ is the disturbance. The subscripts i and t index the subjects and the time period, respectively. As shown by Verbeek (2004, p. 361), the use of lagged dependent variables on the right-hand side of the model introduces estimation problems, more specifically, with the fixed effect model estimator becoming biased regardless of whether α_i is treated as fixed or random. This section focuses on methods based on generalized methods of moments estimation (GMM) that can be used to estimate the parameters in dynamic panel data models.

12.7.1 Dynamic Panel Data Estimation

The estimation technique is based on the class of GMM estimators introduced by Arnello and Bond (1991). They proposed an estimation method based on instrumental variables estimation. As will be seen, the authors took advantage of the independence between the lagged values of the dependent variable and the disturbances. Their general method and formulas can be found in most intermediate–advanced texts on econometrics. I have found Baltagi (2005), Greene (2003), and Verbeek (2004) very useful in understanding the mechanics of the Arellano and Bond estimator. The Proc Panel documentation from SAS Institute, Inc. is also a good reference as it summarizes the key steps and formulas in the estimation process.

A discussion of GMM estimators is beyond the scope of this book (see the above mentioned texts for details). In general, assuming the classical linear model $\mathbf{y} = \mathbf{X}\boldsymbol{\beta} + \boldsymbol{\varepsilon}$ where $\mathbf{X}$ is suspected of being endogenous and where there is a matrix of instruments $\mathbf{Z}$, the GMM estimator takes the form

$$\hat{\boldsymbol{\beta}} = (\mathbf{X}^T \mathbf{Z} \mathbf{W} \mathbf{Z}^T \mathbf{W})^{-1} \mathbf{X}^T \mathbf{Z} \mathbf{W} \mathbf{Z}^T \mathbf{y}$$

where $\mathbf{W}$ is called the weights matrix and is chosen to minimize the asymptotic covariance of $\hat{\boldsymbol{\beta}}$. GMM estimation is usually done in two steps using an initial weight matrix (not optimal) in step 1. The optimal weight matrix is then formed using the residuals from step 1 to calculate the second step GMM estimator.

The estimation steps are best understood by considering a basic dynamic panel data model without exogenous variables. Consider the following simple autoregressive random effects panel model (Baltagi, 2005, p. 136; Verbeek, 2004, p. 361):

$$y_{i,t} = \rho y_{i,t-1} + \alpha_i + \varepsilon_{i,t}, \quad i = 1, \dots n; \, t = 1, \dots, T,$$

where α_i and $\varepsilon_{i,t}$ are independently and identically distributed disturbances with variances σ_α^2 and σ_ε^2, respectively. Furthermore, assume that α_i and $\varepsilon_{i,t}$ are independent of each other.

An initial approach is to take the first differences since this "sweeps" the unobserved individual effects α_i from the model resulting in

$$y_{i,t} - y_{i,t-1} = \rho \lfloor y_{i,t-1} - y_{i,t-2} \rfloor + \lfloor \varepsilon_{i,t} - \varepsilon_{i,t-1} \rfloor .$$

However, the OLS estimator of ρ will still be biased and inconsistent since $Cov(y_{i,t-1}, \varepsilon_{i,t-1}) \neq 0$. To see how instrumental variable estimation can be used to estimate ρ, consider the difference model at $t = 1, 2, \dots$. Obviously, this model is valid for the first time when $t = 3$. The difference model is given by

$$y_{i,3} - y_{i,t-2} = \rho \lfloor y_{i,2} - y_{i,1} \rfloor + \lfloor \varepsilon_{i,3} - \varepsilon_{i,2} \rfloor .$$

Here, $y_{i,1}$ is a valid instrument since it is highly correlated to $y_{2,i} - y_{1,i}$ but is uncorrelated to $\varepsilon_{i,3} - \varepsilon_{i,2}$. Further, we can see that at $t = 4$, the differenced model is

$$y_{i,4} - y_{i,3} = \rho \lfloor y_{i,3} - y_{i,2} \rfloor + \lfloor \varepsilon_{i,4} - \varepsilon_{i,3} \rfloor$$

and now both $y_{i,1}$ and $y_{i,2}$ are valid instruments as both are uncorrelated with $\varepsilon_{i,4} - \varepsilon_{i,3}$. In this fashion, we see that the set of instrumental variables for a given time period t is $y_{i,1}, \dots, y_{i,t-2}, i = 1, \dots, n$ (Baltagi, 2005, p. 137).

If $\mathbf{Z}_i$ denotes the $p \times (T-2)$ matrix of instruments for the i th subject, then it is easy to see that

$$\mathbf{Z}_i^T = \begin{bmatrix} y_{i,1} & 0 & 0 & \dots & \dots & 0 \\ 0 & y_{i,1} & 0 & \dots & \dots & 0 \\ 0 & y_{i,2} & 0 & \dots & \dots & 0 \\ 0 & 0 & \dots & \dots & \dots & 0 \\ \vdots & \vdots & \ddots & \ddots & 0 & y_{i,1} \\ \vdots & \vdots & \ddots & \ddots & 0 & y_{i,2} \\ \vdots & \vdots & \ddots & \ddots & 0 & \vdots \\ 0 & 0 & \dots & \dots & \dots & y_{i,T-2} \end{bmatrix} .$$

The number of rows in $\mathbf{Z}_i$ equals $p = \sum_{t=1}^{T-2} t$. If we combine the instrument variables matrix for all cross sections, we have $\mathbf{Z}^T = \begin{bmatrix} \mathbf{Z}_1^T & \mathbf{Z}_2^T & \dots & \mathbf{Z}_n^T \end{bmatrix}$.

The weight matrix for the first-step estimator is given by $\mathbf{Z}^T\mathbf{H}\mathbf{Z}$ with $\mathbf{H}_i = diag[\mathbf{H}_1, \dots, \mathbf{H}_n]$ where

$$\mathbf{H}_i = \begin{bmatrix} 2 & -1 & 0 & \dots & 0 \\ -1 & 2 & -1 & \ddots & \vdots \\ 0 & \ddots & \ddots & \ddots & 0 \\ \vdots & \ddots & -1 & 2 & -1 \\ 0 & \dots & 0 & -1 & 2 \end{bmatrix}.$$

The $(T-2) \times (T-2)$ matrix $\mathbf{H}_i$ can be constructed as follows:

1. For the ith cross section, denote the vector of differenced residuals, $\Delta\boldsymbol{\varepsilon}_i$ as

$$\Delta\boldsymbol{\varepsilon}_i = (\varepsilon_{i,3}-\varepsilon_{i,2}, \dots, \varepsilon_{i,T}-\varepsilon_{i,T-1}).$$

2. The diagonal elements of the matrix are given by $E(\Delta\varepsilon_{i,t}^2) = 2\sigma_\varepsilon^2$ while the off-diagonal elements are given by $E(\Delta\varepsilon_{i,t}\Delta\varepsilon_{i,t-1}) = -\sigma_\varepsilon^2$, and $E(\Delta\varepsilon_{i,t}\Delta\varepsilon_{i,t-s}) = 0$ for $S \geq 2$.

Notice that σ_ε^2 cancels out in the subsequent steps and is therefore not included in $\mathbf{H}_i$. If we denote the $T-2$ difference terms for all n cross sections as $\Delta\mathbf{y}$, $\Delta\mathbf{y}_{-1}$, and $\Delta\boldsymbol{\varepsilon}$, we can write the difference model as $\Delta\mathbf{y} = \rho\Delta\mathbf{y}_{-1} + \Delta\boldsymbol{\varepsilon}$, then Arellano and Bond's first-step estimator is given by

$$\hat{\rho}_{GMM1} = \left[\Delta\mathbf{y}_{-1}^T\mathbf{Z}(\mathbf{Z}^T\mathbf{H}\mathbf{Z})^{-1}\mathbf{Z}^T\Delta\mathbf{y}_{-1}\right]^{-1} \times \left[\Delta\mathbf{y}_{-1}^T\mathbf{Z}(\mathbf{Z}^T\mathbf{H}\mathbf{Z})^{-1}\mathbf{Z}^T\Delta\mathbf{y}\right].$$

The residuals from this step are given by

$$\Delta\hat{\boldsymbol{\varepsilon}}_i^1 = \Delta\mathbf{y}_i - \hat{\rho}_{GMM1}\Delta\mathbf{y}_{i,-1} \quad \text{for } i = 1, \dots, n$$

and are used to construct the weight matrix for the second-step estimator. That is, if we let $\hat{\boldsymbol{\Omega}} = \Delta\hat{\boldsymbol{\varepsilon}}\Delta\hat{\boldsymbol{\varepsilon}}^T$, then the weight matrix for the second-step estimator is given by $\mathbf{Z}^T\hat{\boldsymbol{\Omega}}\mathbf{Z}$. Arellano and Bond's second-step estimator is given by

$$\hat{\rho}_{GMM2} = \left[\Delta\mathbf{y}_{-1}^T\mathbf{Z}(\mathbf{Z}^T\hat{\boldsymbol{\Omega}}\mathbf{Z})^{-1}\mathbf{Z}^T\Delta\mathbf{y}_{-1}\right]^{-1} \times \left[\Delta\mathbf{y}_{-1}^T\mathbf{Z}(\mathbf{Z}^T\hat{\boldsymbol{\Omega}}\mathbf{Z})^{-1}\mathbf{Z}^T\Delta\mathbf{y}\right].$$

The following computational formulas for the various terms in the GMM calculations are useful when programming the estimation method in Proc IML.

$$\mathbf{Z}^T\mathbf{H}\mathbf{Z} = \sum_{i=1}^{n} \mathbf{Z}_i^T\mathbf{H}_i\mathbf{Z}_i,$$

$$\Delta\mathbf{y}_{-1}^T\mathbf{Z} = \sum_{i=1}^{n} \Delta\mathbf{y}_{i,-1}^T\mathbf{Z}_i,$$

and

$$\Delta\mathbf{y}^T\mathbf{Z} = \sum_{i=1}^{n} \Delta\mathbf{y}_i^T\mathbf{Z}_i.$$

We will illustrate the GMM method for this simple model on the cigar.txt panel data used by Baltagi and Levin (1992). Consider the model (see Baltagi, 2005, pp. 156–158)

$$\ln C_{i,t} = \rho\ln C_{i,t-1} + \alpha_i + \varepsilon_{i,t} \quad i = 1, \dots, 46; \; t = 1, \dots, 30.$$

As discussed by the author, the data set consists of real per capita sales of cigarettes ($C_{i,t}$) in 46 states ($n = 46$) between 1963 and 1992 ($T = 30$). The model contains the lag of this endogenous variable and therefore an OLS estimate of ρ will be inconsistent. The authors fit a more complex dynamic panel model, which will be discussed subsequently. For the moment, we are interested in estimating the basic model given above.

Program 14 in Appendix E contains the complete Proc IML code for estimating both a one- and a two-step GMM estimate for ρ. We leave it as an exercise to verify that the coefficients from the two GMM procedures give values of 1.031457 and 1.031563, respectively. It is very easy to extract the standard errors of the first- and second-step estimators and we leave the details as an exercise for the reader.

12.7.2 Dynamic Panel Data Models with Explanatory Variables

We now turn our attention to dynamic panel data models with explanatory variables. The general form of the model was given in the earlier section. The Arellano and Bond (1991) GMM estimator used in the simple model with no explanatory variables can easily be modified to the case involving explanatory variables.

As before, note that the presence of lagged variables on the right-hand side leads to biased and inconsistent estimators of the parameters. We proceed by taking the first difference of the model and observing the following relationship:

$$y_{i,t} - y_{i,t-1} = \rho \left[y_{i,t-1} - y_{i,t-2} \right] + \left[\mathbf{x}_{i,t}^T - \mathbf{x}_{i,t-1}^T \right] \boldsymbol{\beta} + \left[\varepsilon_{i,t} - \varepsilon_{i,t-1} \right].$$

OLS cannot be used here to estimate the parameters because $Cov(y_{i,t-1}, \varepsilon_{i,t-1}) \neq 0$. Also note that the difference relationship is observed for the first time when $t = 3$ and is given by

$$y_{i,3} - y_{i,2} = \rho[y_{i,2} - y_{i,1}] + [\mathbf{x}_{i,3}^T - \mathbf{x}_{i,2}^T]\boldsymbol{\beta} + [\varepsilon_{i,3} - \varepsilon_{i,2}].$$

Earlier, we saw that $y_{i,1}$ can be used as an instrument for $y_{i,2} - y_{i,1}$ because it is highly correlated with it and is uncorrelated to $\varepsilon_{i,3} - \varepsilon_{i,2}$. Assuming that the explanatory variables are strictly exogenous, we can use all of them as additional instruments for estimating the parameters. That is, the instrumental variables matrix $\mathbf{Z}_i$ is given by

$$\mathbf{Z}_i = \begin{bmatrix} \left[y_{i,1}, \mathbf{x}_{i,1}^T, \ldots, \mathbf{x}_{i,T}^T \right] & 0 & \cdots & 0 \\ 0 & \left[y_{i,1}, y_{i,2}, \mathbf{x}_{i,1}^T, \ldots, \mathbf{x}_{i,T}^T \right] & \cdots & 0 \\ \vdots & \vdots & \ddots & \vdots \\ 0 & & \cdots & \left[y_{i,1}, \ldots, y_{i,T-2}, \mathbf{x}_{i,1}^T, \ldots, \mathbf{x}_{i,T}^T \right] \end{bmatrix}.$$

Under the assumption that the explanatory variables are predetermined, we know that $E(\mathbf{x}_{i,s}\varepsilon_{i,t}) = 0$ for $s < t$ and 0 otherwise. Therefore, we can use $\mathbf{x}_{i,t}$ as instruments up to the same time period as the error term. That is, at time s, only $\mathbf{x}_{i,1}^T, \ldots, \mathbf{x}_{i,s-1}^T$ are valid instruments in the first differenced equation. The matrix of instruments $\mathbf{Z}_i$ in the predetermined case is given by Arellano and Bond (1991) and Baltagi (2005, p. 140):

$$\mathbf{Z}_i = \begin{bmatrix} \left[y_{i,1}, \mathbf{x}_{i,1}^T, \mathbf{x}_{i,2}^T \right] & 0 & \cdots & 0 \\ 0 & \left[y_{i,1}, y_{i,2}, \mathbf{x}_{i,1}^T, \mathbf{x}_{i,2}^T, \mathbf{x}_{i,3}^T \right] & \cdots & 0 \\ \vdots & \vdots & \ddots & \vdots \\ 0 & & \cdots & \left[y_{i,1}, \ldots, y_{i,T-2}, \mathbf{x}_{i,1}^T, \ldots, \mathbf{x}_{i,T-1}^T \right] \end{bmatrix}.$$

The following formulation of the GMM estimator assumes that the explanatory variables are predetermined. The Arellano–Bond estimation method in the more general case involves first constructing $\mathbf{Z}_i$ as defined above along with $\mathbf{H}_i$ (see Section 12.7.1). The weight matrix for the first-step GMM estimator is given by $\mathbf{Z}^T \mathbf{H} \mathbf{Z}$ where $\mathbf{Z}$ and $\mathbf{H}$ are as defined in

Section 12.7.1. Next, for the i th deviation, stack all the deviations and construct $\mathbf{X}_i$ and $\mathbf{Y}_i$ as follows:

$$\mathbf{X}_i = \begin{bmatrix} \Delta y_{i,2} & \Delta \mathbf{x}_{i,3}^T \\ \Delta y_{i,3} & \Delta \mathbf{x}_{i,4}^T \\ \vdots & \vdots \\ \Delta y_{i,T-1} & \Delta \mathbf{x}_{i,T}^T \end{bmatrix} \text{ and } \mathbf{Y}_i = \begin{bmatrix} \Delta y_{i,3} \\ \Delta y_{i,4} \\ \vdots \\ \Delta y_{i,T} \end{bmatrix}.$$

Stacking $\mathbf{X}_i$, $\mathbf{Y}_i$ for all cross sections, we have

$$\mathbf{Z} = \begin{bmatrix} \mathbf{Z}_1 \\ \mathbf{Z}_2 \\ \vdots \\ \mathbf{Z}_n \end{bmatrix}, \mathbf{X} = \begin{bmatrix} \mathbf{X}_1 \\ \mathbf{X}_2 \\ \vdots \\ \mathbf{X}_n \end{bmatrix}, \text{ and } \mathbf{Y} = \begin{bmatrix} \mathbf{Y}_1 \\ \mathbf{Y}_2 \\ \vdots \\ \mathbf{Y}_n \end{bmatrix}.$$

Therefore, the Arellano–Bond first-step estimator is given by

$$\hat{\boldsymbol{\delta}}_{GMM1} = [\mathbf{X}^T \mathbf{Z}(\mathbf{Z}^T \mathbf{H} \mathbf{Z})^{-1} \mathbf{Z}^T \mathbf{X}]^{-1} [\mathbf{X}^T \mathbf{Z}(\mathbf{Z}^T \mathbf{H} \mathbf{Z})^{-1} \mathbf{Z}^T \mathbf{Y}].$$

The residuals from the first-step GMM analysis is given by

$$\Delta \hat{\boldsymbol{\varepsilon}}_i^1 = \mathbf{Y}_i - \mathbf{X}_i \hat{\boldsymbol{\delta}}_{GMM1}$$

and is used to create the optimal weight matrix $\mathbf{Z}^T \hat{\boldsymbol{\Omega}} \mathbf{Z}$ where $\hat{\boldsymbol{\Omega}} = \Delta \hat{\boldsymbol{\varepsilon}} \Delta \hat{\boldsymbol{\varepsilon}}^T$ and is used in place of $\mathbf{Z}^T \mathbf{H} \mathbf{Z}$ to generate the two-step Arellano–Bond GMM estimator

$$\hat{\boldsymbol{\delta}}_{GMM2} = [\mathbf{X}^T \mathbf{Z}(\mathbf{Z}^T \hat{\boldsymbol{\Omega}} \mathbf{Z})^{-1} \mathbf{Z}^T \mathbf{X}]^{-1} [\mathbf{X}^T \mathbf{Z}(\mathbf{Z}^T \hat{\boldsymbol{\Omega}} \mathbf{Z})^{-1} \mathbf{Z}^T \mathbf{Y}].$$

We illustrate the Arellano and Bond GMM estimation method in the general case on the cigar.txt panel data used by Baltagi and Levin (1992). Consider the full model (Baltagi, 2005, p. 156) given by

$$\ln C_{i,t} = \rho \ln C_{i,t-1} + \beta_1 \ln P_{i,t} + \beta_2 \ln Y_{i,t} + \beta_3 \ln Pn_{i,t} + \alpha_i + \varepsilon_{i,t}$$

with $i = 1, \ldots, 46$ and $t = 1, \ldots, 30$. Here, i and t index the states and the time periods, respectively. As described by the author, $C_{i,t}$ is the average number of packs of cigarette sales per person over the age of 14, $P_{i,t}$ is the average retail price of a pack of cigarettes, $Y_{i,t}$ is the disposable income, and $Pn_{i,t}$ is the minimum price of cigarettes in the adjoining states.

Program 15 in Appendix E contains the complete Proc IML code for estimating the cigarette panel data model using the Arellano–Bond method. The reader is asked to verify that the first-step estimates are $\rho = 0.799$ $\beta_1 = -0.259$, $\beta_2 = 0.138$, and $\beta_3 = 0.065$ and that the second-step estimates are $\rho = 0.79$, $\beta_1 = -0.26$, $\beta_2 = 0.139$, and $\beta_3 = 0.033$. It is very easy to extract the standard errors of the first- and second-step GMM estimators. We leave the details as an exercise for the reader.

12.8 HETEROGENEITY AND AUTOCORRELATION IN PANEL DATA MODELS

This section deals with analytical methods where the subjects may be correlated with each other and where heterogeneity is due to significant differences between the within subject (cross section) variances. We will also look at the case of autocorrelation where the correlation is across the time periods. As discussed in Greene (2003, p. 320), a formulation of the model, where the conditional mean is assumed to be the same across the cross sections, can be written as

$$y_{it} = \mathbf{x}_{it}^T \boldsymbol{\beta} + \varepsilon_{it} \quad \text{or} \quad \mathbf{y}_i = \mathbf{X}_i \boldsymbol{\beta} + \boldsymbol{\varepsilon}_i,$$

where i indexes the subjects and t indexes time and we assume that each $\mathbf{X}_i$ is exogenous. Stacking the equations across the n subjects yields $\mathbf{Y} = \mathbf{X}\boldsymbol{\beta} + \boldsymbol{\varepsilon}$ where $\mathbf{Y}$ is $nT \times 1$, $\mathbf{X}$ is $nT \times \mathrm{k}$, $\boldsymbol{\beta}$ is $k \times 1$, and $\boldsymbol{\varepsilon}$ is $nT \times 1$.

If we assume that the subjects are correlated with each other and that there is correlation across the time periods as well, then for the (i,j)th subjects, $E(\boldsymbol{\varepsilon}_i \boldsymbol{\varepsilon}_j^T | \mathbf{X}) = \boldsymbol{\Omega}_{ij}$. The cross-sectional variance across groups can therefore be written as

$$E(\boldsymbol{\varepsilon}\boldsymbol{\varepsilon}^T | \mathbf{X}) = \boldsymbol{\Omega} = \begin{bmatrix} \boldsymbol{\Omega}_{11} & \boldsymbol{\Omega}_{12} & \cdots & \boldsymbol{\Omega}_{1n} \\ \boldsymbol{\Omega}_{21} & \boldsymbol{\Omega}_{22} & \cdots & \boldsymbol{\Omega}_{2n} \\ \vdots & \vdots & \ddots & \vdots \\ \boldsymbol{\Omega}_{n1} & \boldsymbol{\Omega}_{n2} & \cdots & \boldsymbol{\Omega}_{nn} \end{bmatrix}.$$

As shown in Greene (2003), each $\boldsymbol{\Omega}_{ij}$ is a $T \times T$ matrix that incorporates both the cross-sectional and the cross-period correlations. For the case where there is no correlation across the time periods, the above can be rewritten as

$$E(\boldsymbol{\varepsilon}\boldsymbol{\varepsilon}^T | \mathbf{X}) = \boldsymbol{\Omega} = \begin{bmatrix} \sigma_{11}\mathbf{I} & \sigma_{12}\mathbf{I} & \cdots & \sigma_{1n}\mathbf{I} \\ \sigma_{21}\mathbf{I} & \sigma_{22}\mathbf{I} & \cdots & \sigma_{2n}\mathbf{I} \\ \vdots & \vdots & \ddots & \vdots \\ \sigma_{n1}\mathbf{I} & \sigma_{n2}\mathbf{I} & \cdots & \sigma_{nn}\mathbf{I} \end{bmatrix}$$

where the σ_{ij}'s capture the cross-sectional correlations.

12.8.1 GLS Estimation

As stated in Greene (2003, p. 321), the full generalized linear regression model using $\boldsymbol{\Omega}$ consists of $nT(nT + 1)/2$ unknown parameters. Estimation is not possible with nT observations unless restrictions are placed on these parameters. A simple restriction is to assume that there is no correlation across time periods that gives us the simplified version of $\boldsymbol{\Omega}$ given above. If we let $\boldsymbol{\Sigma} = \lfloor \sigma_{ij} \rfloor$, then we can rewrite the variance–covariance matrix as $\boldsymbol{\Omega} = \boldsymbol{\Sigma} \otimes \mathbf{I}$. Using the methods from Chapter 5, we can write the GLS estimator of $\boldsymbol{\beta}$ as

$$\hat{\boldsymbol{\beta}}_{GLS} = (\mathbf{X}^T \boldsymbol{\Omega}^{-1} \mathbf{X})^{-1} \mathbf{X}^T \boldsymbol{\Omega}^{-1} \mathbf{y}.$$

As shown by Greene (2003), if we let $\boldsymbol{\Omega}^{-1} = \boldsymbol{\Sigma}^{-1} \otimes \mathbf{I} = \lfloor \sigma^{ij} \rfloor \otimes \mathbf{I}$, then the GLS estimator is

$$\hat{\boldsymbol{\beta}}_{GLS} = \left[\sum_{i=1}^{n} \sum_{j=1}^{n} \sigma^{ij} \mathbf{X}_i^T \mathbf{X}_j \right]^{-1} \left[\sum_{i=1}^{n} \sum_{j=1}^{n} \sigma^{ij} \mathbf{X}_i^T y_j \right]$$

with asymptotic variance given by

$$Asy.Var(\hat{\boldsymbol{\beta}}_{GLS}) = (\mathbf{X}^T \boldsymbol{\Omega}^{-1} \mathbf{X})^{-1}.$$

12.8.2 Feasible GLS Estimation

In practice, $\boldsymbol{\Omega}$ is unknown and has to be estimated using FGLS methods estimators. The analysis is done in two steps. In step 1, OLS is used on the stacked model to obtain the residuals. Estimates of σ_{ij} are given by $\hat{\sigma}_{ij} = \mathbf{e}_i^T \mathbf{e}_j / T$ (Greene, 2003, p. 322). With $\hat{\sigma}_{ij}$ in hand, the FGLS estimators can easily be calculated.

The groupwise heteroscedasticity estimator that was discussed in Chapter 5 is a special case of the FGLS estimator here with the off-diagonal elements of $\boldsymbol{\Sigma}$ equal to 0. Here, we are assuming that there is no cross-sectional correlation and no correlation across the time periods. However, the cross-sectional variances are significantly different from each other.

Using the methods from Chapter 5, the groupwise heteroscedasticity estimator is given by Greene (2003, p. 323):

$$\hat{\boldsymbol{\beta}}_{GLS} = \left[\sum_{i=1}^{n} \frac{1}{\sigma_i^2} \mathbf{X}_i^T \mathbf{X}_i \right]^{-1} \left[\sum_{i=1}^{n} \frac{1}{\sigma_i^2} \mathbf{X}_i^T \mathbf{y}_i \right].$$

Estimating σ_i as before and using it in the above equation gives the FGLS estimator. We will now illustrate the steps involved in the Grunfeld data set. The Proc IML code is given below.

```
        Proc IML;
*Read the data into appropriate matrices;
        Use SUR;
        Read all var{'F' 'C'} into X;
        Read all var{'I'} into Y;
*Store the dimensions of X;
        r=nrow(X);c=ncol(X);
*Append a column of 1's to X;
        X=J(r,1,1)||X;
*Conduct OLS to get pooled OLS model and calculate the residuals;
*This is step 1 of the procedure;
        BHAT1=inv(X'*X)*X'*Y;
        E1=Y-X*BHAT1;
*Conduct the groupwise heteroscedastic analysis.;
*This is step 2 of the procedure;
        compt=1;
    M=5;
    T=20;
    Temp0=0;Temp1=shape(0,3,3);Temp2=shape(0,3,1);
    do i=1 to M;
        Temp0=E1[compt:compt+t-1,1]'*E1[compt:compt+t-1,1]/T;
        Temp1=Temp1+1/Temp0*X[compt:compt+t-
        1,1:3]'*X[compt:compt+t-1,1:3];
        compt=compt+t;
    end;
    compt=1;
    do i=1 to M;
        Temp0=E1[compt:compt+t-1,1]'*E1[compt:compt+t-1,1]/T;
        Temp2=Temp2+1/Temp0*X[compt:compt+t-
        1,1:3]'*Y[compt:compt+t-1,1];
        compt=compt+t;
    end;
    BHAT_GRP=inv(Temp1)*Temp2;
    Print 'The Groupwise Heteroscedastic Parameter Estimates
    Are';
    Print BHAT_GRP;
*Now, calculate the asymptotic covariance matrix;
    Grp_Sig=Shape(0,5,1);
    compt=1;
    do i=1 to M;
        Grp_Sig[i,1]=E1[compt:compt+t-1,1]'*E1[compt:compt+t-
        1,1]/T;
        compt=compt+T;
    end;
    Cap_Sigma=diag(Grp_Sig);
```

```
    ID=I(T);
    Omega=Cap_Sigma@ID;
    Asy_Var=inv(X'*inv(Omega)*X);
    Print 'The Asymptotic Covariance Matrix Is';
    Print Asy_Var;
    SE=sqrt(vecdiag(Asy_Var));
    Print 'The Asymptotic Standard Errors Are';
    Print SE;
*Now, we get the residuals from the Groupwise Heterogeneity Regression;
    E2=Y-X*BHAT_GRP;
*The final step is to calculate the SUR Pooled estimator.;
*The standard errors are also calculated.;
    temp1=e2[1:20,1]||e2[21:40,1]||e2[41:60,1]||e2[61:80,1]||e2
    [81:100,1];
    countt=1;
    temp1=E2[countt:t,1];
    do i=2 to M;
        countt=countt+t;
        c=E2[countt:countt+t-1,1];
        temp1=temp1||c;
    end;
    temp2=1/t*temp1'*temp1;
    I=I(20);
    Temp3=temp2@I;
    BETA_FGLS=inv(X'*inv(Temp3)*X)*X'*inv(Temp3)*Y;
    Print 'The FGLS Parameter Estimates Are';
    Print BETA_FGLS;
    Asy_Var=inv(X'*inv(Temp3)*X);
    Print 'The Asymptotic Covariance Matrix Is';
    Print Asy_Var;
    SE=sqrt(vecdiag(Asy_Var));
    Print 'The Asymptotic Standard Errors Are';
    Print SE;
*Now, calculate the cross-equation covariance for the SUR pooled model;
    E3=Y-X*BETA_FGLS;
    countt=1;
    temp1=E3[countt:t,1];
    do i=2 to M;
        countt=countt+t;
        c=E3[countt:countt+t-1,1];
        temp1=temp1||c;
    end;
    temp2=1/t*temp1'*temp1;
    Print 'The Cross-Equation Covariance Matrix is';
    Print temp2;
run;
```

The analysis results are given in Output 12.11. Notice that the parameters have the same signs and similar magnitudes as the ones obtained by OLS.

12.9 AUTOCORRELATION IN PANEL DATA

We will now deal with estimation methods when the disturbances are correlated within cross sections and across cross sections. The simplest case is to assume that there is no correlation between the disturbances across cross sections. That is, $Corr(\varepsilon_{it}, \varepsilon_{js}) = 0$ if $i \neq j$.

The Groupwise Heteroscedastic Estimators are

TABLE1		
	BHAT	SE
INTERCEPT	-36.2537	6.1244
F	0.0950	0.0074
C	0.3378	0.0302

The Asymptotic Covariance Matrix is

ASY_VAR		
37.507827	-0.026926	0.0095716
-0.026926	0.0000549	-0.000149
0.0095716	-0.000149	0.0009136

The FGLS Parameter Estimates for Estimator are

TABLE2		
	BHAT	SE
INTERCEPT	-28.2467	4.8882
F	0.0891	0.0051
C	0.3340	0.0167

The Asymptotic Covariance Matrix is

ASY_VAR		
23.894871	-0.017291	0.0011391
-0.017291	0.0000257	-0.000047
0.0011391	-0.000047	0.0002793

The Cross-Equation Covariance Matrix is

COV				
10050.525	-4.805227	-7160.667	-1400.747	4439.9887
-4.805227	305.61001	-1966.648	-123.9205	2158.5952
-7160.667	-1966.648	34556.603	4274.0002	-28722.01
-1400.747	-123.9205	4274.0002	833.35743	-2893.733
4439.9887	2158.5952	-28722.01	-2893.733	34468.976

OUTPUT 12.11. FGLS pooled estimators of the Grunfeld data.

Next, define the AR(1) process $\varepsilon_{it} = \rho_i \varepsilon_{i,t-1} + u_{it}$ for the general linear model where (see Chapter 6 and Greene, 2003, p. 325)

$$Var(\varepsilon_{it}) = \sigma_i^2 = \frac{\sigma_{u_i}^2}{1-\rho_i^2}.$$

We can use the methods of Chapter 6 to estimate β. That is,

1. Use OLS to regress **Y** versus **X** and save the residuals $\mathbf{e} = \mathbf{Y} - \mathbf{X}\boldsymbol{\beta}$. Notice that we will have nT residuals here.
2. Estimate the within cross-section correlation ρ_i by using

$$\hat{\rho}_i = \frac{\sum\limits_{t=2}^{T} e_{it} e_{i,t-1}}{\sum\limits_{t=1}^{T} e_{it}^2}.$$

3. Transform the original data by using the Prais–Winsten approach (see Chapter 6).
4. Conduct an OLS regression of the transformed data to get

$$\hat{\sigma}_{ui}^2 = \frac{\mathbf{e}_{*i}^T \mathbf{e}_{*i}}{T} = \frac{(\mathbf{y}_{*i} - \mathbf{x}_{*i}\boldsymbol{\beta})^T (\mathbf{y}_{*i} - \mathbf{x}_{*i}\boldsymbol{\beta})}{T}.$$

5. Use $\hat{\sigma}_{ui}^2$ to get

$$\hat{\sigma}_i^2 = \frac{\sigma_{ui}^2}{1 - \hat{\rho}_i^2}.$$

6. The FGLS estimator in the presence of within cross-section correlation is given by

$$\hat{\boldsymbol{\beta}}_{FGLS} = \left[\sum_{i=1}^{n} \frac{1}{\hat{\sigma}_i^2} \mathbf{X}_i^T \mathbf{X}_i \right]^{-1} \left[\sum_{i=1}^{n} \frac{1}{\hat{\sigma}_i^2} \mathbf{X}_i^T \mathbf{y}_i \right].$$

7. The covariance matrix can be calculated in the usual way.

We analyze Grunfeld's data using the steps just discussed. The complete IML code and output (Output 12.12) are given below. The estimates of the coefficients along with their standard errors are given below:

$$I_{it} = -26.94 + 0.095 F_{it} + 0.30 C_{it}$$
$$\quad\;\; (6.89) \quad\;\; (0.008) \quad\;\; (0.31)$$

```
Proc IML;
* Read the data into matrices and calculate some constants.;
     Use SUR;
     Read all var{'F' 'C'} into X;
     Read all var{'I'} into Y;
     r=nrow(X);c=ncol(X);
     X=J(r,1,1)||X;
* This is step 1 where the OLS estimates and residuals are
calculated.;
     BHAT1=inv(X'*X)*X'*Y;
     E1=Y-X*BHAT1;
* This is the start of step 2 where the cross correlation vector
is calculated.;
     compt=1;
     M=5;
     T=20;
     rho=shape(0,M,1);
     do i=1 to M;
          Temp0=0;Temp1=0;
          do j= compt+1 to compt+T-1;
```

The least squares estimator from the first stage regression is

BETA_FIRST
-21.93625
0.1083051
0.3987988

The asymptotic variance covariance matrix is

ASY_VAR		
17.9554	-0.018843	-0.023393
-0.018843	0.0000775	-0.000158
-0.023393	-0.000158	0.0012404

The standard errors from the first stage regression is

SE
4.2373813
0.0088058
0.0352189

The FGLS Parameter Estimates are

BETA_FGLS
-26.93677
0.0946555
0.2999458

The Asymptotic Covariance Matrix is

ASY_VAR		
47.525774	-0.032327	0.0021717
-0.032327	0.000062	-0.000156
0.0021717	-0.000156	0.0009455

The Standard Errors are given by

SE
6.893894
0.0078736
0.0307492

OUTPUT 12.12. FGLS estimation of the Grunfeld data under the assumption of cross-correlation.

```
                Temp0=Temp0+E1[j,1]*E1[j-1,1];
        end;
        do j2= compt to compt+T-1;
                Temp1=Temp1+E1[j2,1]*E1[j2,1];
        end;
        rho[i,1]=Temp0/Temp1;
        compt=compt+t;
end;
Print 'The autocorrelation vector is';
Print rho;
* This is step 3 where the data is transformed using the Prais
Winsten Method.;
compt=1;
new_y=shape(0,100,1);
do i=1 to M;
        new_y[compt,1]=y[compt,1]*sqrt(1-rho[i,1]**2);
        do j= compt+1 to compt+T-1;
                new_y[j,1]=y[j,1]-rho[i,1] * y[j-1,1];
        end;
        compt=compt+T;
end;
compt=1;
new_x=shape(0,100,2);
do i=1 to M;
        new_x[compt,1]=x[compt,2]*sqrt(1-rho[i,1]**2);
        new_x[compt,2]=x[compt,3]*sqrt(1-rho[i,1]**2);
        do j= compt+1 to compt+T-1;
                new_x[j,1]=x[j,2]-rho[i,1] * x[j-1,2];
                new_x[j,2]=x[j,3]-rho[i,1] * x[j-1,3];
        end;
        compt=compt+T;
end;
new_x=J(r,1,1)||new_x;
* OLS is now conducted on the transformed data.;
* The standard errors are also calculated.;
beta_first=inv(new_x'*new_x)*new_x'*new_y;
Print 'The least squares estimator from the first stage
regression is';
Print beta_first;
sigma1=shape(0,M,1);
E2=new_y-new_x*beta_first;
compt=1;
do i=1 to M;
        sigma1[i,1]=E2[compt:compt+t-1,1]'*E2[compt:compt+t-
        1,1]/T;
        compt=compt+t;
end;
var_cov=diag(sigma1);
ID=I(T);
Omega=var_cov@ID;
Asy_Var=inv(new_X'*inv(Omega)*new_X);
Print 'The asymptotic variance covariance matrix is';
Print Asy_Var;
SE=sqrt(vecdiag(Asy_Var));
```

```
Print 'The standard errors from the first stage regression
is';
Print SE;
* The second stage FGLS estimates are now calculated.;
compt=1;
temp1=E2[compt:t,1];
do i=2 to M;
     compt=compt+t;
     c=E2[compt:compt+t-1,1];
     temp1=temp1||c;
end;
sigma2=shape(0,M,M);
do i=1 to M;
do j=1 to M;
     sigma2[i,j]=temp1[,i]'*temp1[,i]/T;
     sigma2[i,1]=sigma2[i,1]/(1-rho[i,1]**2);
end;
end;
I=I(T);
Temp3=sigma2@I;
BETA_FGLS=inv(new_X'*inv(Temp3)*new_X)*
new_X'*inv(Temp3)*new_Y;
Print 'The FGLS Parameter Estimates Are';
Print BETA_FGLS;
Asy_Var=inv(new_X'*inv(Temp3)*new_X);
Print 'The Asymptotic Covariance Matrix Is';
Print Asy_Var;
SE=sqrt(vecdiag(Asy_Var));
Print 'The Standard Errors are given by';
Print SE;
run;
```

We will now extend this analysis to the case where we cannot assume that $Corr(\varepsilon_{it},\varepsilon_{js})=0$ if $i\neq j$, thus allowing cross-sectional correlation between subjects. The steps are outlined below. See Greene (2003, pp. 324–326) for more details.

1. Follow steps 1 through 5 from the previous case.
2. Construct $\Sigma = \left[\hat{\sigma}_{ij}^2\right]$ where $i,j=1,\ldots,n$.
3. The FGLS estimator in the presence of cross-sectional correlation is given by

$$\hat{\boldsymbol{\beta}}_{GLS} = (\mathbf{X}^T\hat{\boldsymbol{\Omega}}^{-1}\mathbf{X})^{-1}\mathbf{X}^T\hat{\boldsymbol{\Omega}}^{-1}\mathbf{y}$$

where $\boldsymbol{\Omega}=\boldsymbol{\Sigma}\otimes\mathbf{I}$.

4. The covariance matrix is given by

$$Var(\hat{\boldsymbol{\beta}}_{GLS}) = (\mathbf{X}^T\hat{\boldsymbol{\Omega}}^{-1}\mathbf{X})^{-1}.$$

We will analyze Grunfeld's data using the above-mentioned steps. The IML code and output (Output 12.13) are given below. Note that the code is almost identical to the code previously given. The estimates of the coefficients along with their standard errors are given below. Note that the parameter estimates are comparable to the ones obtained when the cross sections are assumed to be

The autocorrelation vector is

RHO
0.4735903
0.704354
0.8977688
0.5249498
0.8558518

The least squares estimator from the first stage regression is

BETA_FIRST
-16.84981
0.0944753
0.3780965

The asymptotic variance covariance matrix is

ASY_VAR		
13.188884	-0.010868	-0.02407
-0.010868	0.0000552	-0.000131
-0.02407	-0.000131	0.001104

The standard errors from the first stage regression is

SE
3.6316504
0.0074294
0.0332267

The FGLS Parameter Estimates are

BETA_FGLS
-16.36591
0.0895486
0.3694549

OUTPUT 12.13. FGLS estimation for the case when the correlation among the errors is not zero.

uncorrelated.

$$I_{it} = -16.37 + 0.09F_{it} + 0.37C_{it}$$
$$(4.77) \quad (0.009) \quad (0.036)$$

```
Proc IML;
* Read the data into matrices and calculate some constants.;
    Use SUR;
    Read all var{'F' 'C'} into X;
    Read all var{'I'} into Y;
```

The Asymptotic Covariance Matrix is

ASY_VAR		
22.780423	-0.023583	-0.015614
-0.023583	0.0000783	-0.000177
-0.015614	-0.000177	0.0013193

The Standard Errors are given by

SE
4.7728841
0.0088489
0.0363224

OUTPUT 12.13. (*Continued*)

```
      r=nrow(X);c=ncol(X);
      X=J(r,1,1)||X;
* This is step 1 where the OLS estimates and residuals are
calculated.;
      BHAT1=inv(X'*X)*X'*Y;
      E1=Y-X*BHAT1;
* This is the start of step 2 where the cross correlation vector
is calculated.;
      compt=1;
      M=5;
      T=20;
      rho=shape(0,M,1);
      do i=1 to M;
           Temp0=0;Temp1=0;
           do j= compt+1 to compt+T-1;
                Temp0=Temp0+E1[j,1]*E1[j-1,1];
           end;
           do j2= compt to compt+T-1;
                Temp1=Temp1+E1[j2,1]*E1[j2,1];
           end;
           rho[i,1]=Temp0/Temp1;
           compt=compt+t;
      end;
      Print 'The autocorrelation vector is';
      Print rho;
      * This is step 3 where the data is transformed using the Prais
      Winsten Method.;
      compt=1;
      new_y=shape(0,100,1);
      do i=1 to M;
           new_y[compt,1]=y[compt,1]*sqrt(1-rho[i,1]**2);
           do j= compt+1 to compt+T-1;
                new_y[j,1]=y[j,1]-rho[i,1] * y[j-1,1];
           end;
           compt=compt+T;
```

```
end;
compt=1;
new_x=shape(0,100,2);
do i=1 to M;
      new_x[compt,1]=x[compt,2]*sqrt(1-rho[i,1]**2);
      new_x[compt,2]=x[compt,3]*sqrt(1-rho[i,1]**2);
      do j= compt+1 to compt+T-1;
            new_x[j,1]=x[j,2]-rho[i,1] * x[j-1,2];
            new_x[j,2]=x[j,3]-rho[i,1] * x[j-1,3];
      end;
      compt=compt+T;
end;
new_x=J(r,1,1)||new_x;
* OLS is now conducted on the transformed data.;
* The standard errors are also calculated.;
beta_first=inv(new_x'*new_x)*new_x'*new_y;
Print 'The least squares estimator from the first stage
regression is';
Print beta_first;
sigma1=shape(0,M,1);
E2=new_y-new_x*beta_first;
compt=1;
do i=1 to M;
      sigma1[i,1]=E2[compt:compt+t-1,1]'*E2[compt:compt+t-
      1,1]/T;
      compt=compt+t;
end;
var_cov=diag(sigma1);
ID=I(T);
Omega=var_cov@ID;
Asy_Var=inv(new_X'*inv(Omega)*new_X);
Print 'The asymptotic variance covariance matrix is';
Print Asy_Var;
SE=sqrt(vecdiag(Asy_Var));
Print 'The standard errors from the first stage regression
is';
Print SE;
* The second stage FGLS estimates are now calculated.;
compt=1;
temp1=E2[compt:t,1];
do i=2 to M;
      compt=compt+t;
      c=E2[compt:compt+t-1,1];
      temp1=temp1||c;
end;
sigma2=shape(0,M,M);
do i=1 to M;
do j=1 to M;
      sigma2[i,j]=temp1[,i]'*temp1[,i]/T;
      sigma2[i,1]=sigma2[i,1]/(1-rho[i,1]**2);
end;
end;
I=I(T);
```

```
    Temp3=sigma2@I;
    BETA_FGLS=inv (new_X'*inv(Temp3)*new_X)*
    new_X'*inv (Temp3) *new_Y;
    Print 'The FGLS Parameter Estimates Are';
    Print BETA_FGLS;
    Asy_Var=inv (new_X'*inv (Temp3)*new_X);
    Print 'The Asymptotic Covariance Matrix Is';
    Print Asy_Var;
    SE=sqrt (vecdiag (Asy_Var));
    Print 'The Standard Errors are given by';
    Print SE;
run;
```

Appendix A

BASIC MATRIX ALGEBRA FOR ECONOMETRICS

A.1 MATRIX DEFINITIONS

A.1.a Definitions

An $m \times n$ matrix is a rectangular array of elements arranged in m rows and n columns. A general layout of a matrix is given by

$$
\begin{bmatrix}
a_{11} & a_{12} & \cdots & a_{1n} \\
a_{21} & a_{22} & \cdots & a_{2n} \\
\vdots & \vdots & \ddots & \vdots \\
a_{m1} & a_{m2} & \cdots & a_{mn}
\end{bmatrix}.
$$

In this general form, we can easily index any element of the matrix. For instance, the element in the ith row and jth column is given by a_{ij}. It is straightforward to create matrices in Proc IML. For example, the Proc IML command $A = \{2\,4, 3\,1\}$ will create the 2×2 matrix

$$
A = \begin{bmatrix} 2 & 4 \\ 3 & 1 \end{bmatrix}.
$$

A row vector of order n is a matrix with one row and n columns. The general form of a row vector is $y = \begin{bmatrix} y_1 & y_2 & \cdots & y_n \end{bmatrix}$. A column vector of order m is a matrix with m rows and one column. The general form of a column vector is

$$
c = \begin{bmatrix} c_1 \\ c_2 \\ \vdots \\ c_m \end{bmatrix}.
$$

It is straightforward to create row and column vectors in Proc IML. For example, the Proc IML command $y = \{2\ 4\}$ will create the row vector $y = \begin{bmatrix} 2 & 4 \end{bmatrix}$, while the Proc IML command $c=\{-3, 4\}$ will create the column vector

$$
c = \begin{bmatrix} -3 \\ 4 \end{bmatrix}.
$$

Applied Econometrics Using the SAS® System, by Vivek B. Ajmani
Copyright © 2009 John Wiley & Sons, Inc.

Of course, these definitions can easily be extended to matrices of any desired dimension and consequently the Proc IML code can be adjusted to accommodate these changes.

A.1.b   Other Types of Matrices

  i. A **square matrix** is a matrix with equal number of rows and columns. That is, if $A_{m \times n}$ is a square matrix, then $m = n$.

  ii. A **symmetric matrix** is a square matrix where the (ij)th element is the same as the (ji)th element for all i and j. That is, $a_{ij} = a_{ji}, \forall i, j$.

  iii. A **diagonal matrix** is a square matrix where all off-diagonal elements are zero. That is, $a_{ij} = 0, \forall i \neq j$.

  iv. An **identity matrix** (denoted by I) is a diagonal matrix where $a_{ii} = 1, \forall i$. The Proc IML command Id=I(5) will create a 5×5 identity matrix stored under the name *Id*.

  v. The **J** matrix is one where every element equals 1. This matrix frequently occurs in econometric analysis. The Proc IML command J=J(1,5,5) will create a 5×5 matrix of 1's. The size of the matrix can be adjusted by changing the number of rows and/or the number of columns. We can replace the third element in the Proc IML command if we require all the elements to have a different value. For instance, using J(5,5,0) will yield a 5×5 matrix of zeros.

A.2   MATRIX OPERATIONS

A.2.a   Addition and Subtraction

These two operations are defined only on matrices of the same dimension. The operations are themselves very elementary and involve element-by-element addition or subtraction. As an example, consider the following matrices:

$$A = \begin{bmatrix} 2 & 3 \\ 1 & 1 \end{bmatrix}, \qquad B = \begin{bmatrix} 1 & 0 \\ 2 & 1 \end{bmatrix}.$$

Addition is denoted by $A + B$ and is given by

$$A + B = \begin{bmatrix} 3 & 3 \\ 3 & 2 \end{bmatrix}.$$

Similarly, subtraction is denoted by $A - B$ and is given by

$$A - B = \begin{bmatrix} 1 & 3 \\ -1 & 0 \end{bmatrix}.$$

The Proc IML commands C = A + B and D = A − B can be used to carry out these operations.

A.2.b   Scalar Multiplication

For any scalar $r \in \Re$ and any matrix $A \in M(\Re)$, we can define scalar multiplication as rA. Here, each element of the matrix A is multiplied by r. For example, if

$$A = \begin{bmatrix} 2 & 3 \\ 1 & 1 \end{bmatrix},$$

then

$$rA = \begin{bmatrix} 2r & 3r \\ r & r \end{bmatrix}.$$

Let $r = 2$. Then, the Proc IML command C=2*A will yield the result

$$C = \begin{bmatrix} 4 & 6 \\ 2 & 2 \end{bmatrix}.$$

A.2.c   Matrix Multiplication

Assume that matrix A is of order $(k \times m)$ and B is of order $(m \times n)$. That is, the number of rows of B equals the number of columns of A. We say that A and B are conformable for matrix multiplication. Given two conformable matrices, A and B, we define their product C as $C_{k \times n} = A_{k \times m} B_{m \times n}$, where C is of order $(k \times n)$. In general, the (i, j)th element of C is written as

$$
c_{ij} = \begin{pmatrix} a_{i1} & \cdots & \cdots & a_{im} \end{pmatrix} \begin{pmatrix} b_{1j} \\ \vdots \\ \vdots \\ b_{mj} \end{pmatrix}
$$

$$
= a_{i1}b_{1j} + a_{i2}b_{2j} + \cdots + a_{im}b_{mj}
$$

$$
= \sum_{h=1}^{m} a_{ih}b_{hj}.
$$

The Proc IML command C=A*B can be used to carry out matrix multiplications. For instance, if

$$
A = \begin{bmatrix} 1 & 2 \\ 3 & 4 \end{bmatrix}
$$

and

$$
B = \begin{bmatrix} -1 & 6 \\ 4 & 5 \end{bmatrix},
$$

then

$$
C = \begin{bmatrix} 7 & 16 \\ 13 & 38 \end{bmatrix}.
$$

A.3   BASIC LAWS OF MATRIX ALGEBRA

A.3.a   Associative Laws

$$
(A + B) + C = A + (B + C),
$$
$$
(AB)C = A(BC).
$$

A.3.b   Commutative Laws of Addition

$$
A + B = B + A.
$$

A.3.c   Distributive Laws

$$
A(B + C) = AB + AC,
$$
$$
(A + B)C = AC + BC.
$$

The commutative law of addition does not apply to multiplication in general. That is, for two conformable matrices A and B, AB is not necessarily equal to BA.

A.4 IDENTITY MATRIX

A.4.a Definition

The identity matrix is an $n \times n$ matrix with entries satisfying

$$a_{ij} = \begin{cases} 1 & \text{if } i = j, \\ 0 & \text{otherwise.} \end{cases}$$

That is,

$$I = \begin{bmatrix} 1 & 0 & \cdots & 0 \\ 0 & 1 & \cdots & 0 \\ \vdots & \vdots & \ddots & \vdots \\ 0 & 0 & \cdots & 1 \end{bmatrix}.$$

As discussed earlier, it is very easy to create identity matrices in Proc IML. For instance, the command I = I(5) will create an identity matrix of order 5 and store it in the variable I.

A.4.b Properties of Identity Matrices

For an $n \times n$ identity matrix I, the following holds:

   i. For any $k \times n$ matrix A, $AI = A$.
  ii. For any $n \times k$ matrix B, $IB = B$.
 iii. For any $n \times n$ matrix C, $CI = IC = C$.

A.5 TRANSPOSE OF A MATRIX

A.5.a Definition

A transpose matrix of the original matrix, A, is obtained by replacing all elements a_{ij} with a_{ji}. The transpose matrix A^T (or A') is a matrix such that $a_{ji}^T = a_{ij}$, where a_{ij} is the (i, j)th element of A and a_{ji}^T is the (j, i)th element of A^T. For example,

$$\begin{bmatrix} 1 & 2 \\ 3 & 4 \\ 5 & 6 \end{bmatrix}^T = \begin{bmatrix} 1 & 3 & 5 \\ 2 & 4 & 6 \end{bmatrix}.$$

It is straightforward to create transpose of matrices using Proc IML. The command $B = A'$ will store the transpose of the matrix A in B.

A.5.b Properties of Transpose Matrices

   i. $(A + B)^T = A^T + B^T$.
  ii. $(A - B)^T = A^T - B^T$.
 iii. $(A^T)^T = A$.
 iv. $(rA)^T = rA^T$ for any scalar r.
  v. $(AB)^T = B^T A^T$.

A.6 DETERMINANTS

A.6.a Definition

Associated with any square matrix A, there is a scalar quantity called the determinant of A, denoted det(A) or $|A|$. The simplest example involves $A \in M_{2\times2}(\Re)$, where

$$\det \begin{bmatrix} a & b \\ c & d \end{bmatrix} = ad - bc.$$

To define the determinant of a matrix in general form (that is, for any $n \times n$ matrix), we can use the notions of minors and cofactors. Let A be an $n \times n$ matrix and let $\hat{A}_{ij}$ be the $(n-1) \times (n-1)$ submatrix obtained by deleting the ith row and the jth column of A. Then the scalar $M_{ij} = \det(\hat{A}_{ij})$ is called the (i, j)th minor of A. The sign-adjusted scalar

$$C_{ij} = (-1)^{i+j}M_{ij} = (-1)^{i+j}\det(\hat{A}_{ij})$$

is called the (i, j)th cofactor of A. Given this definition, $|A|$ can be expressed in terms of the elements of the ith row (or jth column) of their cofactors as (Greene, 2003, p. 817; Searle, 1982, pp. 84–92)

$$|A| = \sum_{i=1}^{n} a_{ij}C_{ij} = \sum_{i=1}^{n} a_{ij}(-1)^{i+j}|\hat{A}_{ij}|.$$

A.6.b Properties of Determinants

For any $A, B \in M_{n\times n}(\Re)$, we have the following:

i. $|A^T| = |A|$.
ii. $|AB| = |A||B|$.
iii. If every element of a row (or column) of A is multiplied by a scalar r to yield a new matrix B, then $|B| = r|A|$.
iv. If every element of an nth order matrix A is multiplied by a scalar r, then $|rA| = r^n|A|$.
v. The determinant of a matrix is nonzero if and only if it has full rank.

Determinants of matrices can easily be computed in Proc IML by using the command det(A) (Searle, 1982, pp. 82–112).

A.7 TRACE OF A MATRIX

A.7.a Definition

The trace of a $n \times n$ matrix A is the sum of its diagonal elements. That is,

$$tr(A) = \sum_{i=1}^{n} a_{ii}.$$

Note that for any $m \times n$ matrix A,

$$tr(A^TA) = tr(AA^T) = \sum_{i=1}^{m}\sum_{j=1}^{n} a_{ij}^2 \quad \text{(Searle, 1982, pp. 45–46)}.$$

A.7.b   Properties of Traces

    i. $tr(rA) = r \times tr(A)$ for any real number r.

    ii. $tr(A + B) = tr(A) + tr(B)$.

    iii. $tr(AB) = tr(BA)$

    iv. $tr(ABCD) = tr(BCDA) = tr(CDAB) = tr(DABC)$.

    v. $tr(A) = rank(A)$ if A is symmetric and idempotent (Baltagi, 2008, p. 172). As an example, consider

$$A = \begin{bmatrix} 1 & 2 & 3 \\ 4 & 1 & 5 \\ 6 & 7 & 1 \end{bmatrix}.$$

Here, $tr(A) = 3$. The Proc IML command trace(A) will easily calculate the trace of a matrix.

A.8   MATRIX INVERSES

A.8.a   Definition

If, for an $n \times n$ matrix A, there exists a matrix A^{-1} such that $A^{-1}A = AA^{-1} = I_n$, then A^{-1} is defined to be the inverse of A.

A.8.b   Construction of an Inverse Matrix

Let $A \in M_{n \times n}(\mathfrak{R})$ be a nonsingular matrix.

    i. Recall that for any $n \times n$ matrix A, the (i, j)th cofactor of A is $C_{ij} = (-1)^{i+j}\det(\hat{A}_{ij})$.

    ii. From the matrix A, construct a cofactor matrix in which each element of A, a_{ij}, is replaced by its cofactor, c_{ij}. The transpose of this matrix is called the adjoint matrix and is denoted by

$$A^* = adj(A) = cofactor(A)^T = [c_{ji}].$$

That is,

$$adj(A) = \begin{bmatrix} c_{11} & c_{21} & \dots & c_{n1} \\ c_{12} & c_{22} & \dots & c_{n2} \\ \vdots & \vdots & \ddots & \vdots \\ c_{1n} & c_{2n} & \dots & c_{nn} \end{bmatrix}.$$

A^{-1} can then be defined as

$$A^{-1} = \frac{1}{|A|} adj(A) \text{(Searle, 1982, p. 129)}.$$

This implies that A^{-1} does not exist if $|A| = 0$. That is, A is nonsingular if and only if its inverse exists.

A.8.c   Properties of Inverse of Matrices

Let A, B, and C be invertible square matrices. Then (Searle, 1982, p. 130),

    i. $(A^{-1})^{-1} = A$.

    ii. $(A^T)^{-1} = (A^{-1})^T$.

iii. AB is invertible and $(AB)^{-1} = B^{-1}A^{-1}$.

iv. ABC is invertible and $(ABC)^{-1} = C^{-1}B^{-1}A^{-1}$.

A.8.d Some More Properties of Inverse of Matrices

If a square matrix A is invertible, then (Searle, 1982, p. 130)

i. $A^m = \underbrace{A \times A \times \ldots \times A}_{m\ \text{times}}$ is invertible for any integer m and

$$(A^m)^{-1} = (A^{-1})^m = \underbrace{A^{-1} \times A^{-1} \times \ldots \times A^{-1}}_{m\text{times}}.$$

ii. For any integer r and s, $A^r A^s = A^{r+s}$.

iii. For any scalar $r \neq 0$, rA is invertible and $(rA)^{-1} = \frac{1}{r}A^{-1}$.

iv. $|A^{-1}| = \frac{1}{|A|}$.

v. If A is symmetric, then A^{-1} is symmetric.

A.8.e Uniqueness of an Inverse Matrix

Any square matrix A can have at most one inverse. Matrix inverses can easily be computed using Proc IML by using the command inv(A).

A.9 IDEMPOTENT MATRICES

A.9.a Definition

A square matrix A is called idempotent if $A^2 = A$.

A.9.b The M^0 Matrix in Econometrics

This matrix is useful in transforming data by calculating a variable's deviation from its mean. This matrix is defined as

$$M^0 = \left[I - \frac{1}{n} i i^T \right] = \begin{bmatrix} 1 - \dfrac{1}{n} & -\dfrac{1}{n} & \cdots & -\dfrac{1}{n} \\[2mm] -\dfrac{1}{n} & 1 - \dfrac{1}{n} & \cdots & -\dfrac{1}{n} \\[2mm] \vdots & \vdots & \ddots & \vdots \\[2mm] -\dfrac{1}{n} & -\dfrac{1}{n} & \cdots & 1 - \dfrac{1}{n} \end{bmatrix}.$$

For an example of how this matrix is used, consider the case when we want to transform a single variable x. In the single variable case, the sum of squared deviations about the mean is given by (Greene, 2003, p. 808; Searle, 1982, p. 68)

$$\sum_{i=1}^{n} (x_i - \bar{x})^2 = (x - \bar{x})^T (x - \bar{x}) = (M^0 x)^T (M^0 x) = x^T M^{0T} M^0 x.$$

It can easily be shown that M^0 is symmetric so that $M^{0T} = M^0$. Therefore,

$$\sum_{i=1}^{n} (x_i - \bar{x})^2 = x^T M^0 x.$$

For two variables x and y, the sums of squares and cross products in deviations from their means is given by (Greene, 2003, p. 809)

$$\sum_{i=1}^{n}(x_i-\bar{x})(y_i-\bar{y}) = (M^0 x)^T (M^0 y) = x^T M^0 y.$$

Two other important idempotent matrices in econometrics are the P and M matrices. To understand these, let X be a $n \times k$ matrix. Then $X^T X$ is a $k \times k$ square matrix. Define $P = (X^T X)^{-1} X^T$. Then, $P^T P = P$. It can be shown that P is symmetric. This matrix is called the projection matrix.

The second matrix is the M matrix and is defined as $M = I - P$. Then, $M^T = M$ and $M^2 = M$. It can also be shown that M and P are orthogonal so that $PM = MP = 0$ (Greene, 2003, pp. 24–25).

A.10 KRONECKER PRODUCTS

Kronecker products are used extensively in econometric data analysis. For instance, computations involving seemingly unrelated regressions make heavy use of these during FGLS estimation of the parameters. Consider the following two matrices:

$$\mathbf{A} = \begin{bmatrix} a_{11} & a_{12} & \ldots & a_{1n} \\ a_{21} & a_{22} & \ldots & a_{2n} \\ \vdots & \vdots & \ddots & \vdots \\ a_{m1} & a_{m2} & \ldots & a_{mn} \end{bmatrix} \quad \text{and} \quad \mathbf{B}_{p \times q}.$$

The Kronecker product of $\mathbf{A}$ and $\mathbf{B}$ defined as $\mathbf{A} \otimes \mathbf{B}$ is given by the $mp \times nq$ matrix:

$$\mathbf{A} \otimes \mathbf{B} = \begin{bmatrix} a_{11}\mathbf{B} & a_{12}\mathbf{B} & \ldots & a_{1n}\mathbf{B} \\ a_{21}\mathbf{B} & a_{22}\mathbf{B} & \ldots & a_{2n}\mathbf{B} \\ \vdots & \vdots & \ddots & \vdots \\ a_{m1}\mathbf{B} & \ldots & \ldots & a_{mn}\mathbf{B} \end{bmatrix}.$$

The following are some properties of Kronecker products (Greene, 2003, pp. 824–825; Searle, 1982, pp. 265–267):

1. $(\mathbf{A} \otimes \mathbf{B})(\mathbf{C} \otimes \mathbf{D}) = \mathbf{AC} \otimes \mathbf{BD}$,
2. $tr(\mathbf{A} \otimes \mathbf{B}) = tr(\mathbf{A})tr(\mathbf{B})$ is $\mathbf{A}$ and $\mathbf{B}$ are square,
3. $(\mathbf{A} \otimes \mathbf{B})^{-1} = \mathbf{A}^{-1} \otimes \mathbf{B}^{-1}$,
4. $(\mathbf{A} \otimes \mathbf{B})^T = \mathbf{A}^T \otimes \mathbf{B}^T$,
5. $\det(\mathbf{A} \otimes \mathbf{B}) = (\det \mathbf{A})^m (\det \mathbf{B})^n$, $\mathbf{A}$ is $m \times m$ and $\mathbf{B}$ is $n \times n$.

The Proc IML code A@B calculates Kronecker products.

A.11 SOME COMMON MATRIX NOTATIONS

a. A system of m simultaneous equations in n variables is given by

$$a_{11}x_1 + a_{12}x_2 + \cdots + a_{1n}x_n = b_1$$
$$\vdots$$
$$a_{m1}x_1 + a_{m2}x_2 + \cdots + a_{mn}x_n = b_m$$

and can be expressed in matrix form as $\mathbf{Ax} = \mathbf{b}$, where $\mathbf{A}$ is an $m \times n$ matrix of coefficients $\lfloor a_{ij} \rfloor$, $\mathbf{x}$ is a column vector of variables $x_1, \ldots, x_n$, and $\mathbf{b}$ is the column vector of constants $b_1, \ldots, b_m$.

b. *Sum of Values:* We can express the sum $\sum_{i=1}^{n} x_i$ as $\mathbf{i}^T\mathbf{x}$, where $\mathbf{i}$ is a column vector of 1's.

c. *Sum of Squares:* We can express the sums of squares $\sum_{i=1}^{n} x_i^2$ as $\mathbf{x}^T\mathbf{x}$, where $\mathbf{x}$ is a column vector of variables.

d. *Sum of Products:* For two variables $\mathbf{x}$ and $\mathbf{y}$, the sum of their product $\sum_{i=1}^{n} x_i y_i$ can be written as $\mathbf{x}^T\mathbf{y}$.

e. *Weighted Sum of Squares:* Given a diagonal $n \times n$ matrix $\mathbf{A}$ of weights $a_{11}, \ldots, a_{nn}$ the sum $\sum_{i=1}^{n} a_{ii} x_i^2$ can be written as $\mathbf{x}^T\mathbf{A}\mathbf{x}$.

f. *Quadratic Forms:* Given an $n \times n$ matrix A with elements $a_{11}, a_{12}, \ldots, a_{22}, \ldots, a_{nn}$, the sum $a_{11}x_1^2 + a_{12}x_1x_2 + \cdots + a_{22}x_2^2 + \cdots + a_{nn}x_n^2$ can be expressed as $\mathbf{x}^T\mathbf{A}\mathbf{x}$.

See Greene, (2003, p. 807) for more details.

A.12 LINEAR DEPENDENCE AND RANK

A.12.a Linear Dependence/Independence

A set of vectors $\mathbf{v}_1, \ldots, \mathbf{v}_k$ is linearly dependent if the equation $a_1\mathbf{v}_1 + \cdots + a_k\mathbf{v}_k = 0$ has a solution where not all the scalars $a_1, \ldots, a_k$ are zero. If the only solution to the above equation is where all the scalars equal zero, then the set of vectors is called a linearly independent set.

A.12.b Rank

The rank of an $m \times n$ matrix $\mathbf{A}$, denoted as $r(\mathbf{A})$, is defined as the maximum number of linearly independent rows or columns of $\mathbf{A}$. Note that the row rank of a matrix always equals the column rank, and the common value is simply called the "rank" of a matrix. Therefore, $r(\mathbf{A}) \leq \max(m, n)$ and $r(\mathbf{A}) = r(\mathbf{A}^T)$.

Proc IML does not calculate ranks of matrices directly. A way around this is to use the concept of generalized inverses as shown in the following statement round(trace(ginv(A)*A)). Here, $\mathbf{A}$ is the matrix of interest, ginv is the generalized inverse of $\mathbf{A}$, and trace is the trace of the matrix resulting from performing the operation ginv(A)*A. The function round simply rounds the trace value. As an example, consider the following 4×4 matrix given by

$$A = \begin{bmatrix} 1 & 2 & 0 & 3 \\ 1 & -2 & 3 & 0 \\ 0 & 0 & 4 & 8 \\ 2 & 4 & 0 & 6 \end{bmatrix}.$$

The rank of $\mathbf{A}$ is 3 since the last row equals the first row multiplied by 2. Proc IML also yields a rank of 3 for this matrix.

A.12.c Full Rank

If the column(row) rank of a matrix equals the number of columns(rows) of the same matrix, then the matrix is said to be of full rank.

A.12.d Properties of Ranks of Matrices

i. For two matrices $\mathbf{A}$ and $\mathbf{B}$, $r(\mathbf{AB}) \leq \min(r(\mathbf{A}), r(\mathbf{B}))$.

ii. If $\mathbf{A}$ is $m \times n$ and $\mathbf{B}$ is a square matrix of rank n, then $r(\mathbf{AB}) = r(\mathbf{A})$.

iii. $r(\mathbf{A}) = r(\mathbf{A}^T\mathbf{A}) = r(\mathbf{AA}^T)$.

See Greene, (2003, pp. 828–829) for more details.

A.12.e Equivalence

For any square matrix $\mathbf{A}$, the following statements are equivalent (Searle, 1982, p. 172):

    i. $\mathbf{A}$ is invertible.

    ii. Every system of linear equations $\mathbf{Ax} = \mathbf{b}$ has a unique solution for $\forall b \in \Re^n$.

    iii. $\mathbf{A}$ is nonsingular.

    iv. $\mathbf{A}$ has full rank.

    v. The determinant of $\mathbf{A}$ is nonzero.

    vi. All the row(column) vectors of $\mathbf{A}$ are linearly independent.

A.13 DIFFERENTIAL CALCULUS IN MATRIX ALGEBRA

A.13.a Jacobian and Hessian Matrices

Consider the vector function $\mathbf{y} = f(\mathbf{x})$, where $\mathbf{y}$ is a $m \times 1$ vector with each element of $\mathbf{y}$ being a function of the $n \times 1$ vector $\mathbf{x}$. That is,

$$
\begin{aligned}
y_1 &= f_1(x_1, x_2, \ldots, x_n)\\
&\vdots\\
y_m &= f(x_1, x_2, \ldots, x_n).
\end{aligned}
$$

Taking the first derivative of $\mathbf{y}$ with respect to $\mathbf{x}$ yields the Jacobian matrix (Greene, 2003, p. 838; Searle, 1982, p. 338)

$$
J = \frac{\partial \mathbf{y}}{\partial \mathbf{x}^T} = \frac{\partial f(\mathbf{x})}{\partial \mathbf{x}^T} =
\begin{bmatrix}
\dfrac{\partial f_1}{\partial x_1} & \dfrac{\partial f_1}{\partial x_2} & \cdots & \dfrac{\partial f_1}{\partial x_n}\\[2mm]
\dfrac{\partial f_2}{\partial x_1} & \dfrac{\partial f_2}{\partial x_2} & \cdots & \dfrac{\partial f_2}{\partial x_n}\\[2mm]
\vdots & \vdots & \ddots & \vdots\\[2mm]
\dfrac{\partial f_m}{\partial x_1} & \dfrac{\partial f_m}{\partial x_2} & \cdots & \dfrac{\partial f_m}{\partial x_n}
\end{bmatrix}.
$$

Taking the second derivative of $f(\mathbf{x})$ with respect to $\mathbf{x}$ yields the Hessian matrix (Greene, 2003, p. 838; Searle, 1982, p. 341)

$$
H = \frac{\partial^2 \mathbf{y}}{\partial \mathbf{x}^T \partial \mathbf{x}} = \frac{\partial^2 f(\mathbf{x})}{\partial \mathbf{x}^T \partial \mathbf{x}} =
\begin{bmatrix}
\dfrac{\partial f_1}{\partial x_1^2} & \dfrac{\partial f_1}{\partial x_1 \partial x_2} & \cdots & \dfrac{\partial f_1}{\partial x_1 \partial x_n}\\[2mm]
\dfrac{\partial f_2}{\partial x_1 \partial x_2} & \dfrac{\partial f_2}{\partial x_2^2} & \cdots & \dfrac{\partial f_2}{\partial x_2 \partial x_n}\\[2mm]
\vdots & \vdots & \ddots & \vdots\\[2mm]
\dfrac{\partial f_m}{\partial x_1 \partial x_n} & \dfrac{\partial f_m}{\partial x_2 \partial x_n} & \cdots & \dfrac{\partial f_m}{\partial x_n^2}
\end{bmatrix}.
$$

A.13.b Derivative of a Simple Linear Function

Consider the function $f(\mathbf{x}) = \mathbf{a}^T \mathbf{x} = \sum_{i=1}^{n} a_i x_i$. The derivative of $f(\mathbf{x})$ with respect to $\mathbf{x}$ is given by

$$
\frac{\partial f(\mathbf{x})}{\partial \mathbf{x}} = \frac{\partial \mathbf{a}^T \mathbf{x}}{\partial \mathbf{x}} = \mathbf{a}^T.
$$

A.13.c Derivative of a Set of m Linear Functions Ax

Consider the derivative of a set of m linear functions $\mathbf{Ax}$, where $\mathbf{A}$ is a $m \times n$ matrix and

$$\mathbf{Ax} = \begin{bmatrix} a_{11} & a_{12} & \cdots & a_{1n} \\ a_{21} & a_{22} & \cdots & a_{2n} \\ \vdots & \vdots & \ddots & \vdots \\ a_{m1} & a_{m2} & \cdots & a_{mn} \end{bmatrix} \begin{bmatrix} x_1 \\ x_2 \\ \vdots \\ x_n \end{bmatrix} = \begin{bmatrix} \mathbf{a}_1\mathbf{x} \\ \mathbf{a}_2\mathbf{x} \\ \vdots \\ \mathbf{a}_m\mathbf{x} \end{bmatrix}.$$

Therefore,

$$\frac{\partial(\mathbf{Ax})}{\partial \mathbf{x}^T} = \begin{bmatrix} \mathbf{a}_1 \\ \mathbf{a}_2 \\ \vdots \\ \mathbf{a}_m \end{bmatrix} = \mathbf{A}.$$

A.13.d Derivative of a Set of m Linear Functions $\mathbf{x}^T\mathbf{A}$

Consider the derivative of a set of m linear functions $\mathbf{x}^T\mathbf{A}$, where $\mathbf{A}$ is an $n \times m$ matrix and $\mathbf{x}$ is an $n \times 1$ column vector so that

$$\mathbf{x}^T\mathbf{A} = \begin{bmatrix} x_1 & x_2 & \cdots & x_n \end{bmatrix} \begin{bmatrix} a_{11} & a_{12} & \cdots & a_{1m} \\ a_{21} & a_{22} & \cdots & a_{2m} \\ \vdots & \vdots & \ddots & \vdots \\ a_{n1} & a_{n2} & \cdots & a_{nm} \end{bmatrix} = \begin{bmatrix} \mathbf{x}^T\mathbf{a}_1 & \mathbf{x}^T\mathbf{a}_2 & \cdots & \mathbf{x}^T\mathbf{a}_m \end{bmatrix}.$$

Therefore,

$$\frac{\partial(\mathbf{x}^T\mathbf{A})}{\partial \mathbf{x}} = \begin{bmatrix} \mathbf{a}_1 & \mathbf{a}_2 & \cdots & \mathbf{a}_m \end{bmatrix} = \mathbf{A}.$$

A.13.e Derivative of a Quadratic Form $\mathbf{x}^T\mathbf{Ax}$

Consider the derivative of a quadratic form $\mathbf{x}^T\mathbf{Ax}$, where $\mathbf{A}$ is a symmetric $n \times n$ matrix and $\mathbf{x}$ is an $n \times 1$ column vector so that

$$\mathbf{x}^T\mathbf{Ax} = \begin{bmatrix} x_1 & x_2 & \cdots & x_n \end{bmatrix} \begin{bmatrix} a_{11} & a_{12} & \cdots & a_{1m} \\ a_{21} & a_{22} & \cdots & a_{2m} \\ \vdots & \vdots & \ddots & \vdots \\ a_{n1} & a_{n2} & \cdots & a_{nm} \end{bmatrix} \begin{bmatrix} x_1 \\ x_2 \\ \vdots \\ x_n \end{bmatrix}$$
$$= a_{11}x_1^2 + 2a_{12}x_1x_2 + \cdots + 2a_{1n}x_1x_n + a_{22}x_2^2 + 2a_{23}x_2x_3 + \cdots + a_{nn}x_n^2.$$

Taking the partial derivatives of $\mathbf{x}^T\mathbf{Ax}$ with respect to $\mathbf{x}$, we get

$$\frac{\partial(\mathbf{x}^T\mathbf{Ax})}{\partial x_1} = 2(a_{11}x_1 + a_{12}x_2 + \cdots + a_{1n}x_n) = 2\mathbf{a}_1\mathbf{x}$$

$$\vdots$$

$$\frac{\partial(\mathbf{x}^T\mathbf{Ax})}{\partial x_n} = 2(a_{n1}x_1 + a_{n2}x_2 + \cdots + a_{nm}x_n) = 2\mathbf{a}_n\mathbf{x},$$

which is $2\mathbf{Ax}$.

See Greene (2003, pp. 838–840) and Searle (1982, pp. 327–329) for more details.

A.14   SOLVING A SYSTEM OF LINEAR EQUATIONS IN PROC IML

Consider the following linear system of equations in three unknowns:

$$
\begin{aligned}
x + y + z &= 0, \\
x - 2y + 2z &= 4, \\
x + 2y - z &= 2.
\end{aligned}
$$

We will use Proc IML to calculate the value of x, y, and z that satisfies these equations. Let $\mathbf{X}$ be the "data" matrix, $\mathbf{b}$ the vector of unknown coefficients, and let $\mathbf{c}$ be the vector of constants. Then,

$$
\mathbf{X} = \begin{bmatrix} 1 & 1 & 1 \\ 1 & -2 & 2 \\ 1 & 2 & -1 \end{bmatrix},
$$
$$
\mathbf{b} = \begin{bmatrix} x & y & z \end{bmatrix}^T,
$$

and

$$
\mathbf{c} = \begin{bmatrix} 0 & 4 & 2 \end{bmatrix}^T.
$$

It is easy to show that $\mathbf{X}$ is invertible so that $\mathbf{b} = (\mathbf{X}^T\mathbf{X})^{-1}\mathbf{X}^T\mathbf{c}$. We can use the following Proc IML statements to solve for $\mathbf{b}$.

```
proc iml;
     X={1 1 1,1 -2 2,1 2 -1};
     c={0,4,2};
     b=inv(X'*X)*X'*c;
     print b;
run;
```

The program yields a solution set of $x = 4$, $y = -2$, and $z = -2$, which satisfy the original linear system.

Appendix B

BASIC MATRIX OPERATIONS IN PROC IML

B.1 ASSIGNING SCALARS

Scalars can be viewed as 1×1 matrices and can be created using Proc IML by using the statement x=*scalar_value* or x={*scalar_value*}. As an example, the statements x=14.5 and x={14.5} are the same and both store the value 14.5 in x. We can also store character values as the commands name='James' and hello='Hello World' illustrate.

The stored values in the variables can easily be determined by using the print command in Proc IML. For example to view the values in the variables x, *name*, and *hello* use the command Print x name hello.

B.2 CREATING MATRICES AND VECTORS

As mentioned in Appendix A, it is easy to create matrices and vectors in Proc IML. The command A={2 4, 3 1} will create the matrix

$$A = \begin{bmatrix} 2 & 4 \\ 3 & 1 \end{bmatrix}.$$

Each row of the matrix is separated by a comma. That is, each row of the above command yields a row vector. For instance, the command A={1 2 3 4} creates the row vector $A = \begin{bmatrix} 1 & 2 & 3 & 4 \end{bmatrix}$.

If we separate each entry in the row vector by a comma, we will get a column vector. As an example, the command A={1,2,3,4} creates the column vector

$$A = \begin{bmatrix} 1 \\ 2 \\ 3 \\ 4 \end{bmatrix}.$$

Applied Econometrics Using the SAS® System, by Vivek B. Ajmani
Copyright © 2009 John Wiley & Sons, Inc.

These commands can easily be extended to create matrices consisting of character elements. For example, the command A={'a' 'b', 'c' 'd'} will create the matrix

$$A = \begin{bmatrix} a & b \\ c & d \end{bmatrix}.$$

B.3 ELEMENTARY MATRIX OPERATIONS

B.3.a Addition/Subtraction of Matrices

For two conformable matrices, A and B, their sum can be computed by using the command C=A + B, where the sum is stored in C. Changing the addition to a subtraction yields the difference between the two matrices.

B.3.b Product of Matrices

For two conformable matrices, A and B, the element by element product of the two is given by the command C=A#B. For example, consider the two matrices

$$A = \begin{bmatrix} 1 & 2 \\ 3 & 4 \end{bmatrix} \quad \text{and} \quad B = \begin{bmatrix} 5 & 6 \\ 7 & 8 \end{bmatrix}.$$

The element by element product of these two is given by

$$C = \begin{bmatrix} 5 & 12 \\ 21 & 32 \end{bmatrix}.$$

The product of the two matrices is given by using the command C=A*B. In the above example, the product is

$$C = \begin{bmatrix} 19 & 22 \\ 43 & 50 \end{bmatrix}.$$

The square of a matrix is given by either of the following commands C=A##2 or C=A*A. Of course, we can use these commands to raise a matrix to any power (assuming that the product is defined).

B.3.c Kronecker Products

The Kronecker product of two matrices **A** and **B** can be obtained by using the command A@B. For example, let

$$\mathbf{A} = \begin{bmatrix} -1 & 2 \\ 4 & 1 \end{bmatrix} \quad \text{and} \quad \mathbf{B} = \begin{bmatrix} 1 & 0 \\ 0 & 1 \end{bmatrix}.$$

Then, the command C=A@B will produce

$$\mathbf{C} = \begin{bmatrix} -1 & 0 & 2 & 0 \\ 0 & -1 & 0 & 2 \\ 4 & 0 & 1 & 0 \\ 0 & 4 & 0 & 1 \end{bmatrix}.$$

B.3.d Inverses, Eigenvalues, and Eigenvectors

As shown in Appendix A, the inverse of a square matrix A can be computed by using the command C=inv(A). Eigenvalues and eigenvectors can be computed easily by using the commands C=eigval(A) or C=eigvec(A).

B.4   COMPARISON OPERATORS

The max(min) commands will search for the maximum(minimum) element of any matrix or vector. To use these commands, simply type C=max(A) or C=min(A). For two conformable matrices (of the same dimension), we can define the elementwise maximums and minimums. Consider matrices A and B, which were given in (Section B.3.b). The command C=A<>B will find the elementwise maximum between the two matrices. In our example, this will yield

$$C = \begin{bmatrix} 5 & 6 \\ 7 & 8 \end{bmatrix}.$$

The command C=A><B will yield the elementwise minimum between the two matrices. In our example, this is simply A.

B.5   MATRIX-GENERATING FUNCTIONS

B.5.a   Identity Matrix

The command Iden=I(3) will create a 3×3 identity matrix.

B.5.b   The J Matrix

This is a matrix of 1's. The command J=J(3,3) will create a 3×3 matrix of 1's. This command can be modified to create a matrix of constants. For example, suppose that we want a 3×3 matrix of 2's. We can modify the above command as follows J=J(3,3,2).

B.5.c   Block Diagonal Matrices

Often, we will have to work with block diagonal matrices. A block diagonal matrix can be created by using the command C=block $(A_1, A_2, \ldots)$ where $A_1, A_2 \ldots$ are matrices. For example, for the A and B matrices defined earlier, the block diagonal matrix C is given by

$$C = \begin{bmatrix} 1 & 2 & 0 & 0 \\ 3 & 4 & 0 & 0 \\ 0 & 0 & 5 & 6 \\ 0 & 0 & 7 & 8 \end{bmatrix}.$$

B.5.d   Diagonal Matrices

The identity matrix is a matrix with 1's on the diagonal. It is easy to create any diagonal matrix in Proc IML. For instance, the command C=diag({1 2 4}) will create the following diagonal matrix:

$$C = \begin{bmatrix} 1 & 0 & 0 \\ 0 & 2 & 0 \\ 0 & 0 & 4 \end{bmatrix}.$$

Given a square matrix, the *diag* command can be used to extract the diagonal elements. For example, the command C=diag({1 2,3 4}) will create the following matrix

$$C = \begin{bmatrix} 1 & 0 \\ 0 & 4 \end{bmatrix}.$$

B.6   SUBSET OF MATRICES

Econometric analysis using Proc IML often involves extracting specific columns (or rows) of matrices. The command C=A[,1] will extract the first column of the matrix A, and the command R=A[1,] will extract the first row of the matrix A.

B.7   SUBSCRIPT REDUCTION OPERATORS

Proc IML can be used to easily calculate various row- and column-specific statistics of matrices. As an example, consider the 3×3 matrix defined by the command A={0 1 2, 5 4 3, 7 6 8}. Column sums of this matrix can be computed by using the command

Col_Sum=A[+ ,]. Using this command yields a row vector *Col_Sum* with elements 12, 11, and 13. The row sums of this matrix can be computed by using the command Row_Sum=A[, +].

We can also determine the maximum element in each column by using the command Col_Max=A[<>,]. Using this command yields a row vector *Col_Max* with elements 7, 6, and 8. The command Row_Min=A[,><] will yield the minimum of each row of the matrix. The column means can be calculated by using the command Col_Mean=A[:,]. Using this command, yields the row vector *Col_Mean* with elements 4, 3.67, and 4.33. The command Col_Prod=A[#,] results in a row vector *Col_Prod* that contains the product of the elements in each column. In our example, the result is a row vector with elements 0, 24, and 48. We can easily extend this command to calculate the sums of squares of each column. This is calculated by using the command Col_SSQ=A[##,]. The result is a row vector *Col_SSQ* with elements 74, 53, and 77.

B.8 THE Diag AND VecDiag COMMANDS

The Proc IML *Diag* command create a diagonal matrix. For example, if

$$A = \begin{bmatrix} 1 & 3 \\ 2 & 4 \end{bmatrix},$$

then the command B=Diag(A) results in a diagonal matrix *B* whose diagonal elements are the diagonal elements of *A*. That is,

$$B = \begin{bmatrix} 1 & 0 \\ 0 & 4 \end{bmatrix}.$$

This command is useful when extracting the standard errors of regression coefficients from the diagonal elements of the variance–covariance matrices. If a column vector consisting of the diagonal elements of *A* is desirable, then one can use the VecDiag function. As an example, the command B=VecDiag(A) results in

$$B = \begin{bmatrix} 1 \\ 4 \end{bmatrix}.$$

B.9 CONCATENATION OF MATRICES

There are several instances where we have a need to concatenate matrices. A trivial case is where we need to append a column of 1's to a data matrix. Horizontal concatenation can be done by using '||', while vertical concatenation can be done by using '//'. For example, consider the following matrices:

$$\mathbf{A} = \begin{bmatrix} -1 & 2 \\ 2 & 1 \end{bmatrix} \quad \text{and} \quad \mathbf{B} = \begin{bmatrix} 6 & 7 \\ 8 & 9 \end{bmatrix}.$$

The command A||B gives the matrix

$$\begin{bmatrix} -1 & 2 & 6 & 7 \\ 2 & 1 & 8 & 9 \end{bmatrix},$$

whereas the command A//B gives the matrix

$$\begin{bmatrix} -1 & 2 \\ 2 & 1 \\ 6 & 7 \\ 8 & 9 \end{bmatrix}.$$

B.10 CONTROL STATEMENTS

Several Proc IML routines given in this book make use of control statements. For example, we made use of control statements when computing MLE estimates for the parameters. These statements were also used when computing estimates through iterative procedures.

DO-END Statement: The statements following the DO statement are executed until a matching END statement is encountered.

DO Iterative Statement: The DO iterative statements take the form

```
DO Index=start TO end;
IML statements follow
END;
```

For example, the statements

```
DO Index=1 to 5;
Print Index;
END;
```

will print the value of INDEX for each iteration of the DO statement. The output will consist of the values of INDEX starting from 1 through 5.

IF-THEN/ELSE Statement: These statements can be used to impose restrictions or conditions on other statements. The IF part imposes the restriction and the THEN part executes the action to be taken if the restrictions are met. The ELSE portion of the statement execute the action for the alternative. For example, the statements

```
IF MAX(A)<30 then print 'Good Data';
ELSE print 'Bad Data';
```

evaluate the matrix *A*. If the maximum element of the matrix is less than 30, then the statement 'Good Data' is printed, else the statement 'Bad Data' is printed.

B.11   CALCULATING SUMMARY STATISTICS IN PROC IML

Summary statistics on the numeric variables stored in matrices can be obtained in Proc IML by using the SUMMARY command. The summary statistics can be based on subgroups (e.g., Panel Data) and can be saved in matrices for later use. As an example, consider the cost of US airlines panel data set from Greene (2003). The data consist of 90 observations for six firms for 1970–1984. The following SAS statements can be used to summarize the data by airline. The option *opt(save)* saves the summary statistics by airline. The statements will retrieve and save the summary statistics in matrices. The names of the matrices are identical to the names of the variables. The statement 'print LnC' produces the means and standard deviations for the six airlines for the variable LnC. The first column contains the means, whereas the second column contains the standard deviations. We have found this command useful when programming the Hausman–Taylor estimation method for panel data models. The resulting output is given in output B.1.

```
proc import out=airline
      datafile="C:\Temp\airline"
      dbms=Excel Replace;
      getnames=yes;
run;
Data airline;
      set airline;
      LnC=log(C);
      LnQ=Log(Q);
      LnPF=Log(PF);
run;
proc iml;
      use airline;
      summary var {LnC LnQ LnPF} class {i} stat{mean std}
      opt{save};
      print LnC;
run;
```

I	Nobs	Variable	MEAN	STD
1	15	LNC	14.67563	0.49462
		LNQ	0.31927	0.23037
		LNPF	12.73180	0.85990
2	15	LNC	14.37247	0.68054
		LNQ	-0.03303	0.33842
		LNPF	12.75171	0.84978
3	15	LNC	13.37231	0.52207
		LNQ	-0.91226	0.24353
		LNPF	12.78972	0.81772
4	15	LNC	13.13580	0.72739
		LNQ	-1.63517	0.43525
		LNPF	12.77803	0.82784
5	15	LNC	12.36304	0.71195
		LNQ	-2.28568	0.49739
		LNPF	12.79210	0.82652
6	15	LNC	12.27441	0.89175
		LNQ	-2.49898	0.67981
		LNPF	12.77880	0.83292
All	90	LNC	13.36561	1.13197
		LNQ	-1.17431	1.15061
		LNPF	12.77036	0.81237

LNC	
14.675633	0.494617
14.37247	0.6805358
13.372309	0.5220658
13.135799	0.7273866
12.363038	0.7119453
12.274407	0.8917487

OUTPUT B.1. Summary statistics of three variables for each airline.

Appendix C

SIMULATING THE LARGE SAMPLE PROPERTIES OF THE OLS ESTIMATORS

In Chapter 1 we saw that under the least squares assumptions, the estimator $\mathbf{b} = (\mathbf{X}^T\mathbf{X})^{-1}\mathbf{X}^T\mathbf{y}$ for the coefficients vector $\boldsymbol{\beta}$ in the model $\mathbf{y} = \mathbf{X}^T\boldsymbol{\beta} + \boldsymbol{\varepsilon}$ was unbiased with variance–covariance matrix given by $Var(\mathbf{b}|\mathbf{X}) = \sigma^2(\mathbf{X}^T\mathbf{X})^{-1}$. Here, $\sigma^2 = Var(\boldsymbol{\varepsilon}|\mathbf{X})$. We also saw that if $\boldsymbol{\varepsilon}|\mathbf{X} \sim N(0,\sigma^2)$, then the asymptotic distribution of $\mathbf{b}|\mathbf{X}$ is normal with mean $\boldsymbol{\beta}$ and variance–covariance $\sigma^2(\mathbf{X}^T\mathbf{X})^{-1}$. That is, $\mathbf{b}|\mathbf{X} \sim N(\boldsymbol{\beta},\sigma^2(\mathbf{X}^T\mathbf{X})^{-1})$. This appendix presents a simple technique for simulating the large sample properties of the least squares estimator.

Consider the simple linear regression model $y_i = 4 + 10x_i + \varepsilon_i$ with one dependent and one explanatory variable. For simulation purposes, we will assume that $x_i \sim N(10,25)$ and $\varepsilon_i \sim N(0,2.25)$. Note that the random nature of the regressor is simply being used to generate values for the explanatory variable. A single simulation run for this model comprises generating $n = 50$ values of x_i and ε_i, plugging these values into the regression equation to get the corresponding value of the dependent variable, y_i. The simulation ends by running a regression of y_i versus x_i using the 50 simulated values. Proc Reg is used to estimate the values of $\beta_1, \beta_2,$ and σ^2. This simulation is then repeated 10,000 times. Therefore, we end up having 10,000 estimates of the coefficients and of σ^2. Proc Means is then used to generate basic summary statistics for these 10,000 estimates. The output generated can be used to determine how close the means of the 10,000 sample estimates are to the true values of the parameters ($\beta_0 = 10, \beta_1 = 4, \sigma^2 = 2.25$). We conducted the simulation with sample sizes of 50, 100, 500, and 1000. The means for the simulation run with 50 observations is given in Output C.1. Notice that the sample estimates from the simulation runs are almost identical to the true values.

Hypothesis test on the coefficient β_1 was also conducted. We calculated the percentage of times the null hypothesis, $H_0: \beta_1 = 10$, was rejected. This gives us an estimate of the true Type I error rate. We used the confidence interval approach for conducting this test. The *tableout* option of Proc Reg was used to construct the 95% confidence interval for β_1. The null hypothesis was rejected if the confidence interval did not include the value under the null hypothesis (10). The output of the simulation run with 50 observations is given in Output C.2. It indicates that the null hypothesis is rejected 4.89% of the time, which is close to the Type I error rate of 5%.

Finally, we use Proc Univariate with the *histogram* option to generate histograms for different simulation runs to demonstrate the large sample distribution of $\hat{\beta}_1$. The simulation results are given in Figures C.1 – C.4. Notice that the distribution of $\hat{\beta}_1$ is bell-shaped and symmetric even when the sample size is 50. The normality of the distribution becomes more pronounced as the sample size increases. Also notice that the spread of the distribution for the estimate reduces as the sample size increases. This indicates that the standard error of the estimate becomes smaller with increasing sample sizes.

Applied Econometrics Using the SAS® System, by Vivek B. Ajmani
Copyright © 2009 John Wiley & Sons, Inc.

Obs	n	intercept_e	slope_e	MSE_e
1	10000	4.00117	9.99947	1.49198

OUTPUT C.1. Mean and standard deviation of simulated values of the estimate ($n = 50$).

The following SAS statements can be used to conduct the simulation just described.

```
data simulation;
     sigma=1.5;
     beta1=4;
     beta2=10;
     do index1=1 to 10000;
          seed=12345;
          do index2=1 to 50;
               call rannor(seed,x1);
               x=10+5*x1;
               e=sigma*normal(0);
               y=beta1+beta2*x+e;
               output;
          end;
     end;
run;
proc reg data=simulation noprint outest=estimates tableout;
     model y=x;
     by index1;
run;
data estimates1;
     set estimates;
     if y=-1;
     rename x=slope;
     drop _label_;
run;
proc univariate noprint data=estimates1;
     var intercept slope _rmse_;
     output out=estimates1 n=n mean=intercept_e slope_e MSE_e;
run;
proc print data=estimates1;
run;
data estimates2;
     set estimates;
     if _type_ in ('L95B','U95B');
     keep index1 _type_ x;
run;
```

The MEANS Procedure

Analysis Variable : reject				
N	Mean	Std Dev	Minimum	Maximum
10000	0.0497000	0.2173353	0	1.0000000

OUTPUT C.2. Simulated Type 1 error rate ($n = 50$).

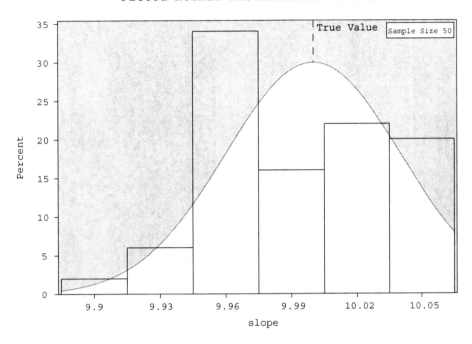

The UNIVARIATE Procedure
Fitted Distribution for slope

Parameters for Normal Distribution		
Parameter	Symbol	Estimate
Mean	Mu	10
Std Dev	Sigma	0.039948

Goodness-of-Fit Tests for Normal Distribution				
Test	Statistic		p Value	
Cramer-von Mises	W-Sq	0.20979495	Pr > W-Sq	0.223
Anderson-Darling	A-Sq	1.21657290	Pr > A-Sq	0.210

Quantiles for Normal Distribution		
	Quantile	
Percent	Observed	Estimated
1.0	9.89767	9.90707
5.0	9.93906	9.93429
10.0	9.94859	9.94881
25.0	9.96272	9.97306
50.0	9.99712	10.00000
75.0	10.02268	10.02694
90.0	10.05022	10.05119
95.0	10.05629	10.06571
99.0	10.05983	10.09293

FIGURE C.1. Histogram of the simulated estimates ($n = 50$).

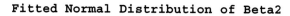

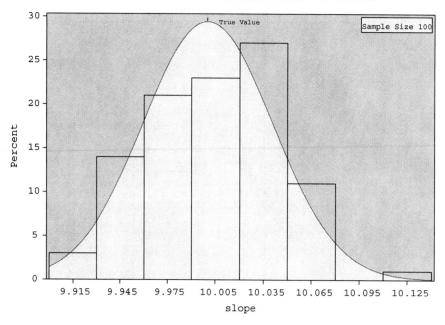

The UNIVARIATE Procedure
Fitted Distribution for slope

Parameters for Normal Distribution		
Parameter	Symbol	Estimate
Mean	Mu	10
Std Dev	Sigma	0.040671

Goodness-of-Fit Tests for Normal Distribution				
Test	Statistic		p Value	
Cramer-von Mises	W-Sq	0.12817105	Pr > W-Sq	>0.250
Anderson-Darling	A-Sq	0.69744815	Pr > A-Sq	>0.250

Quantiles for Normal Distribution		
	Quantile	
Percent	Observed	Estimated
1.0	9.91077	9.90539
5.0	9.93591	9.93310
10.0	9.94713	9.94788
25.0	9.97455	9.97257
50.0	10.00157	10.00000
75.0	10.03171	10.02743
90.0	10.05431	10.05212
95.0	10.06417	10.06690
99.0	10.10230	10.09461

FIGURE C.2. Histogram of the simulated estimates ($n = 100$).

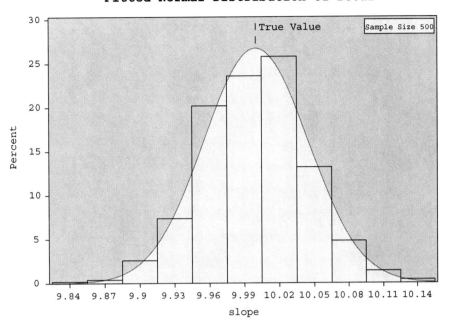

The UNIVARIATE Procedure
Fitted Distribution for slope

Parameters for Normal Distribution		
Parameter	Symbol	Estimate
Mean	Mu	10
Std Dev	Sigma	0.044755

Goodness-of-Fit Tests for Normal Distribution				
Test		Statistic	p Value	
Cramer-von Mises	W-Sq	0.04634466	Pr > W-Sq	>0.250
Anderson-Darling	A-Sq	0.29383767	Pr > A-Sq	>0.250

Quantiles for Normal Distribution		
	Quantile	
Percent	Observed	Estimated
1.0	9.89798	9.89588
5.0	9.92721	9.92638
10.0	9.94281	9.94264
25.0	9.96843	9.96981
50.0	9.99864	10.00000
75.0	10.02923	10.03019
90.0	10.05702	10.05736
95.0	10.07428	10.07362
99.0	10.10539	10.10412

FIGURE C.3. Histogram of the simulated estimates ($n = 500$).

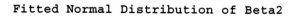

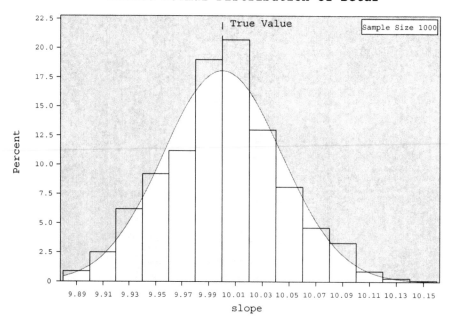

The UNIVARIATE Procedure
Fitted Distribution for slope

Parameters for Normal Distribution		
Parameter	Symbol	Estimate
Mean	Mu	10
Std Dev	Sigma	0.04417

Goodness-of-Fit Tests for Normal Distribution				
Test	Statistic		p Value	
Cramer-von Mises	W-Sq	0.14559887	Pr > W-Sq	>0.250
Anderson-Darling	A-Sq	0.82700450	Pr > A-Sq	>0.250

Quantiles for Normal Distribution		
	Quantile	
Percent	Observed	Estimated
1.0	9.90069	9.89725
5.0	9.92595	9.92735
10.0	9.94106	9.94339
25.0	9.97233	9.97021
50.0	10.00115	10.00000
75.0	10.02828	10.02979
90.0	10.05665	10.05661
95.0	10.07772	10.07265
99.0	10.10588	10.10275

FIGURE C.4. Histogram of the simulated estimates ($n = 1000$).

```
proc transpose data=estimates2 out=estimates2(keep=L95B U95B);
     var x;
     id _type_;
     by index1;
run;
data estimates2;
     set estimates2;
     beta2=10;
     if L95B<beta2 <U95B then reject=0;else reject=1;
run;
proc means data=estimates2;
     var reject;
run;
title 'Fitted Normal Distribution of Beta2';
proc univariate noprint data=estimates0;
     histogram slope /
     normal(mu=10 color=blue fill)
     cfill = ywh
     cframe = ligr
     href = 10
     hreflabel = 'True Value'
     lhref = 2
     vaxis = axis1
     name = 'MyHist';
     axis1 label=(a=90 r=0);
     inset n = 'Sample Size'
     beta / pos=ne cfill=ywh;
run;
```

Appendix D

INTRODUCTION TO BOOTSTRAP ESTIMATION

D.1 INTRODUCTION

Bootstrapping is a general, distribution-free method that is used to estimate parameters of interest from data collected from studies or experiments. It is often referred to as a resampling method because it is carried out by repeatedly drawing samples from the original data that were gathered. This section introduces the basics of bootstrapping and extends it to bootstrapping in regression analysis. For a discussion on calculating bias or calculating confidence intervals using bootstrapping, see Efron and Tibshirani (1993).

Bootstrapping is a useful estimation technique when:

1. The formulas that are to be used for calculating estimates are based on assumptions that may not hold or may not be understood well, or cannot be verified, or are simply dubious.
2. The computational formulas hold only for large samples and are unreliable for small samples or simply not valid for small samples.
3. The computational formulas do not exist.

To begin the discussion of bootstrapping techniques, assume that a study or experiment was conducted resulting in a data set $x_1, \ldots, x_n$ of size n. This is a trivial case where the data are univariate in nature. Most studies involve collection of data on several variables as in the case of regression analysis studies. However, we use the simple example to lay the groundwork for the elements of bootstrapping methods.

Assume that the data set was generated by some underlying distribution $f(\theta)$. Here, $f(\theta)$ is the probability density function and may be either continuous or discrete. It may be the case that the true density function is unknown and the functional form of $f(\theta)$ is, therefore, unknown also. We are interested in estimating the parameter θ, which describes some feature of the population from which the data were collected. For instance, θ could be the true mean, median, the proportion, the variance, or the standard deviation of the population. Assume for the moment that we have a well-defined formula to calculate an estimate, $\hat{\theta}$, of θ. However, no formulas exist for calculating the confidence interval for θ. Under the ideal setting where we have unlimited resources, we could draw a large number of samples from the population. We could then estimate θ by calculating $\hat{\theta}$ for each sample. The calculated values of $\hat{\theta}$ can then be used to construct an empirical distribution of $\hat{\theta}$ that could then be used to construct a confidence interval for θ. However, in reality we just have a single sample that is a justification for the use of bootstrapping method.

The general idea behind bootstrapping is as follows (assuming that a study/experiment resulted in a data set of size n):

1. A sample of size n is drawn with replacement from the data set in hand.
2. An estimate, $\hat{\theta}$, of θ is calculated.

Applied Econometrics Using the SAS® System, by Vivek B. Ajmani
Copyright © 2009 John Wiley & Sons, Inc.

3. Steps 1 and 2 are repeated several times (sometimes thousands of repetitions are used) to generate a (simulated) distribution of $\hat{\theta}$. This simulated distribution is then used for making inferences about θ.

As an example, suppose that we want to construct a 95% confidence interval for θ. However, we do not have formulas that can be used for calculating the interval. We can therefore use bootstrapping to construct the confidence interval. The steps are as follows (Efron and Tibshirani, 1993):

1. Draw 1000 (as an example) bootstrap samples from the original data and calculate $\hat{\theta}_1, \ldots, \hat{\theta}_{1000}$, the estimates from each of the 1000 samples.
2. Next, sort these estimates in increasing order.
3. Calculate the 2.5th and 97.5th percentile from the 1000 simulated values of $\hat{\theta}$. The 2.5th percentile will be the average of the 25th and 26th observation while the 97.5th percentile will be the average of the 975th and 976th observation. That is,

$$\text{Lower confidence limit} = \frac{\hat{\theta}_{25} + \hat{\theta}_{26}}{2},$$

$$\text{Upper confidence limit} = \frac{\hat{\theta}_{975} + \hat{\theta}_{976}}{2}.$$

Notice that we took the lower 2.5% and the upper 2.5% of the simulated distribution of $\hat{\theta}$ out to achieve the desired 95% confidence. Also note that we did not make any assumptions about the underlying distribution that generated the original data set. We will now formalize the general bootstrapping method presented so far. Consider a random variable x with cumulative distribution $F(x; \theta)$. Here, θ is a vector of unknown parameters. For example, if the distribution of x is normal, then $\theta = (\mu, \sigma^2)$. Assume that we are interested in estimating θ or some element of θ that describes some aspect of $f(x; \theta)$, the distribution of x. That is, we may be interested in estimating the mean, or the standard deviation, or the standard error of the mean. As we did before, we will assume that a study/experiment resulted in a random sample $x_1, \ldots, x_n$ of size n. We can use this sample to approximate the cumulative distribution, $F(x; \theta)$, with the empirical distribution function, $\hat{F}(x; \theta)$. The estimate, $\hat{F}(x; \theta)$, can be written as

$$\hat{F}(x; \theta) = \frac{1}{n} \sum_{i=1}^{n} I_{(-\infty, x)}(x_i),$$

where I is an indicator function that counts the number of x's in the original sample that fall in the interval $(-\infty, x)$. This is better illustrated in Figure D.1.

In Figure D.1, the true distribution, $F(x; \theta)$, is given by the smooth line while the estimated function, $\hat{F}(x; \theta)$, is given by the stepwise representation. The parameter vector θ or elements of it could be calculated exactly if the form of $F(x; \theta)$ were known.

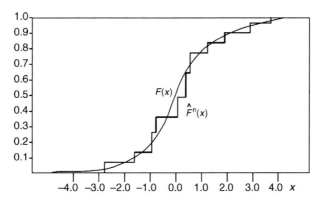

FIGURE D.1. Plot comparing actual cumulative versus simulated cumulative distributions. (Graph reproduced with permission from Paul Glewwe, University of Minnesota.)

That is, if we knew the exact form of $F(x; \boldsymbol{\theta})$, then we could derive the probability density function, $f(x; \boldsymbol{\theta})$, or a function $t(F)$ to calculate θ. However, assume that the functional form of $F(x; \boldsymbol{\theta})$ is unknown and that it was approximated with $\hat{F}(x; \boldsymbol{\theta})$. Therefore, one option we have is to replace $F(x; \boldsymbol{\theta})$ with $\hat{F}(x; \boldsymbol{\theta})$ to get the estimated function $t(\hat{F})$. We can then use $t(\hat{F})$ to calculate an estimate, $\hat{\theta}$, of θ. The estimator $\hat{\theta}$ in this instance is called the plug-in estimator of θ (Efron and Tibshirani, 1993, p. 35). As an example, the plug-in estimator of the population mean μ_x,

$$\mu_x = \int_{-\infty}^{\infty} x f(x)\, dx,$$

is the sample mean

$$\bar{x} = \frac{1}{n} \sum_{i=1}^{n} x_i.$$

Notice that calculating the mean of x was trivial and did not require bootstrapping methods. In general, bootstrapping techniques are used to calculate standard errors and for constructing confidence intervals without making any assumption about the underlying distribution from which the samples are drawn.

D.2   CALCULATING STANDARD ERRORS

We will now discuss how bootstrapping methods can be used to calculate an estimate of the standard error of the parameter of interest. Assume then that we have an estimate of θ. That is, $\hat{\theta}$ was calculated from the original data set without the use of bootstrapping. Bootstrapping, however, will be used to calculate an estimate of the standard error of $\hat{\theta}$. The general method for doing this is as follows (again assume that we have a data set of size n) (Efron and Tibshirani, 2004, p. 45):

1. Draw B samples of size n with replacement from the original data set.
2. Calculate $\hat{\theta}$ for each of the samples from step 1. That is, we now have $\hat{\theta}_1, \ldots, \hat{\theta}_B$.
3. We calculate the standard error from the B estimates of θ by using the standard formulas for standard errors. That is,

$$se_B(\hat{\theta}) = \sqrt{\frac{1}{B-1} \sum_{i=1}^{B} (\hat{\theta}_i - \bar{\hat{\theta}})^2},$$

where $\bar{\hat{\theta}} = B^{-1} \sum_{i=1}^{B} \hat{\theta}_i$ is simply the mean of the $\hat{\theta}_1, \ldots, \hat{\theta}_B$. In practice, B is set to a very large number. Most practitioners use 200–500 bootstrapped samples.

D.3   BOOTSTRAPPING IN SAS

Bootstrapping can easily be programmed in SAS by using simple routines. SAS macros to calculate bootstrapped estimates are available for download from the SAS Institute. The macros can be used to calculate bootstrapped and jackknife estimates for the standard deviation and standard error, and they are also used to calculate the bootstrapped confidence intervals. The macros can also be used to calculate bootstrapped estimates of coefficients in regression analysis. These macros need to be invoked from within SAS. We will illustrate the use of these macros a bit later. For now, we show how a simple program can be written to compute bootstrap estimates.

Consider a data set that consists of 10 values: 196, −12, 280, 212, 52, 100, −206, 188, −100, 202. We will calculate bootstrap estimates of the standard error for the mean. The following SAS statements can be used:

```
data age_data;
     input age;
     cards;
     45
```

```
             40
             9
             7
             17
             16
             15
             11
             10
             8
             54
             76
             87

;
data bootstrap;
     do index=1 to 500;
           do i=1 to nobs;
                 x=round(ranuni(0)*nobs);
                 set age_data
                 nobs=nobs
                 point=x;
                 output;
           end;
     end;
     stop;
run;
```

The following Proc Univariate statements will calculate the mean of the bootstrapped samples.

```
proc univariate data=bootstrap noprint;
     var age;
     by index;
     output out=out1 mean=mean n=n;
run;
```

Finally, the following Proc Univariate statements will calculate the standard deviation of the 500 bootstrapped means.

```
proc univariate data=out1 noprint;
     var mean;
     output out=out2 n=n mean=mean std=se;
run;
proc print data=out2;
run;
```

The analysis results in a mean and standard error of 27.6 and 6.8, respectively.

D.4   BOOTSTRAPPING IN REGRESSION ANALYSIS

Consider the standard linear regression model $y_i = \mathbf{x}_i^T \boldsymbol{\beta} + \varepsilon_i$, where $\mathbf{x}_i$ and $\boldsymbol{\beta}$ are $k \times 1$ column vectors and ε_i is random error. Assume that we have a data set comprising n pairs of observations $(y_1, \mathbf{x}_1), \ldots, (y_n, \mathbf{x}_n)$. Assume that the conditional expectation $E(\varepsilon_i|\mathbf{x}_i) = 0$. Furthermore, assume that we do not know $F(\varepsilon|\mathbf{x})$, the cumulative distribution of ε. In general, F is assumed to be normal.

We will make use the standard least squares estimator for $\boldsymbol{\beta}$, namely, $\hat{\boldsymbol{\beta}} = (\mathbf{X}^T\mathbf{X})^{-1}\mathbf{X}^T\mathbf{y}$, to calculate bootstrapped estimates. That is, as was the case, with the mean being calculated without the use of bootstrapping, we will assume that the least squares estimate can be calculated without any need of bootstrapping. However, we are interested in calculating the standard errors of $\hat{\boldsymbol{\beta}}$. That is, we assume that the formulas for calculating the standard errors are either unknown, unreliable, or simply do not work for small samples.

As shown in Chapter 1, the estimate of the variance of $\hat{\boldsymbol{\beta}}$ is $Var(\hat{\boldsymbol{\beta}}|\mathbf{X}) = \hat{\sigma}^2(\mathbf{X}^T\mathbf{X})^{-1}$, where $\hat{\sigma}^2$ is estimated as

$$\hat{\sigma}^2 = \frac{1}{n}\sum_{i=1}^{n}(y_i - \mathbf{x}_i^T\hat{\boldsymbol{\beta}})^2$$

or

$$\hat{\sigma}^2 = \frac{1}{n-k-1}\sum_{i=1}^{n}(y_i - \mathbf{x}_i^T\hat{\boldsymbol{\beta}})^2.$$

Notice that the first version is not an unbiased estimator for σ^2 whereas the second version is. These versions are often referred to as the "not bias-corrected" and the "bias-corrected" versions, respectively. There are two bootstrapped methods (pairs method, residuals method) that are employed to estimate the standard error of $\hat{\boldsymbol{\beta}}$ (Glewwe, 2006; Efron and Tibshirani, 1993, p. 113).

The bootstrapped pairs method randomly selects pairs of y_i and $\mathbf{x}_i$ to calculate an estimate of ε_i, while the bootstrapped residuals method takes each $\mathbf{x}_i$ just once but then links it with a random draw of an estimate of $\boldsymbol{\varepsilon}$. The next section outlines both methods.

D.4.1 Bootstrapped Residuals Method

As before, we assume that a study or experiment resulted in n observations $(y_1, \mathbf{x}_1), \ldots, (y_n, \mathbf{x}_n)$. The general method for the bootstrapped residuals method is

1. For each i, calculate an estimate, e_i of ε_i. That is, $e_i = y_i - \mathbf{x}_i^T\hat{\boldsymbol{\beta}}$ where $\hat{\boldsymbol{\beta}}$ is the usual OLS estimator calculated from the original data.
2. Randomly draw n values of e_i (from step 1) with replacement. Denote the residuals in the sample as $e_1^*, e_2^*, \ldots, e_n^*$. Notice that the subscripts of the residuals in the selected sample are not the same as the subscripts for the residuals, e_i, which were calculated from the original sample. That is, in general $e_i^* \neq e_i$ for $i = 1, \ldots, n$.
3. With the values of e_i^* (from step 2), compute $y_i^* = \mathbf{x}_i^T\hat{\boldsymbol{\beta}} + e_i^*$. Notice that the subscripts for $\mathbf{x}_i$ here match the subscripts of $\mathbf{x}_i$ in the original data set. That is, we are using each $\mathbf{x}_i$ only once. Notice also that by construction of e_i^*, $y_i \neq y_i^*$.
4. Using the calculated values of y_i^* (from step 3), construct the vector $\mathbf{y}^*$. Finally, use $\mathbf{X} = \begin{bmatrix} \mathbf{x}_1 & \cdots & \mathbf{x}_n \end{bmatrix}^T$ and $\mathbf{y}^* = \begin{bmatrix} y_1^* & \cdots & y_n^* \end{bmatrix}^T$ to calculate $\boldsymbol{\beta}_1^*$, the first bootstrapped estimate of $\boldsymbol{\beta}$. That is, $\boldsymbol{\beta}_1^* = (\mathbf{X}^T\mathbf{X})^{-1}\mathbf{X}^T\mathbf{y}^*$.
5. Steps 2 through 4 are repeated B (typically $B = 200$–500) times to get B estimates of $\boldsymbol{\beta}$.
6. Use the B estimates (from step 5) to calculate the sample standard deviation of $\hat{\boldsymbol{\beta}}$ using the formula

$$s.e.^* = \sqrt{\frac{\sum_{i=1}^{B}(\hat{\boldsymbol{\beta}}_i^* - \hat{\boldsymbol{\beta}}^*)^2}{B-1}},$$

where

$$\hat{\boldsymbol{\beta}}^* = \frac{1}{B}\sum_{i=1}^{B}\hat{\boldsymbol{\beta}}_i^*$$

is the mean of the B residuals method bootstrapped estimates of $\boldsymbol{\beta}$.

D.4.2 Bootstrapped Pairs Method

As before, we assume that a study or experiment resulted in n observations $(y_1, \mathbf{x}_1), \ldots, (y_n, \mathbf{x}_n)$. The general method for the bootstrapped pairs method is

1. Randomly draw n pairs of values, y_i and $\mathbf{x}_i$, with replacement. Denote these as $y_1^*, y_2^*, \ldots, y_n^*$ and $\mathbf{x}_1^*, \mathbf{x}_2^*, \ldots, \mathbf{x}_n^*$. As discussed earlier, the subscripts here do not necessarily match the subscripts in the original data set.
2. Using these values of y_i^* and $\mathbf{x}_i^*$, calculate the first bootstrapped estimate of $\boldsymbol{\beta}$ by using standard OLS techniques. That is, $\mathbf{b}_1^* = (\mathbf{X}^{*T}\mathbf{X}^*)^{-1}\mathbf{X}^{*T}\mathbf{y}^*$.
3. Steps 1 and 2 are repeated B times (typically $B = 200\text{--}500$) to get B estimates of $\boldsymbol{\beta}$.
4. Use the B estimates $\mathbf{b}_i^*$, $i = 1, \ldots, B$, to calculate the sample standard deviation of $\hat{\boldsymbol{\beta}}$ using the formula

$$s.e.^* = \sqrt{\frac{\sum_{i=1}^{B}(\mathbf{b}_i^* - \hat{\boldsymbol{\beta}}^*)^2}{B-1}},$$

where

$$\hat{\boldsymbol{\beta}}^* = \frac{1}{B}\sum_{i=1}^{B}\mathbf{b}_i^*$$

is the mean of the B pairs method bootstrapped estimates of $\boldsymbol{\beta}$. Computationally, the bootstrapped pairs method is more straightforward. As discussed in Efron and Tibshirani (1993, p. 113), the bootstrapped residuals method imposes homoscedasticity because it "delinks" $\mathbf{x}_i$ with e_i. Therefore, if the homoscedasticity assumption is violated, then we should use the bootstrapped pairs method, which does not impose this. On the other hand, if we are very confident of homoscedasticity, then we can use the bootstrapped residuals method to get more precise estimates of the standard error of $\hat{\boldsymbol{\beta}}$. In fact, it can be shown that as $B \to \infty$ the standard errors of the least squares estimates calculated using the bootstrapped residuals method converge to the diagonal elements of the variance–covariance matrix $\hat{\sigma}^2(\mathbf{X}^T\mathbf{X})^{-1}$.

D.4.3 Bootstrapped Regression Analysis in SAS

We will now illustrate the residuals and the pairs methods by using the %BOOT macro that can be downloaded from the SAS Institute website at www.sas.com. We will make use of the gasoline consumption data given in Table F2.1 of Greene (2003). We need the bootstrapped macros (labeled JACKBOOT.SAS here) to be called from within the program. The %include statement can be used for this purpose. The following statements can be used:

```
%include "C:\Temp\jackboot.sas";
```

The data set is then read into SAS and stored into a temporary SAS data set called gasoline. Notice that the raw data are stored in Excel format.

```
proc import out=gasoline
     datafile="C:\Temp\gasoline"
     dbms=Excel Replace;
     getnames=yes;
run;
```

The following SAS data step statements simply transform the variables in the raw data by using the log transformations:

```
data gasoline;
     set gasoline;
     Ln_G_Pop=log(G/Pop);
```

```
        Ln_pg=Log(Pg);
        Ln_Income=Log(Y/Pop);
        Ln_Pnc=Log(Pnc);
        Ln_Puc=Log(Puc);
run;
```

The following Proc Reg statements are used to run OLS regression on the original data set. The residuals from this are stored in a temporary SAS data set called gasoline. The residuals are labeled as *resid*.

```
proc reg data=gasoline;
        model Ln_G_Pop=Ln_pg Ln_Income Ln_Pnc Ln_Puc;
        output out=gasoline r=resid p=pred;
run;
```

The following macro is required before invoking the bootstrapped macros in the program jackboot.sas. The only inputs that require changes are the variable names in the model statement. The remaining statements can be used as is. See Sample 24982-JackKnife and Bootstrap Analyses from the SAS Institute for more details. The following code has been adapted from this publication and has been used with permission from the SAS Institute.

```
%macro analyze(data=,out=);
        options nonotes;
        proc reg data=&data noprint
        outest=&out(drop=Y _IN_ _P_ _EDF_);
        model Ln_G_Pop=Ln_pg Ln_Income Ln_Pnc Ln_Puc;
        %bystmt;
run;
options notes;
%mend;
```

This portion of the code invokes the %boot macro within jackboot.sas and conducts a bootstrapped analysis by using the pairs method. Note that the root mean square error (_RMSE_) is not a plug-in estimator for σ, and therefore the bias correction is wrong. In other words, even though the mean square error is unbiased for σ^2, the root mean square error is not unbiased for σ. However, we choose to ignore this because the bias is minimal.

```
title2 'Resampling Observations-Pairs Method';
title3 '(Bias correction for _RMSE_ is wrong)';
%boot(data=gasoline, random=123);
```

This portion of the code invokes the %boot macro and conducts the bootstrapped analysis by using the residuals method.

```
title2 'Resampling Residuals-Residual Method';
title3 '(bias correction for _RMSE_ is wrong)';
%boot(data=gasoline, residual=resid, equation=y=pred+resid,
random=123);
```

The analysis results are given in Outputs D.1 and D.2. The first part of the output is from the analysis of the original data. We will skip any discussion of this portion of the output as we have already discussed OLS regression output from SAS in detail in Chapter 2. The OLS output is followed by the output where bootstrapping is done by resampling pairs (Output D.1) and where the analysis was done using the residuals method (Output D.2).

Resampling Observations
(bias correction for _RMSE_ is wrong)

The REG Procedure
Model: MODEL1
Dependent Variable: Ln_G_Pop

Number of Observations Read	36
Number of Observations Used	36

Analysis of Variance					
Source	DF	Sum of Squares	Mean Square	F Value	Pr > F
Model	4	0.78048	0.19512	243.18	<0.0001
Error	31	0.02487	0.00080237		
Corrected Total	35	0.80535			

Root MSE	0.02833	R-Square	0.9691
Dependent Mean	-0.00371	Adj R-Sq	0.9651
Coeff Var	-763.79427		

Parameter Estimates					
Variable	DF	Parameter Estimate	Standard Error	t Value	Pr > |t|
Intercept	1	-7.78916	0.35929	-21.68	<0.0001
Ln_pg	1	-0.09788	0.02830	-3.46	0.0016
Ln_Income	1	2.11753	0.09875	21.44	<0.0001
Ln_Pnc	1	0.12244	0.11208	1.09	0.2830
Ln_Puc	1	-0.10220	0.06928	-1.48	0.1502

Resampling Observations
(bias correction for _RMSE_ is wrong)

Name	Observed Statistic	Bootstrap Mean	Approximate Bias	Approximate Standard Error	Approximate Lower Confidence Limit	Bias-Corrected Statistic
Intercept	-7.78916	-7.83880	-0.049634	0.35056	-8.42662	-7.73953
Ln_G_Pop	-1.00000	-1.00000	0.000000	0.00000	-1.00000	-1.00000
Ln_Income	2.11753	2.13101	0.013484	0.09596	1.91597	2.10405
Ln_Pnc	0.12244	0.13541	0.012969	0.13611	-0.15729	0.10947
Ln_Puc	-0.10220	-0.11723	-0.015028	0.08629	-0.25631	-0.08718
Ln_pg	-0.09788	-0.09167	0.006209	0.02904	-0.16101	-0.10409
RMSE	0.02833	0.02623	-0.002093	0.00256	0.02540	0.03042

Name	Approximate Upper Confidence Limit	Confidence Level (%)	Method for Confidence Interval	Minimum Resampled Estimate	Maximum Resampled Estimate	Number of Resamples	LABEL OF FORMER VARIABLE
Intercept	-7.05244	95	Bootstrap Normal	-8.98495	-7.03140	200	Intercept
Ln_G_Pop	-1.00000	95	Bootstrap Normal	-1.00000	-1.00000	200	
Ln_Income	2.29212	95	Bootstrap Normal	1.90256	2.44562	200	
Ln_Pnc	0.37624	95	Bootstrap Normal	-0.23416	0.42458	200	
Ln_Puc	0.08196	95	Bootstrap Normal	-0.32506	0.10206	200	
Ln_pg	-0.04717	95	Bootstrap Normal	-0.16467	0.00233	200	
RMSE	0.03544	95	Bootstrap Normal	0.01844	0.03362	200	Root mean squared error

OUTPUT D.1. Bootstrapped regression analysis (pairs method) of the gasoline consumption data.

```
The REG Procedure
Model: MODEL1
Dependent Variable: Ln_G_Pop
```

Number of Observations Read	36
Number of Observations Used	36

Analysis of Variance

Source	DF	Sum of Squares	Mean Square	F Value	Pr > F
Model	4	0.78048	0.19512	243.18	<0.0001
Error	31	0.02487	0.00080237		
Corrected Total	35	0.80535			

Root MSE	0.02833	R-Square	0.9691
Dependent Mean	-0.00371	Adj R-Sq	0.9651
Coeff Var	-763.79427		

Parameter Estimates

Variable	DF	Parameter Estimate	Standard Error	t Value	Pr > |t|
Intercept	1	-7.78916	0.35929	-21.68	<0.0001
Ln_pg	1	-0.09788	0.02830	-3.46	0.0016
Ln_Income	1	2.11753	0.09875	21.44	<0.0001
Ln_Pnc	1	0.12244	0.11208	1.09	0.2830
Ln_Puc	1	-0.10220	0.06928	-1.48	0.1502

```
Resampling Residuals
(bias correction for _RMSE_ is wrong)
```

Name	Observed Statistic	Bootstrap Mean	Approximate Bias	Approximate Standard Error	Approximate Lower Confidence Limit	Bias-Corrected Statistic
Intercept	-7.78916	-7.78916	2.30926E-14	0	-7.78916	-7.78916
Ln_G_Pop	-1.00000	-1.00000	0	0	-1.00000	-1.00000
Ln_Income	2.11753	2.11753	4.44089E-15	0	2.11753	2.11753
Ln_Pnc	0.12244	0.12244	2.22045E-16	0	0.12244	0.12244
Ln_Puc	-0.10220	-0.10220	1.38778E-16	0	-0.10220	-0.10220
Ln_pg	-0.09788	-0.09788	-1.6653E-16	0	-0.09788	-0.09788
RMSE	0.02833	0.02833	4.85723E-17	0	0.02833	0.02833

Name	Approximate Upper Confidence Limit	Confidence Level (%)	Method for Confidence Interval	Minimum Resampled Estimate	Maximum Resampled Estimate	Number of Resamples	LABEL OF FORMER VARIABLE
Intercept	-7.78916	95	Bootstrap Normal	-7.78916	-7.78916	200	Intercept
Ln_G_Pop	-1.00000	95	Bootstrap Normal	-1.00000	-1.00000	200	
Ln_Income	2.11753	95	Bootstrap Normal	2.11753	2.11753	200	
Ln_Pnc	0.12244	95	Bootstrap Normal	0.12244	0.12244	200	
Ln_Puc	-0.10220	95	Bootstrap Normal	-0.10220	-0.10220	200	
Ln_pg	-0.09788	95	Bootstrap Normal	-0.09788	-0.09788	200	
RMSE	0.02833	95	Bootstrap Normal	0.02833	0.02833	200	Root mean squared error

OUTPUT D.2. Bootstrapped regression analysis (residuals method) of the gasoline consumption data.

The output consists of the OLS estimates in the first column, followed by the mean of the coefficients estimated from the 200 bootstrap samples. The third column gives the bias, which is simply the bootstrap mean minus the observed statistic. The standard errors calculated from the bootstrapped samples are given next. This is followed by the 95% confidence intervals, the bias-corrected statistics, and the minimum and maximum of the estimated coefficient values from the bootstrap samples. Notice that the bootstrap estimates of the coefficients and the standard errors are very similar to the OLS estimates.

There is a remarkable similarity between the bootstrap estimates of the coefficients and the standard errors obtained from the residual method and the OLS estimates. This is not surprising since under the homoscedastic assumption, it can be shown that as the number of bootstrapped samples increases, the estimated values of the standard errors converge to the diagonal elements of $\hat{\sigma}(\mathbf{X}^T\mathbf{X})^{-1}$, where $\hat{\sigma}^2$ is the estimate that is not corrected for bias.

Appendix E

COMPLETE PROGRAMS AND PROC IML ROUTINES

E.1 PROGRAM 1

This program was used in Chapter 2. It is used to analyze Table F3.1 of Greene (2003). In the following data step, we read in the raw data, create a trend variable, *T*, divide GNP and Invest by CPI, and then scale the transformed GNP and Invest time series so that they are measured in trillions of dollars.

```
proc import out=invst_equation
    datafile="C:\Temp\Invest_Data"
    dbms=Excel Replace;
    getnames=yes;
run;
data invst_equation;
    set invst_equation;
    T=_n_;
    Real_GNP=GNP/(CPI*10);
    Real_Invest=Invest/(CPI*10);
run;
/* The start of Proc IML routines.
*/proc iml;
/* Invoke Proc IML and create the X and Y matrices using the variables T, Real_GNP, and
Real_Invest from the SAS data set invst_equation. */
    use invst_equation;
    read all var {'T' 'Real_GNP'} into X;
    read all var {'Real_Invest'} into Y;
/* Define the number of observations and the number of independent variables. */
    n=nrow(X);
    k=ncol(X);
/* Create a column of ones to the X matrix to account for the intercept term. */
    X=J(n,1,1)||X;
```

Applied Econometrics Using the SAS® System, by Vivek B. Ajmani
Copyright © 2009 John Wiley & Sons, Inc.

```
/* Calculate the inverse of X'X and use this to compute B_Hat */
    C=inv(X`*X);
    B_Hat=C*X`*Y;
/* Compute SSE, the residual sum of squares, and MSE, the residual mean square. */
    SSE=y`*y-B_Hat`*X`*Y;
    DFE=n-k-1;
    MSE=sse/DFE;
/* Compute SSR, the sums of squares due to the model; MSR, the sums of squares due to random
error; and the F ratio. */
    Mean_Y=Sum(Y)/n;
    SSR=B_Hat`*X`*Y-n*Mean_Y**2;
    MSR=SSR/k;
    F=MSR/MSE;
/* Compute R-Square and Adj-RSquare. */
    SST=SSR+SSE;
    R_Square=SSR/SST;
    Adj_R_Square=1-(n-1)/(n-k) * (1-R_Square);
/* Compute the standard error of the parameter estimates, their T statistic and P-values. */
    SE=SQRT(vecdiag(C)#MSE);
    T=B_Hat/SE;
    PROBT=2*(1-CDF('T', ABS(T), DFE));
/* Concatenate the results into one matrix to facilitate printing. */
    Source=(k||SSR||MSR||F)//(DFE||SSE||MSE||{.});
    STATS=B_Hat||SE||T||PROBT;
    Print 'Regression Results for the Investment Equation';
    Print Source (|Colname={DF SS MS F} rowname={Model Error} format=8.4|);
    Print 'Parameter Estimates';
    Print STATS (|Colname={BHAT SE T PROBT} rowname={INT T G _R_ P}format=8.4|);
    Print '';
    Print 'The value of R-Square is ' R_Square;
    Print 'The value of Adj R-Square is ' Adj_R_Square;
run;
```

E.2 PROGRAM 2

This program was used in Chapter 3. It analyzes the quarterly data on investment as found in Table 5.1 of Greene (2003). This program is used to conduct the general linear hypothesis—the global F test. We have omitted the data step statements with the hope that users will be able to recreate it with ease. The Proc IML code follows.

```
proc iml;
/* Invoke Proc IML and create the X and Y matrices using the variables Invest, Interest,
delta_p, output, and T from the SAS data set real_invst_eq (Notice we have omitted the data
step) */
    use real_invst_eq;
    read all var {'interest' 'delta_p' 'output' 'T'} into X;
    read all var {'Invest'} into Y;
/* Define the number of observations and the number of independent variables. */
    n=nrow(X);
    k=ncol(X);
/* Create a column of ones to the X matrix to account for the intercept term. */
    X=J(n,1,1)||X;
/* Calculate the inverse of X'X and use this to compute B_Hat. */
    C=inv(X`*X);
```

```
      B_Hat=C*X`*Y;
/* Construct the R Matrix and q matrix. */
      R={0 1 1 0 0,0 0 0 1 0,0 0 0 0 1};
      q={0,1,0};
      j=nrow(R);
/* Compute SSE, the residual sum of squares, and MSE, the residual mean square. */
      SSE=y`*y-B_Hat`*X`*Y;
      DFE=n-k-1;
      MSE=sse/DFE;
/* Calculate the F Statistic. */
      DisVec=R*B_Hat-q;
      F=DisVec`*inv(R*MSE*C*R`)*DisVec/j;
      P=1-ProbF(F,J,n-k);
      Print 'The value of the F Statistic is ' F;
      Print 'The P-Value associated with this is ' P;
run
```

E.3   PROGRAM 3

This program was used in Chapter 3. It was used to analyze the investment equation data given in Table F5.1 of Greene (2003). This program calculates the restricted least squares estimator and the standard errors of the estimator. We have omitted the data step again.

```
proc iml;
/* Invoke Proc IML and create the X and Y matrices using the variables Invest, Interest,
delta_p, output, and T from the SAS data set real_invst_eq. */
      use real_invst_eq;
      read all var {'interest' 'delta_p' 'output' 'T'} into X;
      read all var {'Invest'} into Y;
/* Define the number of observations and the number of independent variables. */
      n=nrow(X);
      k=ncol(X);
/* Create a column of ones to the X matrix to account for the intercept term. */
      X=J(n,1,1)||X;
/* Calculate the inverse of X'X and use this to compute B_Hat. */
      C=inv(X`*X);
      B_Hat=C*X`*Y;
/* Construct the R matrix and q matrix. */
      R={0 1 1 0 0};
      q={0};
/* Calculate the Restricted Least Squares Estimator. */
      M=R'*inv(R*C*R')*(R*B_Hat-q);
      B_Star=B_Hat - C*M;
      print B_Star;
/* Compute SSE, and MSE. */
      SSE=y`*y-B_Hat`*X`*Y;
      DFE=n-k-1;
      MSE=sse/DFE;
/* Compute SSR, MSR, and the F statistic. */
      Mean_Y=Sum(Y)/n;
      SSR=B_Hat`*X`*Y-n*Mean_Y**2;
      MSR=SSR/k;
```

```
    F=MSR/MSE;
/* Compute R-Square and Adj-RSquare. */
    SST=SSR+SSE;
    R_Square=SSR/SST;
    Adj_R_Square=1-(n-1)/(n-k) * (1-R_Square);
/* Compute the standard error of the parameter estimates, their T statistic and P-values. */
    SE=SQRT(vecdiag(C)#MSE);
    T=B_Hat/SE;
    PROBT=2*(1-CDF('T', ABS(T), DFE));
/* Concatenate the results into one matrix. */
Source=(k||SSR||MSR||F)//(DFE||SSE||MSE||{.});
    STATS=B_Hat||SE||T||PROBT;
    Print 'Regression Results for the Restricted Investment Equation';
    Print Source (|Colname={DF SS MS F} rowname={Model Error} format=8.4|);
    Print 'Parameter Estimates';
    Print STATS (|Colname={BHAT SE T PROBT} rowname={INT Interest
    Delta_P Output T} format=8.4|);
    Print '';
    Print 'The value of R-Square is ' R_Square;
    Print 'The value of Adj R-Square is ' Adj_R_Square;
run;
```

E.4 PROGRAM 4

This program was used in Chapter 3 to conduct general linear hypothesis for the investment equation data given in Table F5.1 of Greene (2003). Note that Program 2 simply conducts the global F test, whereas this program does the individual t tests for each of the linear restrictions.

```
proc iml;
/* Invoke Proc IML and create the X and Y matrices using the variables Invest, Interest,
delta_p, output, and T from the SAS data set real_invst_eq. */
    use real_invst_eq;
    read all var {'interest' 'delta_p' 'output' 'T'} into X;
    read all var {'Invest'} into Y;
/* Define the number of observations and the number of independent variables. */
    n=nrow(X);
    k=ncol(X);
/* Create a column of ones to the X matrix to account for the intercept term. */
    X=J(n,1,1)||X;
/* Calculate the inverse of X'X and use this to compute B_Hat. */
    C=inv(X`*X);
    B_Hat=C*X`*Y;
/* Construct the R Matrix and q matrix. */
    R={0 1 1 0 0,0 0 0 1 0,0 0 0 0 1};
    q={0,1,0};
    j=nrow(R);
    R1=R[1,];q1=q[1,];
    R2=R[2,];q2=q[2,];
    R3=R[3,];q3=q[3,];
/* Compute SSE, and MSE. */
    SSE=y`*y-B_Hat`*X`*Y;
```

```
    DFE=n-k-1;
    MSE=sse/DFE;
/* Calculate the t Statistic. */
    T_NUM1=R1*B_Hat-q1;
    se1=sqrt(R1*MSE*C*R1');
    T1=T_NUM1/se1;
    p1=1-ProbT(T1,n-k);
    Print 'The value of the T Statistic for the first restriction is ' t1;
    Print 'The P-Value associated with this is ' P1;
    T_NUM2=R2*B_Hat-q2;
    se2=sqrt(R2*MSE*C*R2');
    T2=T_NUM2/se2;
    P2=1-ProbT(T2,n-k);
    Print 'The value of the T Statistic for the second restriction is ' t2;
    Print 'The P-Value associated with this is ' P2;
    T_NUM3=R3*B_Hat-q3;
    se3=sqrt(R3*MSE*C*R3`);
    T3=T_NUM3/se3;
    P3=1-ProbT(T3,n-k);
    Print 'The value of the T Statistic for the third restriction is ' t3;
    Print 'The P-Value associated with this is ' P3;
run;
```

E.5   PROGRAM 5

This program was used in Chapter 4 to conduct Hausman's specification test on the consumption data that can be found in Table 5.1 of Greene (2003). We have chosen to omit the data step statements.

```
proc iml;
/* Read the data into appropriate matrices. */
    use hausman;
    read all var {'yt' 'it' 'ct1'} into X;
    read all var {'ct'} into Y;
/* Create the instruments matrix Z and some constants. */
    read all var {'it' 'ct1' 'yt1'} into Z;
    n=nrow(X);
    k=ncol(X);
    X=J(n,1,1)||X;
    Z=J(n,1,1)||Z;
/* Calculate the OLS and IV estimators. */
    CX=inv(X`*X);
    CZ=inv(Z`*Z);
    OLS_b=CX*X`*y;
    Xhat=Z*CZ*Z'*X;
    b_IV=inv(Xhat`*X)*Xhat`*y;
/* Calculate the difference between the OLS and IV estimators. Also, calculate MSE */
    d=b_IV-OLS_b;
    SSE=y`*y-OLS_b`*X`*Y;
    DFE=n-k;
    MSE=sse/DFE;
/* Calculate the GINVERSE of the difference inv(X`*X) - inv(Xhat`*Xhat). */
```

```
/* Calculate the Hausman's test statistic. */
    diff=ginv(inv(Xhat`*Xhat)-CX);
    H=d`*diff*d/mse;
    J=round(trace(ginv(diff)*diff));
    Table=OLS_b||b_IV;
    Print Table (|Colname={OLS IV} rowname={Intercept yt it ct1} format=8.4|);
    Print 'The Hausman Test Statistic Value is ' H;
run;
```

E.6 PROGRAM 6

This program was used in Chapter 5 to calculate the estimates of the robust variance–covariance matrices under heteroscedasticity. The analysis is done on the credit card data found in Table 9.1 of Greene (2003). The code calculates White's estimator and the two alternatives proposed by David and MacKinnon.

```
proc iml;
/* Read the data into matrices and create constants. */
    use Expense;
    read all var {'age' 'ownrent' 'income' 'incomesq'} into X;
    read all var {'avgexp'} into Y;
    n=nrow(X);
    X=J(n,1,1)||X;
    k=ncol(X);
/* Calculate the inverse of X'X.*/
    C=inv(X`*X);
/* Calculate the least squares estimator, beta_hat. */
    beta_hat=C*X`*y;
/* Calculate the residuals and MSE. */
    resid=y-X*beta_hat;
    SSE=y`*y-beta_hat`*X`*Y;
    MSE=sse/(n-k);
/* Calculate the S0 term of White's Estimator. */
/* First, initialize a n by n matrix with zero's. */
    S0=J(k,k,0);
    do i=1 to n;
        S0=S0 + resid[i,]*resid[i,]*X[i,]`*X[i,];
    end;
    S0=S0/n;
/* Now, calculate White's Estimator. */
    White=n*C*S0*C;
/* Now, calculate the first recommendation of David & MacKinnon for White's estimator. */
    DM1=n/(n-k) * White;
/* Now, calculate the second recommendation of David & MacKinnon for White's estimator. */
    S0=J(k,k,0);
    do i=1 to n;
        m_ii=1-X[i,]*C*X[i,]`;
        Temp_Ratio=resid[i,]*resid[i,]/m_ii;
        S0=S0+Temp_Ratio*X[i,]`*X[i,];
    end;
    S0=S0/n;
/* Now, calculate the modified White's Estimator. */
    DM2=n*C*S0*C;
```

```
/* Get the standard errors which are nothing but the square root of the diagonal matrix. */
    SE=SQRT(vecdiag(C)#MSE);
    SE_White=SQRT(vecdiag(White));
    SE_DM1=SQRT(vecdiag(DM1));
    SE_DM2=SQRT(vecdiag(DM2));
/* Calculate the t Ratio based on Homoscedastic assumptions. */
    T=Beta_Hat/SE;
/* Print the results. */
    STATS=beta_hat||SE||T||SE_White||SE_DM1||SE_DM2;
    STATS=STATS`;
    print 'Least Squares Regression Results';
    print STATS (|Colname={Constant Age OwnRent Income IncomeSq}
    rowname={Coefficient SE t_ratio White_Est DM1 DM2} format=8.3|);
run;
```

E.7 PROGRAM 7

This program was used in Chapter 5 to conduct White's test to detect heteroscedasticity in the credit card data, which can be found in Table 9.1 of Greene (2003). The data step statements read the data and create the various cross-product terms that are used in the analysis.

```
proc import out=Expense
    datafile="C:\Temp\TableF91"
    dbms=Excel Replace;
    getnames=yes;
run;
data expense;
    set expense;
    age_sq=age*age;
    incomesq=income*income;
    incomefth=incomesq*incomesq;
    age_or=age*ownrent;
    age_inc=age*income;
    age_incsq=age*incomesq;
    or_income=ownrent*income;
    or_incomesq=ownrent*incomesq;
    incomecube=income*incomesq;
    If AvgExp>0;
run;
proc iml;
/* Read the data into matrices and create constants. */
    use expense;
    read all var {'age' 'ownrent' 'income' 'incomesq'} into X;
    read all var {'age' 'ownrent' 'income' 'incomesq' 'age_sq' 'incomefth'
    'age_or' 'age_inc' 'age_incsq' 'or_income' 'or_incomesq' 'incomecube'} into XP;
    read all var {'avgexp'} into Y;
    n=nrow(X);
    np=nrow(XP);
    X=J(n,1,1)||X;
    XP=J(np,1,1)||XP;
    k=ncol(X);
    kp=ncol(XP);
```

```
/* First get the residuals from OLS. */
    C=inv(X`*X);
    beta_hat=C*X`*y;
    resid=y-X*beta_hat;
/* Square the residuals for a regression with cross product terms in White's test. */
    resid_sq=resid#resid;
/* Regress the square of the residuals versus the 13 variables in X. */
    C_E=inv(XP`*XP);
    b_hat_e=C_E*XP`*resid_sq;
/* Calculate R-Square from this regression. */
    Mean_Y=Sum(resid_sq)/np;
    SSR=b_hat_e`*XP`*resid_sq-np*Mean_Y**2;
    SSE=resid_sq`*resid_sq-b_hat_e`*XP`*resid_sq;
    SST=SSR+SSE;
    R_Square=SSR/SST;
    print R_Square;
/* Calculate and print the test statistic value and corresponding p-value. */
    White=np*R_Square;
    pvalue= 1 - probchi(White, kp);
    print 'The test statistic value for Whites Test is 'White;
    print 'The p-value associated with this test is 'pvalue;
run;
```

E.8   PROGRAM 8

This program was used in Chapter 5 to conduct the Breusch-Pagan Lagrange Multiplier test on the credit card data, which can be found in Table 9.1 of Greene (2003). Note that we have omitted the data step statements.

```
proc iml;
/* Read the data into matrices and prep matrices for analysis. */
    use expense;
    read all var {'age', 'ownrent','income','incomesq' } into X;
    read all var {'income', 'incomesq' } into Z;
    read all var {'avgexp' } into y;
/* Create a few constants. */
    n=nrow(X);
    X=J(n,1,1)||X;
    Z=J(n,1,1)||Z;
/* Calculate the residuals from OLS. */
    bhat_OLS=inv(X'*X)*X`*y;
    SSE=(y-X*bhat_OLS)'*(y-X*bhat_OLS);
    resid=y-X*bhat_OLS;
/* Calculate the LM statistic and associated p value. */
    g=J(n,1,0);
    fudge=SSE/n;
    do index=1 to n;
            temp1=resid[index,1]*resid[index,1];
            g[index,1]=temp1/fudge - 1;
    end;
    LM=0.5*g`*Z*inv(Z`*Z)*Z`*g;
/* Calculate the degrees of freedom and print the results. */
    kz=ncol(Z);
```

```
kz=kz-1;
pval=1-probchi(LM,kz);
if (pval<0.05) then
        do;
                print 'The Breusch Pagan Test Statistic Value is 'LM;
                print 'The p value associated with this is 'pval;
                print 'The null hypothesis of homoscedasticity is rejected';
        end;
else
        do;
                print 'The Breusch Pagan Test Statistic Value is 'LM;
                print 'The p value associated with this is 'pval;
                print 'The null hypothesis of homoscedasticity is not
                rejected';
        end;
run;
```

E.9   PROGRAM 9

This program was used in Chapter 5 to calculate the iterative FGLS estimators for the credit card expenditure data found in Table F9.1 of Greene (2003).

```
proc iml;
/* Read the data into matrices and calculate some constants. */
    Use CCExp;
    read all var{'Age' 'OwnRent' 'Income' 'Income_Sq'} into X;
    read all var{'AvgExp'} into y;
    n=nrow(X);
    k=ncol(X);
    X=J(n,1,1)||X;
/* Calculate the OLS estimates, the residuals and the square of the residuals. */
    bhat_OLS=inv(X`*X)*X`*y;
    e=y-X*bhat_OLS;
    r_e=log(e#e);
/* As we have done with this data, we assume that the issue lies with Income_Sq. */
    zi=log(X[,4]);
    Z=J(n,1,1)||zi;
/* Regression of Z (defined above) with the square of the residuals. */
    alpha_m=inv(Z`*Z)*Z`*r_e;
    alpha_s=alpha_m[2,];
    /* Now initialize the weight matrix Omega. */
    omega=J(n,n,0);
    do i=1 to n;
            do j=1 to n;
                    if i=j then omega[i,j]=X[i,4]**alpha_s;
            end;
    end;
/* Calculate the first pass estimates of the parameter vector. */
    bhat_2S=inv(X`*inv(omega)*X)*X`*inv(omega)*y;
/* Start the iterative process (re-do the steps from above). */
    do iter=1 to 100;
```

```
              s1=bhat_2S[1,1]; s2=bhat_2S[2,1]; s3=bhat_2S[3,1];
              s4=bhat_2S[4,1]; s5=bhat_2S[5,1];
              e=y-X*bhat_2S;
              r_e=log(e#e);
              alpha_m=inv(Z`*Z)*Z`*r_e;
              alpha_s=alpha_m[2,];
              omega=J(n,n,0);
              do i=1 to n;
                    do j=1 to n;
                          if i=j then omega[i,j]=X[i,4]**alpha_s;
                    end;
              end;
/* Calculate the parameter estimates for each iteration. */
/* Calculate the difference between subsequent values of these estimates. */
       bhat_2S=inv(X`*inv(omega)*X)*X`*inv(omega)*y;
       n1=bhat_2S[1,1]; n2=bhat_2S[2,1]; n3=bhat_2S[3,1];
       n4=bhat_2S[4,1]; n5=bhat_2S[5,1];
       diff=abs(n1-s1)+abs(n2-s2)+abs(n3-s3)+abs(n4-s4)+abs(n5-s5);
       if diff<0.00001 then
/* Exit strategy! */
              do;
                          print "The value of alpha is " alpha_s;
                          print "Convergence was obtained in " iter "iterations.";
                          stop;
              end;
       end;
       final_MSE=(e`*inv(omega)*e)/(n-k);
       final_cov=final_mse*inv(X`*inv(omega)*X);
       SE=sqrt(vecdiag(final_conv));
       STAT_Table=bhat_2s||SE;
       Print "The estimates of the coefficients are";
       Print STAT_Table (|Colname={BHAT SE} rowname={INT Age OwnRent
       Income Income2} format=8.4|);
run;
```

E.10 PROGRAM 10

This program was used in Chapter 5 to plot ALPHA versus the Likelihood Value for the credit card data, which is found in Table 9.1 of Greene (2003).

```
proc import out=CCExp
       datafile="C:\Temp\TableF91"
       dbms=Excel Replace;
       getnames=yes;
run;
/* Create temp SAS dataset and transform variables. */
data CCExp;
     set CCExp;
     IncomeSq=Income*Income;
     if AvgExp>0;
run;
/* Invoke Proc IML */
proc iml;
```

```
/* Bring in the SAS data set and create matrices. */
    use ccexp;
    read all var{'age','ownrent','income','incomesq'} into x;
    read all var{'income'} into z;
    read all var{'avgexp'} into y;
/* Prep matrices for analysis. */
    n=nrow(x);
    x=j(n,1,1)||x;
    Storage=J(5000,2,0);
/* Generate a range of alpha values. */
    do alpha_ind=1 to 5000;
            alpha=alpha_ind/1000;
/* Compute the GLS estimator of beta for each alpha. */
            omega=J(n,n,0);
            do i=1 to n;
              do j=1 to n;
                if (i=j) then omega[i,j]=z[i,]**alpha;
              end;
            end;
    beta_GLS=inv(x`*inv(omega)*x)*x`*inv(omega)*y;
/* For these alpha and beta values, calculate the generalized sums of squares. */
    GSQ=0.0;
    do i=1 to n;
            temp1=(y[i,1]-x[i,]*beta_GLS);
            temp2=z[i,1]**alpha;
            temp3=(temp1**2)/temp2;
            GSQ=GSQ+temp3;
    end;
    MSE=1/n * GSQ;
/* Calculate the Log Likelihood Stat. */
    Fudge1=-n/2 * (log(2*constant('pi'))+log(MSE));
    temp_sum=0.0;
    do i=1 to n;
            temp1=log(z[i,1]**alpha);
            temp2=1/mse * z[i,1]**(-alpha);
            temp3=(y[i,1]-x[i,]*beta_GLS)**2;
            temp4=temp1+temp2*temp3;
            temp_sum=temp_sum+temp4;
    end;
            temp_sum=-0.5*temp_sum;
            Ln_L=Fudge1+temp_sum;
            storage[alpha_ind,1]=alpha;
            storage[alpha_ind,2]=Ln_L;
    end;
/* Store the plot data. */
    create plot_data from storage;
    append from storage;
run;
/* Invoke the plotting code and plot the data. */
data plot_data;
        set plot_data;
        rename col1=alpha;
        rename col2=Ln_L;
```

```
run;
proc sort data=plot_data;
        by descending Ln_L;
run;
goptions reset=global gunit=pct border cback=white
colors=(black blue green red)
ftitle=swissb ftext=swiss htitle=3 htext=2;

symbol1 value=dot
height=0.5
cv=red
ci=blue
co=green
width=0.5;
proc gplot data=plot_data;
     plot Ln_L*alpha/haxis=axis1
     vaxis=axis2;
     axis1 label=('Alpha');
     axis2 label=(angle=90 'Ln_L');
run;
```

E.11   PROGRAM 11

This program is used in Chapter 5 to calculate the correct standard errors of the GLS estimator for the credit card data, which is found in Table 9.1 of Greene (2003). The optimal value of alpha was found to be 3.651.

```
proc import out=CCExp
     datafile="C:\Temp\TableF91"
     dbms=Excel Replace;
     getnames=yes;
run;
/* Create temp SAS dataset and transform variables. */
data CCExp;
     set CCExp;
     IncomeSq=Income*Income;
     if AvgExp>0;
run;
/* Invoke Proc IML. */
proc iml;
/* Bring in the SAS data set and create matrices. */
     use ccexp;
     read all var{'age','ownrent','income','incomesq'} into x;
     read all var{'income'} into z;
     read all var{'avgexp'} into y;
/* Prep matrices for analysis. */
     n=nrow(x);
     x=j(n,1,1)||x;
/* Generate a range of alpha values. */
     alpha=3.651;
/* Compute the GLS estimator of beta alpha. */
     omega=J(n,n,0);
     do i=1 to n;
```

```
            do j=1 to n;
                    if (i=j) then omega[i,j]=z[i,]**alpha;
            end;
    end;
    beta_GLS=inv(x`*inv(omega)*x)*x`*inv(omega)*y;
/* For this alpha and beta values, calculate the generalized sums of squares. */
    GSQ=0.0;
    do i=1 to n;
            temp1=(y[i,1]-x[i,]*beta_GLS);
            temp2=z[i,1]**alpha;
            temp3=(temp1**2)/temp2;
            GSQ=GSQ+temp3;
    end;
    MSE=1/n * GSQ;
/* Calculate the covariance matrix now. */
    COV=MSE*inv(X`*inv(Omega)*X);
    print COV;
run;
```

E.12   PROGRAM 12

This program uses the credit card data from Table 9.1 of Greene (2003) to get MLEs of the parameters when ALPHA is multivariate.

```
proc import out=CCExp
    datafile="C:\Temp\TableF91"
    dbms=Excel Replace;
    getnames=yes;
run;
/* Create temp SAS dataset and transform variables. */
data CCExp;
    set CCExp;
    IncomeSq=Income*Income;
    if AvgExp>0;
run;
/* Invoke Proc IML.; */
proc iml;
/* Bring in the SAS data set and create matrices.; */
    use ccexp;
    read all var{'age','ownrent','income','incomesq'} into x;
    read all var{'income','incomesq'} into z;
    read all var{'avgexp'} into y;
    s_alpha=J(3,1,0);
    s_beta=J(5,1,0);
/* Prep matrices for analysis.; */
    n=nrow(x);
    x=j(n,1,1) || x;
    z=j(n,1,1) || z;
    CZ=inv(Z`*Z);
/* Compute OLS estimates of beta and mse.; */
    bhat=inv(X`*X)*X`*y;
    e=y-X*bhat;
```

```
/* Log residual square to be used to get alpha values.;
/* Compute alpha values-First iteration.;
    r_e=log(e#e);
    alpha=inv(z`*z)*z`*r_e;
/* Compute GLS beta values-First iteration.;/
    omega=J(n,n,0);
    do i=1 to n;
        do j=1 to n;
            if (i=j) then omega[i,j]=exp(z[i,]*alpha);
        end;
    end;
    beta_GLS=inv(x`*inv(omega)*x)*x`*inv(omega)*y;
/* Update alpha and beta.;
    do i=1 to 100;
        s_alpha[1,1]=alpha[1,1]; s_alpha[2,1]=alpha[2,1]; s_alpha[3,1]=alpha[3,1];
        s_beta[1,1]=beta_GLS[1,1]; s_beta[2,1]=beta_GLS[2,1]; s_beta[3,1]=beta
        GLS[3,1];
        s_beta[4,1]=beta_GLS[4,1]; s_beta[5,1]=beta_GLS[5,1];
        resp=J(n,1,0);
        e=y-x*beta_gls;
        do j=1 to n;
            resp[j,1]=e[j,1]*e[j,1]/exp(z[j,]*alpha) - 1;
        end;
        alpha=alpha+inv(z`*z)*z`*resp;
/* Get a new value of Beta.;
        omega=J(n,n,0);
        do i1=1 to n;
            do i2=1 to n;
                if (i1=i2) then omega[i1,i2]=exp(z[i1,]*alpha);
            end;
        end;
        beta_GLS=inv(x`*inv(omega)*x)*x`*inv(omega)*y;
/* Compute differences.;
        diff_beta=sum(abs(s_beta-beta_gls));
        diff_alpha=sum(abs(s_alpha-alpha));
        diff=diff_beta+diff_alpha;
/* Exit strategy.;
        if diff<0.00001 then
            do;
            print "The estimates of the coefficients are.";
            print beta_gls;
            print "The value of alpha is " alpha;
            print "Convergence was obtained in " i "iterations.";
            stop;
        end;
    end;
var=exp(alpha[1,1]);
var_cov=var*inv(X`*inv(omega)*X);
se=sqrt(vecdiag(var_conv));
STAT_Table=beta_gls||SE;
Print "The estimates of the coefficients are";
Print STAT_Table (|Colname={BHAT SE} rowname={INT Age OwnRent Income
Income2} format=8.4|);
run;
```

E.13 PROGRAM 13

This program was used in Chapter 5 to analyze the airlines data found in Table F7.1 of Greene (2003). The code computes the parameter estimates assuming groupwise heterogeneity.

```
proc iml;
/* Bring in the SAS data set and create matrices.;
    use airline;
    read all var{'LnQ','LF','LnPf','D2','D3','D4','D5','D6'} into x;
    read all var{'D2','D3','D4','D5','D6'} into z;
    read all var{'LnC'} into y;
    s_alpha=J(6,1,0);
    s_beta=J(9,1,0);
/* Prep matrices for analysis.;
    n=nrow(x);
    x=j(n,1,1)||x;
    z=j(n,1,1)||z;
    CZ=inv(Z`*Z);
/* Compute OLS estimates of beta and mse;
    bhat=inv(X`*X)*X`*y;
    e=y-X*bhat;
/* Log residual square to be used to get alpha values.;
/* Compute alpha values-First iteration.;
    r_e=log(e#e);
    alpha=inv(z`*z)*z`*r_e;
/* Compute GLS beta values-First iteration.;
    omega=J(n,n,0);
    do i=1 to n;
            do j=1 to n;
                    if (i=j) then omega[i,j]=exp(z[i,]*alpha);
            end;
    end;
    beta_GLS=inv(x`*inv(omega)*x)*x`*inv(omega)*y;
/* Update alpha and beta.;
    do i=1 to 100;
            s_alpha=alpha;
            s_beta=beta_GLS;
            resp=J(n,1,0);
            e=y-x*beta_gls;
            sum=J(6,1,0.0);
            do j=1 to n;
                    tem=z[j,]`;
                    resp[j,1]=e[j,1]*e[j,1]/exp(z[j,]*alpha) - 1;
                    sum=sum+tem*resp[j,1];
             end;
            alpha=alpha+inv(z`*z)*sum;
/* Get a new value of Beta.;
            omega=J(n,n,0);
            do i1=1 to n;
                    do i2=1 to n;
                            if (i1=i2) then omega[i1,i2]=exp(z[i1,]*alpha);
            end;
            end;
            beta_GLS=inv(x`*inv(omega)*x)*x`*inv(omega)*y;
```

```
/* Compute differences.;
            diff_beta=sum(abs(s_beta-beta_gls));
            diff_alpha=sum(abs(s_alpha-alpha));
            diff=diff_beta+diff_alpha;
/* Exit strategy.;
            if diff<0.00001 then
                do;
                  print "The estimates of the coefficients are.";
                  print beta_gls;
                  print "The value of alpha is " alpha;
                  print "Convergence was obtained in " i "iterations.";
                  stop;
                end;
          end;
/* Calculate the covariance matrix at the optimal values.;
      omega=J(n,n,0);
      do i=1 to n;
            do j=1 to n;
                  if (i=j) then omega[i,j]=exp(z[i,]*alpha);
            end;
      end;
      var_cov=inv(X`*inv(omega)*x);
      var_cov_alpha=2*CZ;
      se=J(9,1,0);
      se_a=J(6,1,0);
      do index=1 to 9;
            se[index,1]=sqrt(var_cov[index,index]);
      end;
      do index=1 to 6;
            se_a[index,1]=sqrt(var_cov_alpha[index,index]);
      end;
      print se;
      print se_a;
run;
```

E.14 PROGRAM 14

This program was used in Chapter 7 to estimate the parameters of a dynamic panel data for the cigarettes data set with no explanatory variables.

```
* Read the data into SAS;
proc import out=cigar
    datafile="C:\Temp\cigar"
    dbms=Excel Replace;
    getnames=yes;
run;
* Take the Log Transformation;
data cigar;
    set cigar;
    if state=. or year=. then delete;
    Log_C=Log(C);
    keep state year Log_C;
```

```
run;
proc iml;
* Define constants;
N=46;T=30;P=406;
* Read the variable into a matrix;
use cigar;
     read all var{'Log_C'} into Y;
* H is fixed. So, define H here;
     H=shape(0,T-2,T-2);
     do j=1 to T-2;
          do l=1 to T-2;
                if j=l then H[l,j]=2;
                if (j=l+1) then H[l,j]=-1;if (j=l-1) then H[l,j]=-1;
          end;
     end;
* Initialize four sum matrices and the counter;
     ZHZ=shape(0,P,P);YMZ=shape(0,1,P);YZ=shape(0,1,P);ZPZ=shape(0,P,P);
     compt=1;
     do i=1 to N;
* Calculate the diff matrix;
     Y_DIFF=shape(0,T-2,1);
     Y_DIFFS=shape(0,T-2,1);
     do Index=1 to T-2;
          Y_DIFF[Index]=Y[Index+compt+1,1]-Y[Index+Compt,1];
          Y_DIFFS[Index]=Y[Index+compt,1]-Y[Index+compt-1,1];
     end;
* Calculate the BZI matrix;
     j=1;cpt=j;cpt2=compt;
     BZI=shape(0,1,cpt);BZI[1,1]=Y[cpt2,1];
     do j=2 to T-2;
          cpt=j;
          cpt2=compt;
          C=shape(0,1,cpt);
          do k=1 to j;
                C[1,k]=Y[cpt2+k-1,1];
          end;
     BZI=block(BZI,C);
     end;
* Calculate the matrix sums;
     ZHZ=ZHZ+BZI`*H*BZI;
     YMZ=YMZ+Y_DIFF`*BZI;
     YZ=YZ+Y_DIFFS`*BZI;
     compt=compt+T;
     end;
* Calculate the first step coefficient estimate;
     Delta_Est1=inv(YZ*inv(ZHZ)*YZ`)*YZ*inv(ZHZ)*YMZ`;
     print Delta_Est1;
* Calculate the Residual Vector;
     compt=1;
     do i=1 to N;
* Calculate the diff matrix;
     Y_DIFF=shape(0,T-2,1);
     Y_DIFFS=shape(0,T-2,1);
     E=shape(0,T-2,1);
```

```
    do Index=1 to T-2;
            Y_DIFF[Index,1]=Y[Index+compt+1,1]-Y[Index+Compt,1];
            Y_DIFFS[Index,1]=Y[Index+compt,1]-Y[Index+compt-1,1];
    end;
    E=Y_DIFF-Delta_Est1*Y_DIFFS;
* Calculate the BZI matrix;
    j=1;cpt=j;cpt2=compt;
    BZI=shape(0,1,cpt);BZI[1,1]=Y[cpt2,1];
    do j=2 to T-2;
            cpt=j;cpt2=compt;
            C=shape(0,1,cpt);
            do k=1 to j;
    C[1,k]=Y[cpt2+k-1,1];
    end;
    BZI=block(BZI,C);
    end;
* Calculate the weight matrix for the second step;
    ZPZ=ZPZ+BZI`*E*E`*BZI;
    compt=compt+T;
    end;
* Calculate the second step Arellano and Bond Estimator;
    Delta_Est2=inv(YZ*ginv(ZPZ)*YZ`)*YZ*ginv(ZPZ)*YMZ';
    print Delta_Est2;
run;
```

E.15   PROGRAM 15

This program was used in Chapter 7 to estimate the cigarettes data set dynamic panel model with explanatory variables. This code calculates the Anderson–Hso estimator.

```
* Read the data from Excel;
proc import out=cigar
    datafile="C:\Temp\cigar"
    dbms=Excel Replace;
    getnames=yes;
run;
* Create the log transformations;
data cigar;
    set cigar;
    if state=. or year=. then delete;
    Log_C=Log(C);
    Log_MIN=Log(MIN);
    Log_NDI=Log(NDI);
    Log_Price=Log(Price);
RUN;
proc iml;
* This program will calculate the Anderson-Hso Estimator for the Cigar.TXT dataset;
* Define constants;
N=46;T=30;P=4;
* Read the variables into a matrix;
use cigar;
    read all var {'Log_C','Log_Min','Log_NDI','Log_Price'} into Y;
```

```
* Initialize three sum matrices;
    ZHZ=shape(0,P,P);YMZ=shape(0,1,P);YZ=shape(0,4,P);
* Initialize the counter;
    compt=1;
* Begin the loop for calculating the first step Arellano-Bond Estimator;
    do i=1 to N;
* Calculate the diff matrix relating to the Y's;
    Y_DIFF=shape(0,T-2,1);
    Y_DIFFS=shape(0,T-2,1);
    do Index=1 to T-2;
            Y_DIFF[Index,1]=Y[Index+compt+1,1]-Y[Index+Compt,1];
            Y_DIFFS[Index,1]=Y[Index+compt,1]-Y[Index+compt-1,1];
    end;
* Calculate the diff matrix relating to the X's;
    X_DIFF1=shape(0,T-2,1);
    X_DIFF2=shape(0,T-2,1);
    X_DIFF3=shape(0,T-2,1);
    do index=1 to T-2;
            X_DIFF1[Index,1]=Y[Index+compt+1,2]-Y[Index+Compt,2];
            X_DIFF2[Index,1]=Y[Index+compt+1,3]-Y[Index+Compt,3];
            X_DIFF3[Index,1]=Y[Index+compt+1,4]-Y[Index+Compt,4];
    end;
* Create the XI matrix;
    XI=shape(0,T-2,4);
    XI=Y_DIFFS||X_DIFF1||X_DIFF2||X_DIFF3;
* Calculate the BZI matrix;
    BZI=shape(0,t-2,p);
    do index=1 to t-2;
            BZI[Index,1]=Y[Index+Compt-1,1];
            BZI[Index,2]=Y[Index+compt+1,2]-Y[Index+Compt,2];
            BZI[Index,3]=Y[Index+compt+1,3]-Y[Index+Compt,3];
            BZI[Index,4]=Y[Index+compt+1,4]-Y[Index+Compt,4];
    end;
    ZHZ=ZHZ+BZI`*BZI;
    YMZ=YMZ+Y_DIFF`*BZI;
    YZ=YZ+XI`*BZI;
    compt=compt+T;
    end;
* Calculate the coefficient estimate;
    Delta_Est1=inv(YZ*inv(ZHZ)*YZ)*YZ`*inv(ZHZ)*YMZ`;
    print Delta_Est1;
run;
```

E.16   PROGRAM 16

This program was used in Chapter 7 to estimate the cigarettes data set dynamic panel model with explanatory variables. This code calculates the Arnello–Bond estimator.

```
proc import out=cigar
    datafile="C:\Documents and Settings\E81836\Desktop\Economics Book\cigar"
    dbms=Excel Replace;
    getnames=yes;
```

```
run;
data cigar;
     set cigar;
     Log_C=Log(C);
     Log_MIN=Log(MIN);
     Log_NDI=Log(NDI);
     Log_Price=Log(Price);
run;
proc iml;
* This program will calculate the Arellano-Bond estimator for the Cigar.TXT dataset;
* We will assume that all X's are predetermined;
* Define constants;
N=1;T=30;P=1708;
* Read the variables into a matrix;
     use cigar;
     read all var {'Log_C','Log_Min','Log_NDI','Log_Price'} into Y;
* H is fixed. So, define H here;
     H=shape(0,T-2,T-2);
     do j=1 to T-2;
          do l=1 to T-2;
               if j=l then H[l,j]=2;
               if (j=l+1) then H[l,j]=-1;
               if (j=l-1) then H[l,j]=-1;
          end;
     end;
* Initialize Four sum matrices;
     ZHZ=shape(0,P,P); YMZ=shape(0,1,P); YZ=shape(0,4,P); ZPZ=shape(0,P,P);
* Initialize the counter;
     compt=1;
* Begin the loop for calculating the first step Arellano-Bond estimator;
     do i=1 to N;
* Calculate the diff matrix relating to the Y's;
          Y_DIFF=shape(0,T-2,1);
          Y_DIFFS=shape(0,T-2,1);
          do Index=1 to T-2;
               Y_DIFF[Index,1]=Y[Index+compt+1,1]-Y[Index+Compt,1];
               Y_DIFFS[Index,1]=Y[Index+compt,1]-Y[Index+compt-1,1];
          end;
* Calculate the diff matrix relating to the X's;
          X_DIFF1=shape(0,T-2,1);
          X_DIFF2=shape(0,T-2,1);
          X_DIFF3=shape(0,T-2,1);
     do index=1 to T-2;
          X_DIFF1[Index,1]=Y[Index+compt+1,2]-Y[Index+Compt,2];
          X_DIFF2[Index,1]=Y[Index+compt+1,3]-Y[Index+Compt,3];
          X_DIFF3[Index,1]=Y[Index+compt+1,4]-Y[Index+Compt,4];
     end;
* Create the XI matrix;
     XI=shape(0,T-2,4);
     XI=Y_DIFFS||X_DIFF1||X_DIFF2||X_DIFF3;
* Calculate the BZI matrix;
     j=1;
     cpt=j+3*(j+1);
```

```
    cpt2=compt;
    BZI=shape(0,1,cpt);BZI[1,1]=Y[cpt2,1];BZI[1,2]=Y[cpt2,2];
    BZI[1,3]=Y[cpt2+1,2];BZI[1,4]=Y[cpt2,3];BZI[1,5]=Y[cpt2+1,3];
    BZI[1,6]=Y[cpt2,4];BZI[1,7]=Y[cpt2+1,4];
    do j=2 to T-2;
            cpt=j+3*(j+1);
            cpt2=compt;
            C=shape(0,1,cpt);
            do k=1 to j;
                    C[1,k]=Y[cpt2+k-1,1];
            end;
    do k=1 to j+1;
            c[1,j+k]=Y[cpt2+k-1,2];
            c[1,j+j+1+k]=Y[cpt2+k-1,3];
            c[1,j+2*(j+1)+k]=Y[cpt2+k-1,4];
    end;
            BZI=block(BZI,C);
    end;
    ZHZ=ZHZ+BZI`*H*BZI;
    YMZ=YMZ+Y_DIFF`*BZI;
    YZ=YZ+XI`*BZI;
    compt=compt+T;
    end;
* Calculate the coefficient estimate;
    Delta_Est1=inv(YZ*ginv(ZHZ)*YZ`)*YZ*ginv(ZHZ)*YMZ`;
    print Delta_Est1;
* Calculate the Residual Vector;
    compt=1;
    do i=1 to N;
* Calculate the diff matrix;
    Y_DIFF=shape(0,T-2,1);
    Y_DIFFS=shape(0,T-2,1);
    E=shape(0,T-2,1);
    do Index=1 to T-2;
            Y_DIFF[Index,1]=Y[Index+compt+1,1]-Y[Index+Compt,1];
            Y_DIFFS[Index,1]=Y[Index+compt,1]-Y[Index+compt-1,1];
    end;
* Calculate the diff matrix relating to the X's;
    X_DIFF1=shape(0,T-2,1);X_DIFF2=shape(0,T-2,1);X_DIFF3=shape(0,T-2,1);
    do index=1 to T-2;
            X_DIFF1[Index,1]=Y[Index+compt+1,2]-Y[Index+Compt,2];
            X_DIFF2[Index,1]=Y[Index+compt+1,3]-Y[Index+Compt,3];
            X_DIFF3[Index,1]=Y[Index+compt+1,4]-Y[Index+Compt,4];
    end;
* Create the XI matrix;
    XI=shape(0,T-2,4);
    XI=Y_DIFFS||X_DIFF1||X_DIFF2||X_DIFF3;
    E=Y_DIFF-XI*Delta_Est1;
* Calculate the BZI matrix;
    j=1;cpt=j+3*(j+1);cpt2=compt;
    BZI=shape(0,1,cpt);BZI[1,1]=Y[cpt2,1];BZI[1,2]=Y[cpt2,2];
    BZI[1,3]=Y[cpt2+1,2];BZI[1,4]=Y[cpt2,3];BZI[1,5]=Y[cpt2+1,3];
    BZI[1,6]=Y[cpt2,4]; BZI[1,7]=Y[cpt2+1,4];
    do j=2 to T-2;
```

```
            cpt=j+3*(j+1);
            cpt2=compt;
            C=shape(0,1,cpt);
            do k=1 to j;
                    C[1,k]=Y[cpt2+k-1,1];
            end;
            do k=1 to j+1;
                    c[1,j+k]=Y[cpt2+k-1,2];
                    c[1,j+j+1+k]=Y[cpt2+k-1,3];
                    c[1,j+2*(j+1)+k]=Y[cpt2+k-1,4];
            end;
            BZI=block(BZI,C);
    end;
    ZPZ=ZPZ+BZI`*E*E`*BZI;
    compt=compt+T;
    end;
    Delta_Est2=inv(YZ*ginv(ZPZ)*YZ`)*YZ*ginv(ZPZ)*YMZ`;
    print Delta_Est2;
    run;
```

E.17   PROGRAM 17

This code (including the following comments) was written by Thomas Fomby (Department of Economics, Southern Methodist University) in 2005. This SAS IML program conducts a duration analysis of the lengths of strikes as a function of the deviation of output from its trend level, an indicator of the business cycle position of the economy. The data was downloaded from the CD provided in the Greene textbook, *Econometric Analysis*, 4th edn., Table A20.1. The data was originally analyzed by J. Kennan (1985) in his paper "The Duration of Contract Strikes in U.S. Manufacturing," *Journal of Econometrics*, 28, 55–28.

```
data strike;
input dur eco;
datalines;
7.00000   .0113800
9.00000   .0113800
13.0000   .0113800
14.0000   .0113800
26.0000   .0113800
29.0000   .0113800
52.0000   .0113800
130.000   .0113800
9.00000   .0229900
37.0000   .0229900
41.0000   .0229900
49.0000   .0229900
52.0000   .0229900
119.000   .0229900
3.00000  -.0395700
17.0000  -.0395700
19.0000  -.0395700
28.0000  -.0395700
72.0000  -.0395700
99.0000  -.0395700
104.000  -.0395700
```

```
114.000  -.0395700
152.000  -.0395700
153.000  -.0395700
216.000  -.0395700
15.0000  -.0546700
61.0000  -.0546700
98.0000  -.0546700
2.00000  .00535000
25.0000  .00535000
85.0000  .00535000
3.00000  .0742700
10.0000  .0742700
1.00000  .0645000
2.00000  .0645000
3.00000  .0645000
3.00000  .0645000
3.00000  .0645000
4.00000  .0645000
8.00000  .0645000
11.0000  .0645000
22.0000  .0645000
23.0000  .0645000
27.0000  .0645000
32.0000  .0645000
33.0000  .0645000
35.0000  .0645000
43.0000  .0645000
43.0000  .0645000
44.0000  .0645000
100.000  .0645000
5.00000  -.104430
49.0000  -.104430
2.00000  -.00700000
12.0000  -.00700000
12.0000  -.00700000
21.0000  -.00700000
21.0000  -.00700000
27.0000  -.00700000
38.0000  -.00700000
42.0000  -.00700000
117.000  -.00700000
;

data strike;
set strike;
dum = 1;
proc iml;
start mle;
use strike;
read all into t var{dur};
read all into x var{eco};
read all into d var{dum};
/* Calculation of Unrestricted MLE estimates using Newton-Raphson Method */
```

```
theta= {1,4,-9};
crit =1;
n=nrow(t);
ones = j(n,1,1);
result=j(10,9,0);
do iter=1 to 10 while (crit>1.0e-10);
sigma=theta[1,1];
beta1=theta[2,1];
beta2=theta[3,1];
w = (ones/sigma)#(log(t) - ones#beta1 - x#beta2);
lnL = d#(w - log(sigma))- exp(w);
lnL = sum(lnL);
g1 = (ones/sigma)#(w#exp(w) - d#(w + ones));
g2=(ones/sigma)#(exp(w) - d);
g3 = (ones/sigma)#x#((exp(w) - d));
g1=sum(g1);
g2=sum(g2);
g3=sum(g3);
g=g1//g2//g3;
h11= -(ones/sigma**2)#((w##2)#exp(w) + 2#w#exp(w) - 2#w#d - d);
h11= sum(h11);
h12= -(ones/sigma**2)#(exp(w) - d + w#exp(w));
h12 = sum(h12);
h13= -(ones/sigma**2)#x#(exp(w) - d + w#exp(w));
h13 = sum(h13);
h21 = h12;
h31 = h13;
h22 = -(ones/sigma**2)#exp(w);
h22 = sum(h22);
h23 = -(ones/sigma**2)#x#exp(w);
h23 = sum(h23);
h32 = h23;
h33 = -(ones/sigma**2)#(x##2)#exp(w);
h33 = sum(h33);
h=(h11||h12||h13)//(h21||h22||h23)//(h31||h32||h33);
db=-inv(h)*g;
thetanew = theta + db;
crit = sqrt(ssq(thetanew-theta));
theta=thetanew;
result[iter,] = iter||(theta')||g1||g2||g3||crit||lnL;
end;
lnLu = lnL;
cnames = {iter,sigma,beta1,beta2,g1,g2,g3,crit,lnLu};
print "Calculation of Unrestricted MLE estimates using Hessian-Based Newton-Raphson
Method";
print "Iteration steps ", result[colname=cnames];
print , "Unrestricted Log-likelihood = ", lnLu;
/*Covariance matrix from Hessian*/
cov = -inv(h);
se_sigma_h = sqrt(cov[1,1]);
se_beta1_h = sqrt(cov[2,2]);
se_beta2_h = sqrt(cov[3,3]);
z_sigma_h = sigma/se_sigma_h;
```

```
z_beta1_h = beta1/se_beta1_h;
z_beta2_h = beta2/se_beta2_h;
/*Covariance matrix from BHHH*/
g1 = (ones/sigma)#(w#exp(w) - d#(w + ones));
g2 = (ones/sigma)#(exp(w) - d);
g3 = (ones/sigma)#x#((exp(w) - d));
gmat = g1||g2||g3;
bhhh = gmat'*gmat;
covbh3 = inv(bhhh);
se_sigma_b=sqrt(covbh3[1,1]);
se_beta1_b=sqrt(covbh3[2,2]);
se_beta2_b=sqrt(covbh3[3,3]);
z_sigma_b = sigma/se_sigma_b;
z_beta1_b = beta1/se_beta1_b;
z_beta2_b = beta2/se_beta2_b;
pnames = {sigma,beta1, beta2};
print , "The Maximum Likelihood Estimates: Hessian-Based Newton-Raphson Iteration",
theta [rowname=pnames];
print , "Asymptotic Covariance Matrix-From Hessian", cov
[rowname=pnames colname=pnames];
print "Standard errors: ",se_sigma_h,se_beta1_h,se_beta2_h;
print , "Asymptotic Covariance Matrix-From bhhh", covbh3
[rowname=pnames colname=pnames];
print "Standard errors: ",se_sigma_b,se_beta1_b, se_beta2_b;
print "Wald test of hypothesis of constant hazard (sigma=1)";
Wald = (sigma-1)*inv(cov[2,2])*(sigma-1);   * Wald test;
critval = cinv(.95,1);  * calculates the 95th percentile of chi-square 1;
pval = 1 - probchi(wald,1); * calculates the probability value of Wald;
print "Results of Wald test Using Hessian" Wald critval pval;

Wald = (sigma-1)*inv(covbh3[2,2])*(sigma-1);   * Wald test;
critval = cinv(.95,1);   * calculates the 95th percentile of chi-square 1;
pval = 1 - probchi(wald,1); * calculates the probability value of Wald;
print "Results of Wald test Using BHHH" Wald critval pval;
/* ML Estimation of Restricted Model*/
print , "Maximum Likelihood Estimation of Restricted Model";
print "*********************************************";
theta = {4,-9};
crit = 1;
n = nrow(t);
result = j(10,7,0);
do iter = 1 to 10 while (crit > 1.0e-10);
beta1=theta[1,1];
beta2=theta[2,1];
w = (log(t) - ones#beta1 - x#beta2);
lnLr = d#w - exp(w);
lnLr = sum(lnLr);
g1 = -(d - exp(w));
g1 = sum(g1);
g2 = -x#(d - exp(w));
g2 = sum(g2);
g = g1//g2;
h11 = -exp(w);
```

```
h12 = -x#exp(w);
h22 = -(x##2)#exp(w);
h11 = sum(h11);
h12 = sum(h12);
h21 = h12;
h22 = sum(h22);
h = (h11||h12)//(h21||h22);
db = -inv(h)*g;
thetanew = theta + db;
crit = sqrt(ssq(thetanew - theta));
result[iter,] = iter||(theta')||g1||g2||crit||lnLr;
theta = thetanew;
end;
cov = -inv(h);
cnames = {iter,beta1,beta2,g1,g2,crit,lnLr};
print "Iteration steps",result [colname=cnames];
pnames = {beta1,beta2};
print , "The Maximum Likelihood Estimates-Restricted Model", (theta')
[colname=pnames];
print , "Asymptotic Covariance Matrix-From Hessian of Restricted Model", cov
[rowname=pnames colname=pnames];
/* Gradient evaluated at restricted MLE estimates */
sigma = 1;
w = (ones/sigma)#(log(t) - ones#beta1 - x#beta2);
g1 = (ones/sigma)#(w#exp(w) - d#(w + ones));
g2=(ones/sigma)#(exp(w) - d);
g3 = (ones/sigma)#x#((exp(w) - d));
gmat = g1||g2||g3;
g1=sum(g1);
g2=sum(g2);
g3=sum(g3);
g=g1//g2//g3;
/* Hessian evaluated at restricted MLE estimates */
h11= -(ones/sigma**2)#((w##2)#exp(w) + 2#w#exp(w) - 2#w#d - d);
h11= sum(h11);
h12= -(ones/sigma**2)#(exp(w) - d + w#exp(w));
h12 = sum(h12);
h13= -(ones/sigma**2)#x#(exp(w) - d + w#exp(w));
h13 = sum(h13);
h21 = h12;
h31 = h13;
h22 = -(ones/sigma**2)#exp(w);
h22 = sum(h22);
h23 = -(ones/sigma**2)#x#exp(w);
h23 = sum(h23);
h32 = h23;
h33 = -(ones/sigma**2)#(x##2)#exp(w);
h33 = sum(h33);
h=(h11||h12||h13)//(h21||h22||h23)//(h31||h32||h33);
LM = g'*(-inv(h))*g;          * LM test;
critval = cinv(.95,1);
pval = 1 - probchi(LM,1);
print "Results of LM test Using Hessian" LM critval pval;
```

```
/* BHHH evaluated at Restricted MLE*/
bhhh = gmat'*gmat;
covbh3r = inv(bhhh);
LM = g'*covbh3r*g;          * LM test;
critval = cinv(.95,1);
pval = 1 - probchi(LM,1);
print "Results of LM test Using BHHH" LM critval pval;
LR = -2*(lnLr-lnLu);        * Likelihood Ratio test;
pval = 1 - probchi(LR,1);
print "Results of LR test" LR critval pval;
/* Let's see if we get essentially the same maximum likelihood estimates if we use
a BHHH-based Newton-Raphson iteration. */
theta= {1,3.77,-9.35};
crit =1;
n=nrow(t);
ones = j(n,1,1);
result=j(60,9,0);
do iter= 1 to 60 while (crit>1.0e-10);
sigma=theta[1,1];
beta1=theta[2,1];
beta2=theta[3,1];
w = (ones/sigma)#(log(t) - ones#beta1 - x#beta2);
lnL = d#(w - log(sigma))- exp(w);
lnL = sum(lnL);
g1 = (ones/sigma)#(w#exp(w) - d#(w + ones));
g2=(ones/sigma)#(exp(w) - d);
g3 = (ones/sigma)#x#((exp(w) - d));
gmat = g1||g2||g3;
g1 = sum(g1);
g2 = sum(g2);
g3 = sum(g3);
g = g1//g2//g3;
bhhh = gmat'*gmat;
db= inv(bhhh)*g;
thetanew = theta + db;
crit = sqrt(ssq(thetanew-theta));
theta = thetanew;
result[iter,] = iter||(theta')||g1||g2||g3||crit||lnL;
end;
cnames = {iter,sigma,beta1,beta2,g1,g2,g3,crit,lnL};
print "Calculation of Unrestricted MLE estimates using BHHH-Based Newton-Raphson Method";
print "Iteration steps ", result[colname=cnames];
finish;
run mle;
```

REFERENCES

Agresti, A. (1990). *Categorical Data Analysis*, John Wiley & Sons, Inc., New York.

Aigner, D. K. Lovell, and Schmidt, P. (1977). Formulation and Estimation of Stochastic Frontier Production Models. *Journal of Econometrics*, **6**: 21–37.

Allison, P. D. (1995). *Survival Analysis Using SAS: A Practical Guide*, SAS Institute, Inc., Cary, N.C.

Allison, P. D. (2003). *Logistic Regression using the SAS® System: Theory and Application*, SAS Institute Inc., Cary, NC.

Anderson, T., and Hsiao, C. (1981). Estimation of Dynamic Models with Error Components. *Journal of American Statistical Association*, **76**: 598–606.

Anderson, T., Hsiao, C. (1982). Formulation and Estimation of Dynamic Models Using Panel Data. *Journal of Econometrics*, **18**: 67–82.

Arellano, M. (1987). Computing Robust Standard Errors for Within-Groups Estimators. *Oxford Bulletin of Economics and Statistics*, **49**: 431–434.

Arellano, M., and Bond, S. (1991). Some Tests for Specification for Panel Data: Monte Carlo Evidence and an Application to Employment Equations. *Review of Economic Studies*, **58**: 277–297.

Ashenfelter, O., Levine, P. B., and Zimmerman, D. J. (2003). *Statistics and Econometrics: Methods and Applications*, John Wiley & Sons, Inc., New York.

Baltagi, B. H. (2005). *Econometric Analysis of Panel Data*, John Wiley & Sons, Inc., New York.

Baltagi, B. H. (2008). *Econometrics*, Springer, New York.

Balatgi, B. H., and Levin, D. (1992). Cigarette taxation: Raising revenues and reducing consumption, *Structural Change and Economic Dynamics* **3**: 321–335.

Bollerslev, T. (1986). Generalized Autoregressive Conditional Heteroscedasticity. *Journal of Econometrics*, **31**: 307–327.

Breusch, T., and Pagan, A. (1979). A Simple Test for Heteroscedasticity and Random Coefficients Variation. *Econometrica*, **47**: 1287–1294.

Breusch, T., and Pagan, A. (1980). The LM Test and Its Application to Model Specification in Econometrics. *Review of Economic Studies*, **47**: 239–254.

Brocklebank, J. C., and Dickey, D. A. (2003). *SAS® for Forecasting Time Series*, SAS Institute Inc., Cary, NC.

Brown, B., Durbin, J., and Evans, J. (1975). Techniques for Testing the Constancy of Regression Relationships Over Time. *Journal of Royal Statistical Society, Series B*, **37**: 149–172.

Casella, G., and Berger, R. L. (1990). *Statistical Inference*, Wadsworth, Inc., California.

Chow, G. (1960). Tests of Equality Between Sets of Coefficients in Two Linear Regressions. *Econometrica*, **28**: 591–605.

Chung, C. F., Schmidt, P. and Witte, A. D. (1991). Survival Analysis: A Survey, *Journal of Quantitative Criminology* **7**: 59–98.

Cincera, M. (1997). Patents, R&D, and Technological Spillovers at the Firm Level: Some Evidence from Econometric Count Models for Panel Data, *Journal of Applied Econometrics*, **12**: 265–280.

Cornwell, C., and Rupert, P. (1988). Efficient Estimation with Panel Data: An Empirical Comparison of Instrumental Variables Estimators. *Journal of Applied Econometrics*, **3**: 149–155.

Davidson, R., and MacKinnon, J. (1993). *Estimation and Inference in Econometrics*, New York: Oxford University Press.

Efron, B., and Tibshirani, R. J. (1993). *An Introduction to Bootstrap*, Chapman & Hall, London, UK.

Enders, W. (2004). *Applied Econometric Time Series*. John Wiley & Sons, Inc., New York.

Engle, R. (1982). Autoregressive Conditional Heteroscedasticity with Estimates of the Variance of United Kingdom Inflations. *Econometrica*, **50**: 987–1008.

Fomby, T. B. (2007). Department of Economics, Southern Methodist University, Dallas, TX, personal communication, March 31, 2007.

Freund, R., and Littell, R. C. (2000). *SAS® System for Regression*, 3rd Edition, SAS Institute Inc., Cary, NC.

Freund, R. J., and Wilson, W. J. (1998). *Regression Analysis*, San Diego, Academic Press.

Fuller, W. A., and Battese, G. E. (1974). Estimation of Linear Models with Crossed-Error Structure. *Journal of Econometrics*, **2**: 67–78.

Glewwe, P. (2006). Department of Applied Economics, St. Paul, MN, personal communication, January 31, 2006.

Graybill, F. A. (2000). Theory and *Application of Linear Models*, Duxbury Press.

Greene, W. (1992). A Statistical Model for Credit Scoring. Working Paper No. EC-92-29, New York University, Department of Economics, Stern School of Business.

Greene, W. H. (2003). *Econometric Analysis*, Prentice Hall, New Jersey.

Grunfeld, Y. (1958). The Determinants of Corporate Investment. Unpublished Ph.D. thesis, Department of Economics, University of Chicago.

Hallam, A. Unpublished Lecture Notes, Department of Economics, Iowa State University, Ames, Iowa.

Hausman, J. (1978). Specification Tests in Econometrics. *Econometrica*, **46**: 1251–1271.

Hausman, J., and Taylor, W. (1977). Panel Data and Unobservable Individual Effects. *Econometrica*, **45**: 919–938.

Hausman, J., and Taylor W. (1981). Panel Data and Unobservable Individual Effects. *Econometrica*, **49**: 1377–1398.

Heckman, J. (1979). Sample Selection Bias as a Specification Error. *Econometrica*, **47**: 153–161.

Hildebrand, G., and T. Liu (1957). *Manufacturing Production Functions in the United States*. Ithaca, N.Y.: Cornell University Press.

Financial Dictionary, www.investopedia.com, A Forbes Digital Company, 2008.

Jackson, E. J., A *User's Guide to Principal Components*, John Wiley & Sons, NY, 2003.

Jaeger, D. A. (2007). Department of Economics, University of Bonn, Germany, April 30, 2007.

Kiviet, J. (1995). On Bias, Inconsistency, and Efficiency of Some Estimators in Dynamic Panel Data Models. *Journal of Econometrics*, **68**(1): 63–78.

Kennan, J. (1985). The Duration of Contract Strikes in U.S. Manufacturing, *Journal of Econometrics*, **28**: 5–28.

Koenker, R. (1981). A Note on Studentizing a Test for Heteroscedasticity. *Journal of Econometrics*, **17**: 107–112.

Koenker, R., and Bassett, G. (1982). Robust Tests for Heteroscedasticity Based on Regression Quantiles. *Econometrica*, **50**: 43–61.

Lee, E. T. (1992). *Statistical Methods for Survival Data Analysis*, John Wiley & Sons, Inc., New York.

Littell, R. C., Stroup, W. W., and Freund, R. (2002). *SAS® for Linear Models*, SAS Institute Inc., Cary, NC.

Littell, R. C., Milliken, G. A., Stroup, W. W., and Wolfinger, R. D. (2006). *SAS for Mixed Model*, 2nd Edition, SAS Institute Inc., Cary, NC.

Lovell, M. C. (2006). A Simple Proof of the FWL (Frisch–Waugh–Lovell) Theorem. Available at SSRN: http://ssrn.com/abstract=887345.

MacKinnon, J., and White, H. (1985). Some Heteroscedasticity Consistent Covariance Matrix Estimators with Improved Finite Sample Properties. *Journal of Econometrics*, **19**: 305–325.

McCall, B. P. (1995). The Impact of Unemployment Insurance Benefit Levels on Recipiency, *Journal of Business and Economic Statistics*, **13**: 189–198.

McCullough, G. (2005). Department of Applied Economics, St. Paul, MN, personal communication, September 30, 2005.

McLeod, A., and Li, W. (1983). Diagnostic Checking ARMA Time Series Models Using Squared Residual Correlations. *Journal of Time Series Analysis* **4**: 269–273.

Meyers, H. M. (1990). *Classical and Modern Regression with Applications*, PWS-Kent, Massachusetts.

Montgomery, D. C. (1991). *Introduction to Statistical Quality Control*, John Wiley & Sons, New York.

Mroz, T. (1987). The Sensitivity of an Empirical Model of Married Women's Hours of Work to Economic and Statistical Assumptions. *Econometrica*, **55**: 765–799.

Nickell, S. (1981). Biases in Dynamic Models with Fixed Effects. *Econometrica*, **49**: 1417–1426.

NIST/SEMATECH, *e-Handbook of Statistical Methods*, available at http://www.itl.nist.gov/div898/handbook.

Page, E. S. (1954). Continuous Inspection Schemes. *Biometrika*, **41**(1): 100–115.

Park, H. M. (2005). Linear Regression Models for Panel Data Using SAS, Stata, LIMDEP and SPSS. Technical Working Paper. The University Information Technology Services (UITS) Center for Statistics and Mathematics, Indiana University, Indiana.

Sargan, J. D. (1958). The Estimation of Economic Relationships Using Instrumental Variables. *Econometrica*, **26**: 393–415.

Searle, S. R. (1982). *Matrix Algebra Useful for Statistics*, John Wiley & Sons, Inc., New York.

Snedecor, G. W., and Cochran, W. G. (1983). *Statistical Methods*, Iowa State University Press, Iowa.

Stokes, M. E., Davis, C. S., and Koch, G. G. (2001). *Categorical Data Using the SAS® System*, SAS Institute Inc., Cary, NC.

Verbeek, M. (2006). *A Guide to Modern Econometrics*, John Wiley & Sons Ltd., West Sussex, England.

Walter, E. (2004). *Applied Econometric Time Series*, John Wiley & Sons, Inc., New York.

White, H. (1980). A Heteroscedasticity-Consistent Covariance Matrix Estimator and a Direct Test for Heteroscedasticity. *Econometrica*, **48**: 817–838.

Woodall, W. H., and Ncube, M. M. (1985). Multivariate CUSUM Quality Control Procedures. *Technometrics*, **27**(3): 285–292.

Wooldridge, J. M. (2002). *Econometric Analysis of Cross Section and Panel Data*, Massachusetts Institute of Technology, Cambridge, MA.

Zellner, A. (1962). An Efficient Method of Estimating Seemingly Unrelated Regression and Tests of Aggregation Bias. *Journal of the American Statistical Association*, **57**: 500–509.

INDEX

Applied Econometrics Using the SAS® System, by Vivek B. Ajmani
Copyright © 2009 John Wiley & Sons, Inc.

311

CPSIA information can be obtained at www.ICGtesting.com
Printed in the USA
BVOW031315300912

301717BV00003B/4/P